Results and Problems in Cell Differentiation

Volume 64

More information about this series at http://www.springer.com/series/400

Adam Telerman • Robert Amson
Editors

TCTP/tpt1 - Remodeling Signaling from Stem Cell to Disease

Springer

Editors
Adam Telerman
Bâtiment B2M
Institut Gustave Roussy
Villejuif, Val-de-Marne
France

Robert Amson
Bâtiment B2M
Institut Gustave Roussy
Villejuif, Val-de-Marne
France

ISSN 0080-1844 ISSN 1861-0412 (electronic)
Results and Problems in Cell Differentiation
ISBN 978-3-319-67590-9 ISBN 978-3-319-67591-6 (eBook)
DOI 10.1007/978-3-319-67591-6

Library of Congress Control Number: 2017957604

Printed on acid-free paper

This Springer imprint is published by Springer Nature
The registered company is Springer International Publishing AG
The registered company address is: Gewerbestrasse 11, 6330 Cham, Switzerland

Contents

1 Introduction: How We Encountered TCTP and Our Purpose in Studying It . . . 1
Adam Telerman and Robert Amson
1.1 The Initial Years: The Tumor Reversion Project . . . 2
1.2 Learning to Work with High-Throughput Technology and the First Molecular Data . . . 3
1.3 The Year 2000: Giving a Decisive Turn into the Understanding of the Tumor Reversion Program . . . 3
1.4 The P53-TCTP Reciprocal Negative Feedback Loop and the Clinical Significance . . . 4
1.5 Conclusion and Perspectives . . . 5
References . . . 6

2 Structural Insights into TCTP and Its Interactions with Ligands and Proteins . . . 9
Nadine Assrir, Florian Malard, and Ewen Lescop
2.1 Introduction . . . 9
2.2 Sequence and Structure of TCTP . . . 10
2.2.1 Description of the Structure of TCTP . . . 10
2.2.2 Structural Homologues of TCTP . . . 12
2.2.3 Functional Elements Within TCTP Sequence . . . 17
2.3 Binding Properties and Structural Aspects of TCTP in Complex with Ions, Small Molecules, Carbohydrates, Peptides, and Nucleic Acids . . . 21
2.3.1 Calcium Binding . . . 21
2.3.2 Antihistaminic Drugs and the Related Sertraline/Thioridazine . . . 23
2.3.3 Peptides . . . 26
2.3.4 Heme, Artemisinine, and Analogs . . . 26
2.3.5 Nucleic Acids . . . 27

2.3.6 *Bombyx mori* TCTP as a Binding Platform for Saccharides 28
2.4 Structural Aspects of TCTP in Complex with Proteins 29
2.4.1 TCTP Directly Interacts with Dozens of Proteins 29
2.4.2 Structural Information on Native Complexes 35
2.4.3 TCTP Tends to Self-associate 38
2.5 Conclusions 38
References 40

3 **Structure-Function Relationship of TCTP** 47
Beatriz Xoconostle-Cázares and Roberto Ruiz-Medrano
3.1 Introduction 48
3.2 Conserved Functions of TCTP Across Kingdoms 49
3.3 Taxon-Specific Functions of TCTP 50
3.3.1 Fungi 51
3.3.2 Plants 52
3.3.3 Blood-Borne and Other Vertebrate Parasites 55
3.4 Non-cell Autonomous Functions of TCTP 60
3.5 Perspectives 63
References 64

4 **The Translational Controlled Tumour Protein TCTP: Biological Functions and Regulation** 69
Ulrich-Axel Bommer
4.1 Introduction 70
4.1.1 The 'Translationally Controlled Tumour Protein TCTP': Names and History 71
4.1.2 Gene Structure and mRNA 72
4.1.3 Molecular Structure, Conservation and Interactions . . . 74
4.2 Biological Functions of TCTP 75
4.2.1 Maintaining Cell Homoeostasis and Survival 75
4.2.2 Involvement in the Cell Cycle and in Early Development 81
4.2.3 TCTP in Cell Growth Regulation, Protein Synthesis and Degradation 85
4.2.4 Extracellular Functions of TCTP 88
4.3 Involvement of TCTP in Disease Processes 91
4.3.1 TCTP in Human Cancer 91
4.3.2 Participation in Immunological Reactions 98
4.3.3 TCTP in Lower Animals and Parasitic Infections 100
4.3.4 TCTP in Other (Patho)physiological Processes 103
4.4 Regulation of Cellular TCTP Levels 105
4.4.1 Cell Physiologic Conditions That Result in Regulation of TCTP Levels 105
4.4.2 Mechanisms Involved in Regulation of Cellular TCTP Levels 105

4.5 Synopsis 112
References 114

5 Current Understanding of the TCTP Interactome 127
Siting Li and Feng Ge
5.1 Introduction 127
5.2 Global Interactome Profiling Methods 128
5.3 The Current Knowledge of the TCTP Interactome 129
5.3.1 Chaperone Proteins 130
5.3.2 Nucleic Acid-Binding Proteins 130
5.3.3 Cytoskeletal Proteins 130
5.3.4 Other Functions 132
5.4 Concluding Remarks 133
References 133

6 Role and Fate of TCTP in Protein Degradative Pathways 137
Michel Vidal
6.1 Introduction 138
6.2 TCTP as Protein Stabilizer 138
6.2.1 TCTP Masks the Ubiquitination Sites of Its Partners . . 139
6.2.2 TCTP Binding Leads to E3 Ligase Degradation 140
6.2.3 Mmi1/ScTCTP Modulates Proteasome Activity 140
6.3 TCTP as Degradation Inducer 141
6.4 TCTP Degradation 142
6.5 Conclusion 144
References 146

7 Roles of the Translationally Controlled Tumor Protein (TCTP) in Plant Development 149
Léo Betsch, Julie Savarin, Mohammed Bendahmane, and Judit Szecsi
7.1 Introduction 149
7.2 Features of Plant TCTP Genes 150
7.3 TCTP Is Essential for Plant Development 154
7.4 Role of TCTP in Plant Signaling 158
7.4.1 Is TCTP a Component of the TOR Pathway? 158
7.4.2 Role of TCTP in Hormone Signaling 160
7.4.3 Response to Abiotic and Biotic Stresses 162
7.5 Conclusion 166
References 166

8 Function of Translationally Controlled Tumor Protein in Organ Growth: Lessons from Drosophila Studies 173
Kwang-Wook Choi, Sung-Tae Hong, and Thao Phuong Le
8.1 Introduction 173
8.2 Identification of *Drosophila* Tctp Function in Organ Growth . . . 174

8.3 Role of Tctp in TOR Signaling 175
8.4 Regulation of Tctp Function by 14-3-3 178
8.5 Effects of 14-3-3 Isoforms and Tctp on Cyclin E 181
8.6 Tctp Function in DNA Damage Control 181
8.7 Tctp in Chromatin Remodeling and Genome Stability 184
8.8 Concluding Remarks 185
References 187

9 Translationally Controlled Tumor Protein (TCTP/HRF) in Animal Venoms 193
Andrea Senff-Ribeiro
9.1 Introduction 193
9.2 *Loxosceles intermedia* TCTP 194
9.3 Other TCTPs Found in Animal Venoms 196
9.4 Perspectives 198
References 199

10 Tctp in Neuronal Circuitry Assembly 201
Cláudio Gouveia Roque and Christine E. Holt
10.1 Introduction 202
10.1.1 Features of Axon Development 202
10.1.2 Axonal mRNA Localization: One in Thousands 202
10.1.3 Axon Guidance and Cancer: Shared Features 205
10.1.4 Axonal Mitochondria 207
10.2 TCTP in Neuronal Circuitry Assembly 207
10.3 Summary and Future Directions 212
References 213

11 Elusive Role of TCTP Protein and mRNA in Cell Cycle and Cytoskeleton Regulation 217
Jacek Z. Kubiak and Malgorzata Kloc
11.1 TCTP and Cell Cycle 218
11.1.1 The Role of TCTP Protein 218
11.1.2 The Role of TCTP mRNA 220
11.2 TCTP and Cytoskeleton 221
11.2.1 TCTP and Microtubules 222
11.2.2 TCTP and Actin 223
11.3 Conclusions 223
References 223

12 The Translationally Controlled Tumor Protein and the Cellular Response to Ionizing Radiation-Induced DNA Damage 227
Jie Zhang, Grace Shim, Sonia M. de Toledo, and Edouard I. Azzam
12.1 Introduction 228
12.2 Primary Effects of Ionizing Radiation 230
12.2.1 Direct and Indirect Effects of Ionizing Radiation 230

12.3 Endogenous and Radiation-Induced DNA Alterations 232
12.3.1 DNA Damage Response Pathways and DNA Damage Repair Mechanisms 233
12.4 TCTP and the Sensing of Genotoxic Stress 237
12.5 TCTP and the Repair of DNA Damage 240
12.6 TCTP and Control of Cell Cycle Progression Under Normal and Stress Conditions 242
12.7 TCTP and Cell Death 244
12.8 Perspective 245
References 247

13 TCTP Has a Crucial Role in the Different Stages of Prostate Cancer Malignant Progression 255
Virginie Baylot, Sara Karaki, and Palma Rocchi
13.1 TCTP Is Upregulated in Prostate Cancer 256
13.2 TCTP: A Promising Target in Castration-Resistant Prostate Cancer 256
13.2.1 TCTP Mediates Heat Shock Protein 27 Cytoprotective Function in CRPC 256
13.2.2 TCTP and P53 in CRPC: "Neither Can Live While the Other Survives" 257
13.2.3 Development of a TCTP Antisense Oligonucleotide for Clinical Applications 259
13.3 Discussion 260
References 260

14 Role of TCTP for Cellular Differentiation and Cancer Therapy . . . 263
Ean-Jeong Seo, Nicolas Fischer, and Thomas Efferth
14.1 Introduction 265
14.1.1 TCTP in Differentiation Processes 266
14.2 TCTP in Cancer 268
14.2.1 TCTP and Tumor Reversion 268
14.2.2 TCTP as Antiapoptotic Protein 269
14.2.3 Cell Cycle Regulation of TCTP 270
14.2.4 TCTP Reduces Cellular Stress 271
14.3 TCTP for Differentiation Therapy 271
14.3.1 Approaches of Differentiation Therapy in General 271
14.3.2 Retinoids 272
14.3.3 Histone Deacetylase Inhibitors 272
14.3.4 PPARγ Agonists 273
14.3.5 Vitamin D 273
14.3.6 Differentiation Therapy with Antihistaminic Drugs . . . 274
14.4 Conclusions and Perspectives 275
References 276

15 Targeting TCTP with Sertraline and Thioridazine in Cancer Treatment 283
Robert Amson, Christian Auclair, Fabrice André, Judith Karp, and Adam Telerman
15.1 Introduction 283
15.2 Experimental Approach 284
15.3 Conclusion and Therapeutic Perspectives 288
References 289

16 History of Histamine-Releasing Factor (HRF)/Translationally Controlled Tumor Protein (TCTP) Including a Potential Therapeutic Target in Asthma and Allergy 291
Susan M. MacDonald
16.1 Introduction/Cloning 292
16.2 Clinical Relevance of HRF/TCTP 293
16.3 HRF/TCTP Extracellular Functions 294
16.4 Other Functions of HRF/TCTP (Mainly Intracellular) 295
16.5 An Inducible HRF/TCTP Transgenic Mouse 296
16.6 The Importance of Ship-1 on HRF/TCTP Signaling 298
16.7 Additional Intracellular Signaling by HRF/TCTP 300
16.8 HRF/TCTP as a Therapeutic Target 301
16.9 Summary 303
References 304

Conclusion 309

Chapter 1
Introduction: How We Encountered TCTP and Our Purpose in Studying It

Adam Telerman and Robert Amson

Abstract In this brief introduction, we describe our encounter with TCTP. Back in 2000, we discovered TCTP in two quite different ways: first, we looked at protein partners of TSAP6 and one of them was TCTP. Then, in collaboration with Sidney Brenner, we performed a high-throughput differential screening comparing the parental cancer cells with revertants. The results indicated that TCTP was of the most differentially expressed genes. These two approaches were carried out only months apart. They guided our research and led to the discoveries of drugs that inhibit the function of TCTP. Much of the preclinical data on sertraline as an inhibitor of TCTP in cancer were obtained with Judith Karp at Johns Hopkins. This drug is now given in combination with Ara-C to patients in a phase I clinical trial for Acute Myeloid Leukemia. We will here detail how all this happened in our lab while working around one central project: tumor reversion.

It is both fascinating and challenging to edit the very first book on a protein. The implication of Translationally Controlled Tumor Protein (TCTP) in disease was discovered by Susan MacDonald at Johns Hopkins University: she identified it as the histamine-releasing factor (HRF) (MacDonald et al. 1995). Only later its function in cancer and more specifically in tumor reversion was discovered (Tuynder et al. 2001a, b, 2002, 2004; Amson et al. 2013a, b; Telerman and Amson 2009). Today, we know much more about TCTP and the mechanisms by which it controls cell fate. The fact that it is present in all eukaryotes, in stem cells, and that it interacts with the apoptotic machinery—including members of the Bcl2 family as well as p53-mdm2—makes of it a key-protein in regulatory processes (Amson et al. 2012b; Cans et al. 2003; Susini et al. 2008; Thebault et al. 2016).

In this book, we gave voice to some of the scientists that provided the most significant advances in the field. We have chosen not to devote chapters on describing the genetic and biologic studies on TCTP done in our laboratory,

A. Telerman (✉)
Bâtiment B2M, Institut Gustave Roussy, Unité Inserm U981, 114 rue Édouard-Vaillant, 94805 Villejuif, France
e-mail: atelerman@gmail.com

A. Telerman, R. Amson (eds.), *TCTP/tpt1 - Remodeling Signaling from Stem Cell to Disease*, Results and Problems in Cell Differentiation 64,
DOI 10.1007/978-3-319-67591-6_1

which have already been reviewed extensively. Our single chapter concerns TCTP as a target in the treatment of cancer and the clinical study that we initiated together with Judith Karp, from Johns Hopkins.

Our introduction sheds light, for the first time, on how those discoveries were made in our laboratory. Indeed, we have been asked numerous times to describe those events in detail, since this could be relevant for young researchers in planning their work.

1.1 The Initial Years: The Tumor Reversion Project

When we were doing our postdoctoral training at the Weitzman Institute of Science, the vast majority of the investigators in the field of cancer sought to understand how a normal cell becomes a tumor cell. At that time, oncogenes were the main focus of research in almost every oncology laboratory worldwide. When we decided to set up our laboratory, it seemed to us pointless to concentrate our efforts on a project in which some of the strongest intellects in the field of biology had already made such tremendous contributions to answer that question. We thought that there was a different way to proceed in cancer research: not trying to understand how a normal cell becomes malignant, but rather how a malignant cell can quit its malignant phenotype (Telerman et al. 1993a, b, c). This laid the basis of the tumor reversion project (Telerman and Amson 2009). Max Askanazy had already provided at the beginning of the twentieth century the most striking example of tumor reversion (Askanazy 1907; Telerman and Amson 2009). He observed that ovarian carcinoma was composed of a homogeneous tumor cell population at an early stage, and that ultimately these cells differentiate into teeth and hairs. This quite unbelievable observation turned out to be of dramatic importance. If an ovarian carcinoma cell could become hair or teeth, it meant that those cancer cells could be entirely reprogrammed. It is precisely this reprogramming at the genetic and molecular level that became our project for almost 30 years now. In the 1950–1960s, Armin Braun (1951, 1959, 1965) confirmed tumor reversion in plants. Later, a series of investigators found in cellular systems, consisting mostly of in vitro cultures, that in very rare instances cancer cells transformed by oncogenes could lose their malignant phenotype (Bissell and Labarge 2005; Brinster 1974; Ge et al. 2011; Hendrix et al. 2007; Macpherson 1965; Mintz and Illmensee 1975; Pierce and Dixon 1959; Telerman and Amson 2009; Weaver et al. 1997). In most of the cases, this was due to the loss of the transforming oncogene, but not in all cases.

When we started our laboratory we found that that there was a desperate need for the proper biological models to study the molecular pathways of tumor reversion. This is why we sought to obtain parental malignant cells and derive from those the revertant ones. Another laboratory in Brussels studied at that time a quite peculiar virus: the H1 Parvovirus that kills preferentially cancer cells while sparing their normal counterparts (Mousset and Rommelaere 1982; Toolan 1967). We thought that we could use the H1 Parvovirus as a negative selective agent that would kill the malignant cells but spare those that would have reverted and lost some of their malignant properties. With the help of Marcel Tuynder we started the experiments with the human erythroleukemia cell line K562 and after three rounds of infection with the Parvovirus we succeeded in rescuing the cells with a suppressed malignant phenotype, which we called "KS" for "K562 Suppressed" (Telerman et al. 1993a, b, c). In the following years, we expanded the experiment to different types of cancer—leukemia, breast, colon, lung, and melanoma (Tuynder et al. 2004, 2002). The next step was to provide a differential analysis of gene expression between the malignant and the revertant cells (Tuynder et al. 2002).

1.2 Learning to Work with High-Throughput Technology and the First Molecular Data

In 1994 after publishing our first work on tumor reversion we moved to Paris, France, to join Daniel Cohen and Jean Dausset at the Fondation Jean Dausset—Centre d'Etude du Polymorphysme Humain. Daniel Cohen had made a tremendous contribution in creating a human genome center with the highest scientific standards and the most up-to-date technology and we could learn from the way they envisaged the progress in biology. Things had to be fast, precise, efficient, and large scale. We used the method of Liang and Pardee (1992) to make a first differential gene analysis using Moshe Oren's system of M1/LTR6 cells (Yonish-Rouach et al. 1991). This yielded with the first ten differentially expressed genes that have later been proven to be so useful for our studies of tumor reversion (Telerman et al. 1996; Amson et al. 2000, 1996; Linares-Cruz et al. 1998; Nemani et al. 1996; Roperch et al. 1998, 1999). Another inspiring mentor, Georges Charpak, helped us in quantifying these data in such an elegant way with his new developed technology (Amson et al. 1996).

1.3 The Year 2000: Giving a Decisive Turn into the Understanding of the Tumor Reversion Program

We divided our laboratory in several groups. Marcel Tuynder was focused on the biological models of tumor reversion and their characterization. Laurent Susini was working on the differential gene expression analysis, Giusy Fiucci on the murine knockout models, and the crystallography and Brent Passer on the yeast two hybrid analysis.

We teamed up with Sydney Brenner that had just developed the Megasort and MPSS screening strategies (Brenner et al. 2000a, b). Laurie Goodman from Brenner's lab came to Paris with a short list of the ten mostly differentially expressed genes between the U937 cancer cells and their revertants, the US cells (Tuynder et al. 2000, 2001a, b). At the top of the list was Translationally Controlled

Tumor Protein (TCTP) with 248 signals in the parental U937 cancer cells versus 2 in the revertant US cells using Megasort, and this was proportional to the amount of mRNA. Decreasing TCTP by siRNA induced cell death in the parental U937 cells and a reprogramming of breast cancer cells into structures with a similar architecture of normal cells. These results were presented at the Annual Meeting on Oncogenes, Frederick, Maryland, USA, June 2001 and also at the Conference on Programmed Cell Death. Cold Spring Harbor, September 2001. The work on the anti-apoptotic of TCTP has been confirmed by another group a couple of months later; unfortunately, they changed the name of TCTP and invented a new one (Li et al. 2001).

Meanwhile, on the other side of our laboratory, Brent Passer was investigating one of the genes we had previously identified, TSAP6 (Amson et al. 1996; Amzallag et al. 2004; Passer et al. 2003). Among the potential partner proteins of TSAP6 Brent found the Histamine Releasing Factor (HRF) (MacDonald et al. 1995) that was just another name for TCTP. Brent had come to these results before we received the short list from Sydney Brenner. Later, we found that TSAP6 was promoting the secretion of TCTP via the exosomal pathway (Amzallag et al. 2004; Lespagnol et al. 2008). As explained later in the book, it was this HRF function of TCTP that led us to the discovery of the first drugs inhibiting the function of TCTP.

1.4 The P53-TCTP Reciprocal Negative Feedback Loop and the Clinical Significance

It took us a long time to understand how TCTP functions and what are the molecular mechanisms that it regulates (Amson et al. 2013b). We first observed that in different biological models, increasing P53 was decreasing TCTP (Amson et al. 2012a). In contrast, overexpression of TCTP strongly decreased P53. So we tried to understand what was really going on; Alexandra Lespagnol found that the promoter of TCTP has a consensus-binding site for P53 and that this results in a negative regulation of TCTP. On the other side, TCTP promotes the degradation of P53 by stabilizing MDM2. Together with Pier Paolo Di Fiore, Salvatore Pece, and Jean-Christophe Marine, we investigated the details of these mechanisms and most importantly how it applied to stem cell biology and breast cancer, this time in patients. TCTP was highly expressed in normal breast stem cells and in breast cancer like stem cells. Decreasing TCTP inhibited the colony forming efficiency in mammosphere assays. Di Fiore's group also made the observation that in a cohort of 508 breast cancer patients, tumors with high levels of TCTP induced a more aggressive disease and a poor prognosis. Accordingly, low levels of TCTP led to a significantly better survival. TCTP stands as a prognostic marker on its own.

The search for a drug targeting TCTP in cancer treatment is addressed further in this book and deserves a chapter on its own. Briefly, as soon as we saw that decreasing TCTP could be of potential clinical relevance, we searched for compounds that would be able to inhibit its action. The fact that TCTP was also the Histamine Releasing Factor led us to the hypothesis that anti-histaminic agents such as hydroxyzine, promethazine, and dexchlorpheniramine could kill cancer cells or revert them (Tuynder et al. 2004). A couple of months after these initial findings, chemists in our laboratory identified sertraline and thioridazine as having the same structural backbone as some of the anti-histaminic drugs and those last two drugs proved to be efficient in vitro and in vivo against cancer cells. They both bind TCTP in a domain very close to its mobile loop and inhibit this way its function (Amson et al. 2012b, 2013b). After some fascinating discussions, Judith Karp at Johns Hopkins in Baltimore decided to start a clinical trial in combination with Cytosine Arabinoside in refractory adult myeloid leukemia (AML). With Judith we started to apply for grants in 2010. She teemed up with Ivana Gojo also working at Johns Hopkins and Mark Frattini at Columbia University in New York. The study received the support of the American Leukemia & Lymphoma Society in 2014, based on the initial grant that we wrote in 2010 and including this time some of the preliminary work we did with Judith Karp (https://www.lls.org/content/the-clinical-application-of-tumor-reversion-a-phase-i-study-of-sertraline-zoloft-in-combination-with-timed-sequential-cytosine-arabinoside-ara-c-in). For us today, the drugability of TCTP becomes one of the main subjects of our research.

1.5 Conclusion and Perspectives

Working on the tumor reversion project taught us to be very systematic in our research: starting by addressing one biologically and medically relevant question; then build the biological models that would enable us to answer this question, no matter how long this could take. As Kurt Isselbacher told us at the beginning of our work during a seminar we were presenting at the Massachusetts General Hospital Cancer Center: "With biological models you know when you start, but you never know when you will actually have them and this might take so many years." He was so right. It took us almost 10 years to obtain the different biological models. Fortunately, Sydney Brenner invented the Megasort and MPSS and this helped us a lot to extract the molecular information out of the biological models. Then we could go to the biology and genetics of tumor reversion, just focusing on a small number of genes. The genetic models of tumor reversion and cell reprogramming remain a major challenge for the years to come. Throughout the course of our research, we were very lucky to be surrounded by some of the brightest mentors and collaborators: Jean Dausset, Daniel Cohen, Sydney Brenner, Georges Charpak, Moshe Oren, Joseph Schlessinger, Pier Paolo Di Fiore, Salvatore Pece, Jean-Christophe Marine, and Michel Vidal. Our postdoctoral investigators, Marcel Tuynder, Laurent Susini, Giusy Fiuci, Andrea Senff Ribeiro, just to name a few of them, were all outstanding. Marcel Tuynder and Laurent Susini joined us long before they had their PhD; they obtained it while they worked in our lab and they pursued as project directors. Alexandra Lespagnol continued with us as a postdoctoral investigator after obtaining her PhD on tumor reversion. Another of our students, Stéphanie Thebault, obtained the crystal structure of the protein complex between TCTP and Bcl-xL. With Judith Karp a new avenue was opened for the use of Sertraline as a TCTP inhibitor in patients. Without each one of them, this task would have been impossible and there is still so much to do. Ultimately, we are so grateful to Jacek Kubiak and Malgorzata Kloc for convincing us and giving us the opportunity to put together a book on TCTP. Last but not least, we are indebted to Sabine Schwartz for the important suggestions she made in organizing this book, encouragement, patience, and kindness.

TCTP is of wide interest in today's biology and we give now the word to these researchers that made such seminal discoveries.

Acknowledgements This work was supported by grants from the French National Agency for Research (ANR, ANR-09-BLAN-0292) to AT, RA, and MV; European Union Network of Excellence CONTICANET to AT and RA; LabEx LERMIT to AT and RA; INCa Projets libres de Recherche 2013-1-PL BIO-10 "Biologie et Sciences du Cancer" to AT and RA; Ligue Nationale Contre le Cancer to AT and RA; Odyssea fund to RA.

References

Amson RB, Nemani M, Roperch JP, Israeli D, Bougueleret L, Le Gall I, Medhioub M, Linares-Cruz G, Lethrosne F, Pasturaud P et al (1996) Isolation of 10 differentially expressed cDNAs in p53-induced apoptosis: activation of the vertebrate homologue of the drosophila seven in absentia gene. Proc Natl Acad Sci U S A 93:3953–3957

Amson R, Lassalle JM, Halley H, Prieur S, Lethrosne F, Roperch JP, Israeli D, Gendron MC, Duyckaerts C, Checler F et al (2000) Behavioral alterations associated with apoptosis and down-regulation of presenilin 1 in the brains of p53-deficient mice. Proc Natl Acad Sci U S A 97:5346–5350

Amson R, Pece S, Lespagnol A, Vyas R, Mazzarol G, Tosoni D, Colaluca I, Viale G, Rodrigues-Ferreira S, Wynendaele J, Chaloin O, Hoebeke J, Christophe Marine J-C, Paolo Di Fiore P, Telerman A (2012a) Relevance of the repressive feedback loop between p53 and TCTP in stemness and cancer treatment. Mechanisms and models of cancer. Salk Institute for Biological Studies, La Jolla, CA, Aug 2011

Amson R, Pece S, Lespagnol A, Vyas R, Mazzarol G, Tosoni D, Colaluca I, Viale G, Rodrigues-Ferreira S, Wynendaele J et al (2012b) Reciprocal repression between P53 and TCTP. Nat Med 18:91–99

Amson R, Karp JE, Telerman A (2013a) Lessons from tumor reversion for cancer treatment. Curr Opin Oncol 25:59–65

Amson R, Pece S, Marine JC, Di Fiore PP, Telerman A (2013b) TPT1/TCTP-regulated pathways in phenotypic reprogramming. Trends Cell Biol 23:37–46

Amzallag N, Passer BJ, Allanic D, Segura E, Thery C, Goud B, Amson R, Telerman A (2004) TSAP6 facilitates the secretion of translationally controlled tumor protein/histamine-releasing factor via a nonclassical pathway. J Biol Chem 279:46104–46112

Askanazy M (1907) Die Teratome nach ihrem Bau, ihrem Verlauf, ihrer Genese und im Vergleich zum experimentellen Teratoid. Verhandl Deutsch Pathol 11:39–82

Bissell MJ, Labarge MA (2005) Context, tissue plasticity, and cancer: are tumor stem cells also regulated by the microenvironment? Cancer Cell 7:17–23

Braun AC (1951) Recovery of tumor cells from effects of the tumor-inducing principle in crown gall. Science 113:651–653

Braun AC (1959) A demonstration of the recovery of the crown-gall tumor cell with the use of complex tumors of single-cell origin. Proc Natl Acad Sci U S A 45:932–938

Braun AC (1965) The reversal of tumor growth. Sci Am 213:75–83

Brenner S, Williams SR, Vermaas EH, Storck T, Moon K, McCollum C, Mao JI, Luo S, Kirchner JJ, Eletr S et al (2000a) In vitro cloning of complex mixtures of DNA on microbeads: physical separation of differentially expressed cDNAs. Proc Natl Acad Sci U S A 97:1665–1670

Brenner S, Johnson M, Bridgham J, Golda G, Lloyd DH, Johnson D, Luo S, McCurdy S, Foy M, Ewan M et al (2000b) Gene expression analysis by massively parallel signature sequencing (MPSS) on microbead arrays. Nat Biotechnol 18:630–634

Brinster RL (1974) The effect of cells transferred into the mouse blastocyst on subsequent development. J Exp Med 140:1049–1056

Cans C, Passer BJ, Shalak V, Nancy-Portebois V, Crible V, Amzallag N, Allanic D, Tufino R, Argentini M, Moras D et al (2003) Translationally controlled tumor protein acts as a guanine nucleotide dissociation inhibitor on the translation elongation factor eEF1A. Proc Natl Acad Sci U S A 100:13892–13897

Ge F, Zhang L, Tao SC, Kitazato K, Zhang ZP, Zhang XE, Bi LJ (2011) Quantitative proteomic analysis of tumor reversion in multiple myeloma cells. J Proteome Res 10:845–855

Hendrix MJ, Seftor EA, Seftor RE, Kasemeier-Kulesa J, Kulesa PM, Postovit LM (2007) Reprogramming metastatic tumour cells with embryonic microenvironments. Nat Rev Cancer 7:246–255

Lespagnol A, Duflaut D, Beekman C, Blanc L, Fiucci G, Marine JC, Vidal M, Amson R, Telerman A (2008) Exosome secretion, including the DNA damage-induced p53-dependent secretory pathway, is severely compromised in TSAP6/Steap3-null mice. Cell Death Differ 15:1723–1733

Li F, Zhang D, Fujise K (2001) Characterization of fortilin, a novel antiapoptotic protein. J Biol Chem 276:47542–47549

Liang P, Pardee AB (1992) Differential display of eukaryotic messenger RNA by means of the polymerase chain reaction. Science 257:967–971

Linares-Cruz G, Bruzzoni-Giovanelli H, Alvaro V, Roperch JP, Tuynder M, Schoevaert D, Nemani M, Prieur S, Lethrosne F, Piouffre L et al (1998) p21WAF-1 reorganizes the nucleus in tumor suppression. Proc Natl Acad Sci U S A 95:1131–1135

MacDonald SM, Rafnar T, Langdon J, Lichtenstein LM (1995) Molecular identification of an IgE-dependent histamine-releasing factor. Science 269:688–690

Macpherson I (1965) Reversion in hamster cells transformed by Rous sarcoma virus. Science 148:1731–1733

Mintz B, Illmensee K (1975) Normal genetically mosaic mice produced from malignant teratocarcinoma cells. Proc Natl Acad Sci U S A 72:3585–3589

Mousset S, Rommelaere J (1982) Minute virus of mice inhibits cell transformation by simian virus 40. Nature 300:537–539

Nemani M, Linares-Cruz G, Bruzzoni-Giovanelli H, Roperch JP, Tuynder M, Bougueleret L, Cherif D, Medhioub M, Pasturaud P, Alvaro V et al (1996) Activation of the human homologue of the Drosophila sina gene in apoptosis and tumor suppression. Proc Natl Acad Sci U S A 93:9039–9042

Passer BJ, Nancy-Portebois V, Amzallag N, Prieur S, Cans C, Roborel de Climens A, Fiucci G, Bouvard V, Tuynder M, Susini L et al (2003) The p53-inducible TSAP6 gene product regulates apoptosis and the cell cycle and interacts with nix and the Myt1 kinase. Proc Natl Acad Sci U S A 100:2284–2289

Pierce GB, Dixon FJ Jr (1959) Testicular teratomas. I. Demonstration of teratogenesis by metamorphosis of multipotential cells. Cancer 12:573–583

Roperch JP, Alvaro V, Prieur S, Tuynder M, Nemani M, Lethrosne F, Piouffre L, Gendron MC, Israeli D, Dausset J et al (1998) Inhibition of presenilin 1 expression is promoted by p53 and p21WAF-1 and results in apoptosis and tumor suppression. Nat Med 4:835–838
Roperch JP, Lethrone F, Prieur S, Piouffre L, Israeli D, Tuynder M, Nemani M, Pasturaud P, Gendron MC, Dausset J et al (1999) SIAH-1 promotes apoptosis and tumor suppression through a network involving the regulation of protein folding, unfolding, and trafficking: identification of common effectors with p53 and p21(Waf1). Proc Natl Acad Sci U S A 96:8070–8073
Susini L, Besse S, Duflaut D, Lespagnol A, Beekman C, Fiucci G, Atkinson AR, Busso D, Poussin P, Marine JC et al (2008) TCTP protects from apoptotic cell death by antagonizing bax function. Cell Death Differ 15:1211–1220
Telerman A, Amson R (2009) The molecular programme of tumour reversion: the steps beyond malignant transformation. Nat Rev Cancer 9:206–216
Telerman A, Tuynder M, Dupressoir T, Robaye B, Sigaux F, Shaulian E, Oren M, Rommelaere J, Amson R (1993a) A model for tumor suppression using H-1 parvovirus. In: 5th Parvovirus workshop, Crystal River, FL, USA, Nov 1993
Telerman A, Tuynder M, Dupressoir T, Robaye B, Sigaux F, Shaulian E, Oren M, Rommelaere J, Amson R (1993b) Selection of a suppressed phenotype in a human leukemia by means of H-1 parvovirus. In: Ninth annual meeting on oncogenes, Frederick, MD, USA, June 1993
Telerman A, Tuynder M, Dupressoir T, Robaye B, Sigaux F, Shaulian E, Oren M, Rommelaere J, Amson R (1993c) A model for tumor suppression using H-1 parvovirus. Proc Natl Acad Sci U S A 90:8702–8706
Telerman A, Nemani M, Roperch JP, Cruz GL, Giovanelli HB, Alvaro V, Israeli D, Medhioub M, Bougueleret L, Cherif D, Le Paslier D, Le Gall I, Dausset J, Calvo F, Oren M, Cohen D, Amson R (1996) The human homologue of the Drosophila seven in absentia gene in apoptosis and tumor suppression. In: Twelfth annual meeting on oncogenes, Frederick, MD, USA, June 1996
Thebault S, Agez M, Chi X, Stojko J, Cura V, Telerman SB, Maillet L, Gautier F, Billas-Massobrio I, Birck C et al (2016) TCTP contains a BH3-like domain, which instead of inhibiting, activates Bcl-xL. Sci Rep 6:19725
Toolan HW (1967) Lack of oncogenic effect of the H-viruses for hamsters. Nature 214:1036
Tuynder M, Lethrosne F, Prieur S, Roperch JP, Susini L, Oren M, Telerman A (2000) Uncovering tumor suppression pathways by investigationg tumor reversion. In: Conference on cancer genetics and tumor suppressor genes. Cold Spring Harbor, Aug 2000
Tuynder M, Susini L, Richardson J, Goodman L, Bowen B, Géant A, Oren M, Amson R, Telerman A (2001a) Characterization of biological models and molecular pathways of tumor reversion: cross-talk with programmed cell death. In: Conference on programmed cell death. Cold Spring Harbor, Sept 2001
Tuynder M, Susini L, Richardson J, Goodman L, Géant A, Dufour F, Mas E, Zyto S, Tolédano J, Oren M, Brenner S, Amson R, Telerman A (2001b) Characterisation of biological models and molecular pathways of tumor reversion. In: Annual meeting on oncogenes, Frederick, MD, USA, June 2001
Tuynder M, Susini L, Prieur S, Besse S, Fiucci G, Amson R, Telerman A (2002) Biological models and genes of tumor reversion: cellular reprogramming through tpt1/TCTP and SIAH-1. Proc Natl Acad Sci U S A 99:14976–14981
Tuynder M, Fiucci G, Prieur S, Lespagnol A, Geant A, Beaucourt S, Duflaut D, Besse S, Susini L, Cavarelli J et al (2004) Translationally controlled tumor protein is a target of tumor reversion. Proc Natl Acad Sci U S A 101:15364–15369
Weaver VM, Petersen OW, Wang F, Larabell CA, Briand P, Damsky C, Bissell MJ (1997) Reversion of the malignant phenotype of human breast cells in three-dimensional culture and in vivo by integrin blocking antibodies. J Cell Biol 137:231–245
Yonish-Rouach E, Resnitzky D, Lotem J, Sachs L, Kimchi A, Oren M (1991) Wild-type p53 induces apoptosis of myeloid leukaemic cells that is inhibited by interleukin-6. Nature 352:345–347

Chapter 2
Structural Insights into TCTP and Its Interactions with Ligands and Proteins

Nadine Assrir, Florian Malard, and Ewen Lescop

Abstract The 19–24 kDa Translationally Controlled Tumor Protein (TCTP) is involved in a wide range of molecular interactions with biological and nonbiological partners of various chemical compositions such as proteins, peptides, nucleic acids, carbohydrates, or small molecules. TCTP is therefore an important and versatile binding platform. Many of these protein–protein interactions have been validated, albeit only few received an in-depth structural characterization. In this chapter, we will focus on the structural analysis of TCTP and we will review the available literature regarding its interaction network from a structural perspective.

2.1 Introduction

This chapter will focus on the structural aspects of TCTP in the context of its wide interaction network, with the aim of being as comprehensive as possible. First we will describe the available structures of TCTP and compare them with other structurally related proteins. Then in a second part, we will discuss the properties of some amino acid regions of TCTP that are important due to their conservation and/or specific functions. Then the last two parts will describe the large interactome of TCTP involving non-proteic or proteic molecules. Two recent reviews (Kawakami et al. 2012; Amson et al. 2013) also covered part of the topics of this chapter. However, the last 4 years have witnessed astonishing progress in TCTP field, and we felt that an updated description of TCTP interactome was necessary. We believe this chapter will be useful not only for the general reader but also for TCTP experts, to overcome the difficulties associated with the multiple names of TCTP found in literature. Indeed, depending on its intra- or extracellular localization, or on the species, TCTP is also called Histamine-Releasing factor (HRF), p23, p21, Q23, fortilin, Mmi1p (yeast), or Tpt1. This confusing nomenclature

---

N. Assrir • F. Malard • E. Lescop (✉)
Institut de Chimie des Substances Naturelles, CNRS UPR2301, Université Paris-Sud, Université Paris-Saclay, 1 avenue de la Terrasse, 91190 Gif-sur-Yvette, France
e-mail: Ewen.LESCOP@cnrs.fr

A. Telerman, R. Amson (eds.), *TCTP/tpt1 - Remodeling Signaling from Stem Cell to Disease*, Results and Problems in Cell Differentiation 64,
DOI 10.1007/978-3-319-67591-6_2

undoubtedly hinders the diffusion of knowledge on this protein within the community and hence slows down the progress of its characterization.

2.2 Sequence and Structure of TCTP

2.2.1 Description of the Structure of TCTP

The high-resolution structures of TCTP from different organisms have been determined by NMR or X-ray crystallography (see Fig. 2.1). These include the malaria parasites *Plasmodium falciparum* (Eichhorn et al. 2013) and *knowlesi* (Vedadi et al. 2007), the yeast *Schizosaccharomyces pombe* (Thaw et al. 2001), the worm *Caenorhabditis elegans* (Lange et al. 2012), and wild type (Feng et al. 2007; Susini et al. 2008) or E12V mutant (Dong et al. 2009) human TCTP. The structures of TCTP are highly conserved between different organisms. TCTP is a monomeric protein, although covalent and non-covalent TCTP dimers have been observed, as discussed in Sect. 2.4.3. The TCTP fold contains three α-helices (α_1, α_2, α_3) and eleven β-strands arranged in two small β-sheets β_2–β_1–β_{11} and β_5–β_6 and a larger β-sheet β_7–β_8–β_9–β_{10}–β_4–β_3 (Fig. 2.1a). The two β-sheets β_2–β_1–β_{11} and β_7–β_8–β_9–β_{10}–β_4–β_3 are twisted and their relative arrangement forms a β-tent (Lupas et al. 2015). The helices α_2 and α_3 are connected by a short loop that creates a kink to form a helical hairpin. This hairpin sits on one side of the large surface defined by the six-stranded β-sheet. The different TCTP structures differ by the secondary structure elements that slightly vary in length and relative positioning. For example, the β-sheet β_2–β_1–β_{11} is severely distorted in *C. elegans* TCTP compared to the other structures. One conserved feature of TCTP structures is the long ~30–33-amino acid loop connecting strands β_5 and β_6 (between residues T39 and V66 in human TCTP sequence, see sequence alignment in Fig. 2.2). This loop is highly flexible as judged from missing electron density in crystal structures (Eichhorn et al. 2013; Vedadi et al. 2007; Susini et al. 2008; Dong et al. 2009) and from the scarcity of long-range NOE restraints in this region leading to poor structural convergence in NMR structures (Thaw et al. 2001; Lange et al. 2012; Feng et al. 2007). The ^{15}N relaxation NMR study revealed that, in human TCTP, the loop explores a wide conformation range at the pico- to nano-second timescale (Feng et al. 2007). The N- and C-extremities of the loop are more or less rigid and tend to form a short β-sheet β_5–β_6 that projects the loop towards the bulk solution and away from the core structure. This is clearly visible in the NMR structures of human and *S. pombe* TCTP (Fig. 2.1b). The NMR structure ensemble of *C. elegans* TCTP (Fig. 2.1b) is more compact and the loop explores a more restricted conformational space, suggesting that a few long-range NOE-derived distances bring the loop in relative close proximity to the core TCTP structure. It is therefore possible that this loop may have distinct dynamic properties in the different species.

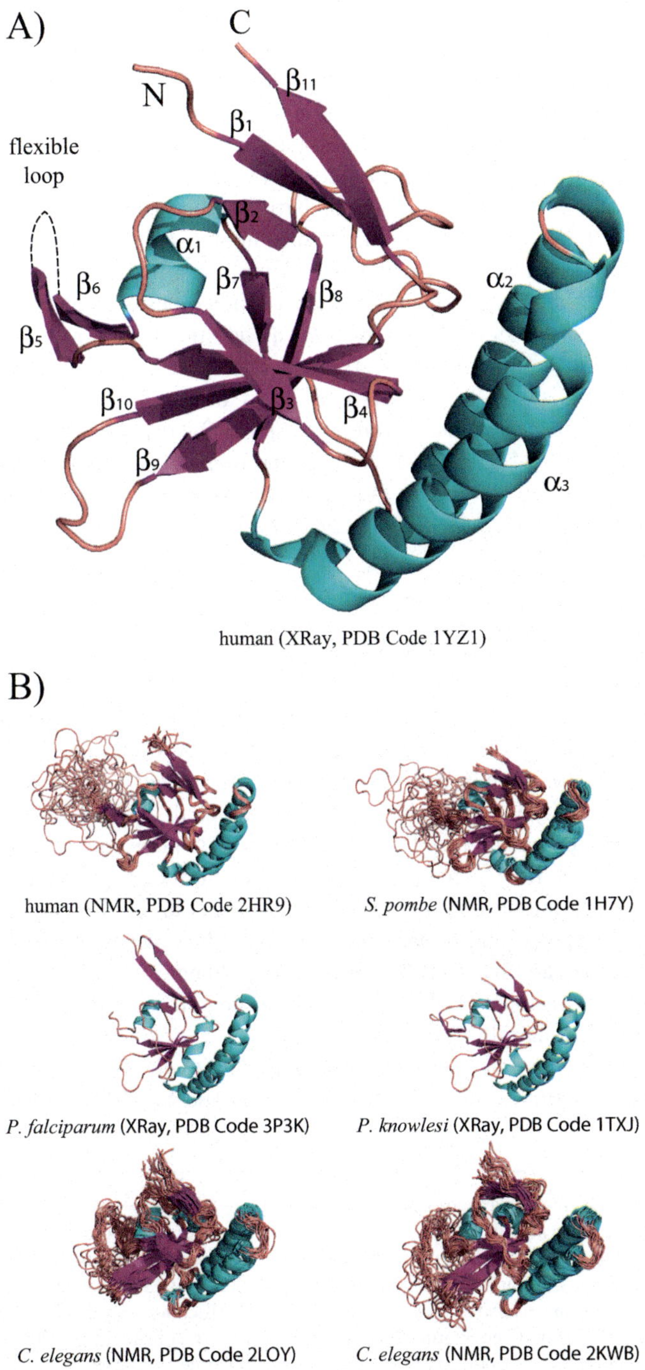

Fig. 2.1 Ribbon representation of the structures of TCTP from different organisms. (**a**) Crystal structure of the human TCTP [PDB Code 1YZ1 (Susini et al. 2008)]. The secondary structure elements are shown using the nomenclature from Fig. 2.2 and the α-helical, β-strand, and coil

TCTP is a highly charged acidic protein with an isoelectric point around 4.5. Accordingly, the 172 amino acid human TCTP contains up to 31 Asp/Glu and 20 Lys/Arg amino acid residues. The vast majority of these charged residues are solvent-exposed at the surface of the protein, making TCTP a highly water-soluble molecule. TCTP is amongst the most abundant proteins in many eukaryotic cells, and the high solubility of TCTP is therefore an important feature. Several charged residues form salt bridges that are partially buried at the surface of the protein. In all TCTP structures, one aspartate residue (D6 in human TCTP sequence) located at the C-terminus of strand β_1 is significantly buried in a hydrophobic environment where it makes hydrogen bond with the main chain amides of I8 and S9 forming an Asx-turn motif on the loop β_1–β_2. D6 forms an additional H-bond with the amide group of M145 at the N-terminus of strand β_9 and, hence, creates contact between the β_2–β_1–β_{11} and β_7–β_8–β_9–β_{10}–β_4–β_3 β-sheets that define the β-tent. Another aspartate (D11), located at the beginning of strand β_2, is also partially buried and form H-bond with the backbone of N139 in the loop β_8–β_9. Both D6 and D11 are strictly conserved, thus revealing their potential roles in the stabilization of the β-tent conformation and consequently of the TCTP fold.

2.2.2 Structural Homologues of TCTP

The long helical hairpin represents a hallmark of TCTP and shares strong structural similarity with other proteins (Susini et al. 2008). These include transmembrane domains of diphtheria toxin and bacterial colicins as well as the helices H5–H6 found in Bcl-2 family proteins, such as Bax (Susini et al. 2008). However, the similarity is restricted to structural features since there is poor amino acid homology between these proteins. The helical hairpin, and in particular residue K102, plays a role in the anti-apoptotic function of TCTP (Susini et al. 2008). Bax has a strong pro-apoptotic property, and remarkably, replacing the essential helices H5–H6 of Bax by helices α_2 and α_3 from TCTP does not change much Bax pro-apoptotic functions suggesting that TCTP helical hairpin can structurally and functionally replace Bax helical hairpin (Susini et al. 2008).

With the determination of the structure of TCTP, it was also realized that TCTP shares strong structural similarities with the MsrB and Mss4/Dss4 families (Thaw

---

Fig. 2.1 (continued) regions are colored *cyan*, *magenta*, and *rose*, respectively. The 30–33 amino acid long loop between strands β_5 and β_6 is not visible in the crystal structure and is indicated as a *dotted line*. Of note, the length of the *dotted line* does not represent the effective length of the loop. (**b**) TCTP structures from human (Feng et al. 2007), fission yeast *Schizosaccharomyces pombe* (Thaw et al. 2001), the parasites *Plasmodium falciparum* (Eichhorn et al. 2013) and *knowlesi* (Vedadi et al. 2007), and the worm *Caenorhabditis elegans* (Lange et al. 2012). For NMR structures, the ensemble of conformations is shown to illustrate the flexibility of the long loop due to the absence or scarcity of experimentally determined long-range distance constraints in this region

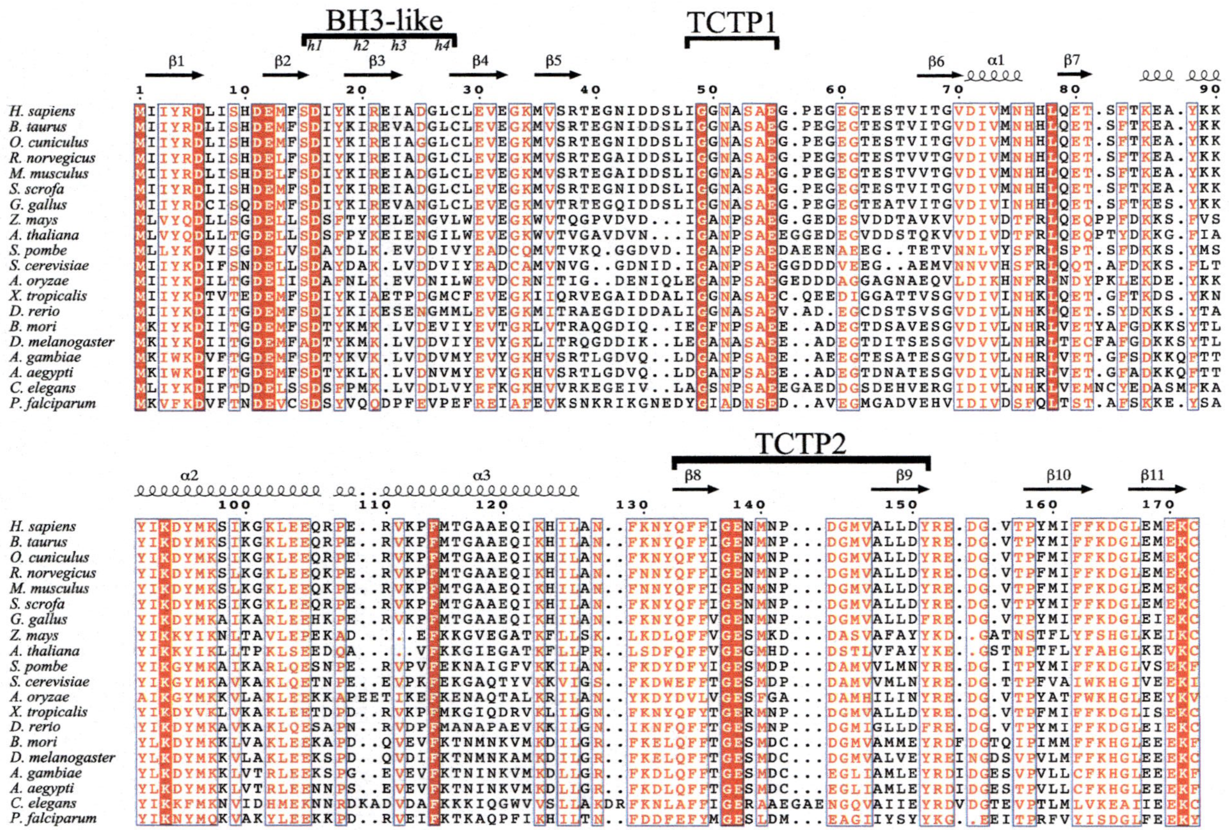

Fig. 2.2 Amino acid alignment of selected TCTP sequences from mammalians, parasites, plants and insects. The amino acid numbering and the secondary structures are from the human sequences (PDB 2HR9) and are shown on the top of the alignment. The picture was prepared using the ESPript webserver (Robert and Gouet 2014) with the Blosum62 color scheme. Specific sequences such as TCTP1 and TCTP2 signatures and the BH3-like domain together with the h1, h2, h3, h4 positions conserved in conventional BH3 domains are highlighted

et al. 2001; Lowther et al. 2002) (see Fig. 2.3f,g). The methionine-R-sulfoxide reductase B (MsrB) is an enzyme involved in the protection of cell against oxidation damages by reducing methionine sulfoxide back to methionine. The Mss4/Dss4 proteins bind the GDP/GTP free form of Rab GTPase proteins and act as a poorly efficient guanine nucleotide exchange factor or guanine nucleotide-free chaperon (Itzen et al. 2006). Despite their different functions, the three protein families share a similar topology. Although the length of the β-strands differs in the different families, they all have in common the two β-sheets forming the β-tent. Nevertheless, some clear variations occur. Firstly, the long flexible loop is absent in MsrB and Mss4/Dss4 proteins and is specific to TCTP. Secondly, the MsrB and Mss4/Dss4 families do not possess the long helical hairpin present in TCTP. In the case of Mss4/Dss4 proteins, an additional two-stranded β-sheet occupies roughly the position where the helical hairpin is located in TCTP structure (see Fig. 2.3d,f, g). In MsrB, a short helical hairpin is present roughly at the same spatial position as the long helical hairpin of TCTP with respect to the β-tent; however, in MsrB the two helices represent insertions at the N-terminus instead of being inserted between strands β_7–β_8 as in TCTP. The size and the position of the helical hairpin in MsrB allows the positioning of the substrate on the solvent accessible surface of the larger β-sheet as seen in Fig. 2.3g. The similarity of TCTP and Mss4 folds has prompted studies to explore the role of TCTP in guanine nucleotide exchange. TCTP was found to be a GDP exchange inhibitor in the elongation step of protein synthesis (Cans et al. 2003). In contrast, it has been proposed to stimulate the GTP/GDP exchange on the Rheb GTP-binding protein to control mTORC1-dependent cell growth and proliferation (Dong et al. 2009; Hsu et al. 2007), although this function has been challenged (Rehmann et al. 2008; Wang et al. 2008).

The Protein DataBank was recently interrogated for structural similarities (Amson et al. 2013; Lupas et al. 2015) and several new TCTP structural homologues were identified, although the proteins shared very little sequence homology: Cereblon, Mis18, RIG-I, and DUF427 (Fig. 2.3). Because structural similarities could reveal hints about yet unknown TCTP function, we will describe those proteins from a structural but also interactome perspective.

Cereblon (CRBN) is a multidomain protein that interacts with the damaged DNA-binding protein 1 (DDB1) and forms one component of the CUL4–RBX1–DDB1–CRBN E3 ubiquitin ligase complex to regulate the selective proteolysis of key proteins in DNA repair, replication, and transcription (Iovine et al. 2011). The N-terminal extremity of Cereblon contains the LON protease domain and the DDB1-binding region whereas the C-terminal extremity contains the CULT domain [Cereblon domain of Unknown activity, binding cellular Ligands and Thalidomide (Lupas et al. 2015)] that shows structural homology to TCTP. Cereblon has been identified as the primary teratogenic target of the well-known thalidomide drug (Ito et al. 2010), and the thalidomide-binding region is located in the CULT domain (Fischer et al. 2014) (see Fig. 2.3a). The binding of the immunomodulatory (IMiD) agents such as thalidomide to Cereblon inhibits ubiquitination of the CUL4–RBX1–DDB1–CRBN E3 ubiquitin ligase substrates and redirects the enzyme towards new protein targets such as the ikaros family of transcription

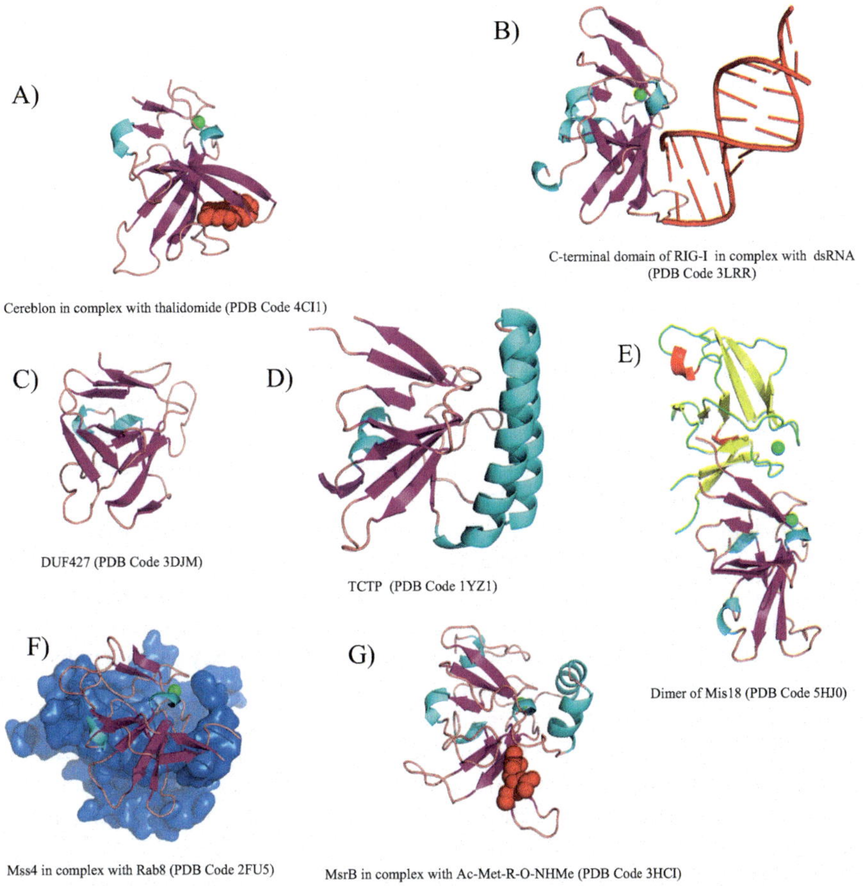

Fig. 2.3 Comparison of structures showing homology to TCTP: TCTP [PDB code 1YZ1 (Susini et al. 2008)], Cereblon in complex with thalidomide [PDB Code 4CI1 (Fischer et al. 2014)], C-terminal domain of RIG-I in complex with dsRNA [PDB 3LRR (Lu et al. 2010)], DUF427 (PDB Code 3DJM, unpublished), the dimer of Mis18 [PDB Code 5HJ0 (Subramanian et al. 2016)], Mss4 in complex with Rab8 [PDB Code 2FU5, Rab8 protein is shown as a *light blue* surface (Itzen et al. 2006)], and MsrB in complex with the substrate Ac-Met-R-O-NHMe [PDB Code 3HCI (Ranaivoson et al. 2009)]. The structures were superimposed on their common β-tent fold. The protein ribbons are colored according to the secondary structure elements with the α-helical, β-strand, and coil regions colored *cyan*, *magenta*, and *rose*, respectively. For the dimeric Mis18, the two monomers are colored with different schemes. When available, zinc ions are represented as *green spheres*, and ligands are shown in *red*

factors IKZF1 and IKZF3 (Fischer et al. 2014) or casein kinase 1α (CK1α) (Kronke et al. 2015). This IMiD-induced reprogramming of CUL4–RBX1–DDB1–CRBN E3 ubiquitin ligase relies on novel interactions at the surface of the IMiD–CULT domain complex (Petzold et al. 2016) that allows new substrate recognition such as CK1α. Cereblon is involved in several protein–protein interactions including

BKCa, ClC-2, AMPK, PSMB4, ikaros and aiolos (IKZF3), and MEIS2 as well as with Ago2 (Xu et al. 2016).

The protein Mis18 is a component of the kinetochore, an essential actor in centromere localization. In *S. pombe*, Mis18 acts as an obligatory homodimeric form mediated by the N-terminal Yippee-like domain that adopts a β-tent conformation (Subramanian et al. 2016) (see Fig. 2.3e). The interface of the dimer is stabilized by strong interactions between the three-stranded β-sheets of the two protomers. An additional α-helix at the C-terminus is involved in tetramerization (not visible in fig 2.3e). In human Mis18, oligomerization is conserved but involves an heterodimer formed between two Mis18 isoforms, Mis18α and Mis18β that share 29% identity (Subramanian et al. 2016). Within the Mis18 complex, the Mis18α and Mis18β have evolved to different functions. Mis18α interacts with the Mis18-Binding Protein 1 (Mis18BP1) through its Yippee-like domain whereas Mis18β interacts with the C terminus of CENP-C also through its Yippee-like domain (Stellfox et al. 2016). The conserved substrate-binding pocket in Mis18 is required for its function although the partners are not known exactly (Subramanian et al. 2016).

The retinoic acid-inducible gene I (RIG-I) and its homologs MDA5 and LGP2 of the RIG-I like receptors (RLR) family recognize replicating viral RNA for the innate antiviral immune response. They possess a helicase domain followed by a C-terminal conserved Yippee-like domain responsible for the binding specificity to double stranded and 5′-triphosphated single stranded RNA. RNA binding induces a major conformational change that releases RLR autoinhibition and results in the activation of type I interferon for the control of viral infection (Leung and Amarasinghe 2012, 2016). The C-terminal region that shows homology to TCTP contributes to RNA recognition through a positively charged groove formed by the large β-sheet and involves interactions essentially with strands β_7–β_8–β_9–β_{10}–β_4–β_3 (Lu et al. 2010; Cui et al. 2008) (see Fig. 2.3b).

The DUF427 protein also belongs to the same structural family as judged from its 3D structure (see Fig. 2.3c). The function of this protein is currently unknown. As proposed before (Lupas et al. 2015), the glutathione-dependent formaldehyde-activating enzyme (GFA) that catalyzes the formation of S-hydroxymethylglutathione from formaldehyde and glutathione also shares some structural similarities with the abovementioned proteins. We did not include this enzyme in the analysis because the similarities are restricted to a structural subdomain.

One common feature of a subset of proteins from the β-tent family is to bind a zinc (Zn^{2+}) ion at the apex of the β-tent. This zinc ion is present in Mis18, Cereblon, RIG-I, and MSS4 and is coordinated by two conserved CXXC motifs. It has been demonstrated that zinc binding is essential for RIG-I in vivo (Cui et al. 2008), and the zinc-binding site may contribute to stabilize the two β-sheets forming the β-tent. In contrast, the cysteines required for zinc binding are largely absent in TCTP and DUF427 and most often are lacking in MsrB. Accordingly, zinc binding has not been reported for these proteins. Consequently, zinc likely does not play any substantial role for these proteins. Alternative processes therefore stabilize the β-tent fold in proteins lacking the zinc-binding site. We proposed that the conserved aspartates D6/D11 could contribute to the TCTP fold (*vide supra*).

Proteins from the β-tent fold family share no detectable sequence homology and have very distinct biological activities and functions. They bind a wide spectrum of compounds ranging from small molecules to nucleic acids and proteins. As noted in a recent survey (Lupas et al. 2015), Cereblon, RIG-I, and MsrB bind partners (small molecules or RNA) through the solvent-exposed large C-terminal β-sheet. This is clearly illustrated in Fig. 2.3 that shows the similar positioning of the ligands (shown in red) sitting on the same face of the large β-sheet. It is likely that the β-tent conformation has converged to expose this binding surface. In TCTP, the binding position is occupied by the helix hairpin, which reinforces the versatility of the C-terminal β-sheet to bind various molecular types, including internal peptide resulting from gene evolution. The presence of the helical hairpin in TCTP clearly hampers binding of TCTP ligands on the C-terminal β-sheet but, at the same time, provides a novel surface formed by the helix hairpin. As shown in Fig. 2.3, other proteins adopting the β-tent fold expose other interfaces for interaction. In Mis18, homo- or heterodimerization occur through the smaller β-sheet. Another example of the versatility of the β-tent to expose binding surface is provided by the structure of the Mss4/Rab8 complex. In this structure, the stretch encompassing helix α_1 and the following strand β_7 from Mss4 is largely involved in the interaction with Rab8. Taken together, these proteins most likely result from a convergent process to adopt the β-tent fold that can accommodate various binding modes and binding partners. Not surprisingly, structural elements that represent extensions when compared to the minimal fold also occupy known binding interface. This is for example illustrated by the helical hairpin of TCTP and the long insertion between strands β_3 and β_4 in Mss4 that both interact with the C-terminal β-sheet. Taken together, TCTP belongs to a large structural family that exposes different binding surfaces and has the ability to interact with molecules of various types. Due to this diversity, it is difficult to predict the interactome and the function for members from this family on the sole basis of the fold.

2.2.3 *Functional Elements Within TCTP Sequence*

2.2.3.1 Conserved Signatures

TCTP amino acid sequences are highly conserved across eukaryotic cells, including in animal and plant kingdom, as well as in yeast (Hinojosa-Moya et al. 2008) (Fig. 2.2). TCTP homologues have also been detected in spider venom, *C. elegans* or various parasites. Two primary regions of high sequence homology were identified and termed TCTP1 and TCTP2 signatures (Thaw et al. 2001). TCTP1 is an eight amino acid sequence (consensus sequence IG[A-G]N[A-P]SAE) located between residues 48 and 55 in the flexible loop (we use here by default human TCTP numbering, see Fig. 2.2) and is largely hydrophilic. Because this region is not involved in the stabilization of the overall fold of the protein, its conservation is most likely related to functional constraints, such as protein–protein interaction or posttranslational modifications. The residue S53 is a predicted phosphorylation site for PI3K kinase and its phosphorylation was observed in human cell line during mitosis (Dephoure et al. 2008), but not confirmed in two other studies (Maeng et al. 2015; Zhang et al. 2012). The conservation of S53 in TCTP1 signature might therefore be related to TCTP regulation. The TCTP2 signature is located between residues 133 and 151 and largely conserved residues are: F134-F135 in strand β_8, G137-E138-M140-D143 in the following β_7–β_8 loop, and Y151 at the C-terminal extremity of strand β_9. Many of these residues are solvent-exposed and do not contribute significantly to the 3D TCTP fold, suggesting that their conservation reveals nonstructural evolutionary pressure. Beyond TCTP1 and TCTP2, other residues are extremely well conserved in TCTP, including D6, D11, E12, D16, L78, K93, F114, V156, and K171. D6 and D11 have already been discussed. E12 has been shown to be essential for protein–protein interaction (Dong et al. 2009; Hsu et al. 2007; Hong and Choi 2016) and for TCTP guanine nucleotide exchange (GEF) activity (Dong et al. 2009; Hsu et al. 2007). The conserved solvent exposed residue K93 in helix α_2 is involved in protein–protein interaction (Wu et al. 2015).

2.2.3.2 Functional Motifs in TCTP

TCTP Contains a Noncanonical Cell-Penetrating Peptide

TCTP is able to spontaneously penetrate cells of various types but also multiple organs (Kim et al. 2011a). This property is associated to a protein transduction domain (PTD) corresponding to the first ten residues of human TCTP (MIIYRDLISH) (Kim et al. 2011a). The internalization seems to involve lipid raft-mediated endocytosis and macropinocytosis (Kim et al. 2011a, 2015). The mechanism is not yet understood and seemingly differs from other known protein transduction domain, in the sense that it does not involve recognition by the cell surface heparin sulfate (Kim et al. 2011a). TCTP–PTD has been advantageously used as a cargo for the internalization of fused peptides or proteins with potential in drug delivery (Bae and Lee 2013; Kim et al. 2011b; Lee et al. 2011). Nevertheless, the biological meaning of the cell-penetrating property of TCTP is not yet clarified. A recent study pointed out that extracellular TCTP is sufficient to reprogram intracellular signaling pathways to promote migration and invasiveness in colorectal cancer cells (Xiao et al. 2016), which strongly supports the idea that TCTP cell penetration may play (patho)physiological roles. In these processes, TCTP import may mirror the exosome-mediated TCTP export process for cell-to-cell communication (Amzallag et al. 2004). From a structural perspective, the PTD domain encompasses the first strand β_1 and the following loop β_1–β_2 that form the central part of the smaller β-sheet. In TCTP structure, most of the side chains of the PTD are accessible for interaction at the surface. However, the 3D conformation of the peptide seems not to be required for cell penetration since the 10-mer TCTP–PTD peptide can efficiently transport various molecules, although it probably lacks stable 3D conformation. Hence, it is not clear whether TCTP remains folded or undergoes severe unfolding during cell internalization.

TCTP Contains a Noncanonical BH3-like Domain

The anti-apoptotic Bcl-xL protein is a partner of TCTP, and the N-terminal fifty residues of TCTP were identified to contribute to TCTP/Bcl-xL interaction (Yang et al. 2005). An in-depth sequence comparison with classical BH3 domains, that are known as Bcl-xL ligands, revealed that TCTP contains a BH3-like domain between residues 16 and 27 (Thebault et al. 2016). BH3 domains usually fold as an α-helix in protein–protein complexes and are characterized by highly conserved residues at positions h1, h2, h3, and h4 that line on one face of the α-helix and that contribute to stabilize the helix in the BH3-binding groove of the partners, such as Bcl-xL. Compared to classical BH3 domains, the TCTP BH3-like domain contains the conserved residues I20, I23, and L27 at h2, h3, and h4 positions, respectively, but lacks the hydrophobic residue commonly found at h1 position in canonical BH3 domains (see Fig. 2.2). In TCTP, the h1 position is occupied by the strictly conserved D16. Accordingly, the structure of Bcl-xL in complex with a peptide derived from $TCTP_{11-31}$ sequence showed that residues 16–27 of TCTP folds into a α-helix that occupies the classical BH3-binding groove of Bcl-xL (Thebault et al. 2016). Surprisingly, instead of decreasing the anti-apoptotic activity of Bcl-xL, as would be expected from competition of TCTP BH3-like domains with canonical BH3 domains at the same binding groove on Bcl-xL, TCTP appears to potentiate the anti-apoptotic activity of Bcl-xL through a yet unknown mechanism (Thebault et al. 2016). Whether TCTP BH3-like domain acts also on other BH3-binding proteins, such as Mcl-1, remains to be investigated.

TCTP Contains an ADP/Cofilin Motif

TCTP interacts with the actin cytoskeleton (Bazile et al. 2009). The comparison of the primary sequences of TCTP and of ADF/cofilin, a family of actin-binding proteins that destabilize actin filaments, unveiled a region of high sequence homology (Tsarova et al. 2010). Indeed, the stretch of residues G69 to E105 that encompasses the helices α_1 and α_2 and the intervening strand β_7 shows significant conservation with the G-actin-binding site of cofilin (Tsarova et al. 2010). Accordingly, TCTP preferentially binds to the globular actin (G-actin) than to filamentous actin (F-actin), but TCTP binding does not alter actin dynamics (Tsarova et al. 2010).

2.2.3.3 Posttranslational Modifications of TCTP

Several posttranslational modifications are predicted on mammalian TCTP. The ELM server (Dinkel et al. 2016) (http://elm.eu.org) predicts for TCTP solvent-exposed regions the following modifications: cleavage sites for caspases 3 and 7, glycosaminoglycan or N-glycosylation attachment site (S53), CK2 phosphorylation sites (residues S9, S37), or Polo-like kinase-1 (Plk1) phosphorylation sites (S46, S64, T65 and S82). Experimentally, only a few posttranslational modifications have been observed.

Biologically important phosphorylations occur at residues S46 and S64. In vivo, the polo-like kinase Plk1 phosphorylates these two serines to detach TCTP from the mitotic spindle for proper mitosis (Yarm 2002). The Plk1-dependent phosphorylation of TCTP contributes to the subcellular localization of TCTP (Yarm 2002; Cucchi et al. 2010; Lucibello et al. 2015). S46 phosphorylation has been proposed to be a biomarker of Plk1 level and kinase activity, with potential interest in antitumor drug design strategy targeting Plk1 (Cucchi et al. 2010) and is observed in mitotic cells (Dephoure et al. 2008). In vitro, the activated Plk1 can phosphorylate TCTP at position S46 but not at position S64 (Johnson et al. 2008). A hierarchical mechanism by which S64 phosphorylation occurs only when S46 is already phosphorylated has also been proposed (Yarm 2002). The serine S46 is conserved in higher eukaryotes whereas S64 is only partially conserved (conserved in mammalian but not in chicken sequences for example, see Fig. 2.2a). Therefore, the impact of S46/S64 phosphorylations is limited to higher eukaryotes (Johnson et al. 2008). Mutations of serines 46 or 64 to glutamate residues abrogate TCTP binding to MDM2 and to the drugs sertraline and thioridazine (Amson et al. 2012). Considering that these mutations mimic phosphoserines, it might indicate that TCTP phosphorylation could also perturb TCTP interactome. Furthermore, it has been proposed that phosphorylated TCTP could be a target of dihydroartemisinin in cancer cells (Lucibello et al. 2015). More recently, the insulin-dependent phosphorylation of S9 and S15 has been reported (Maeng et al. 2015), albeit with yet unknown functional consequences. T39 and S53 phosphorylation have also been observed in mitotic human cell (Dephoure et al. 2008). Phosphorylation of TCTP at definite sites is therefore prone to play important roles in TCTP function.

Beyond phosphorylation, the N-glycosylation of TCTP has been reported (Teshima et al. 1998). The attachment site is not known but S53 is a serious candidate, as judged from ELM predictions. The BioGrid server (http://thebiogrid.org) reports several proteomics studies indicating that TCTP can be ubiquitinated or sumoylated. The Ubc9-mediated sumoylation of TCTP controls its subcellular localization, and the residue K164 was identified as a SUMO-1 substrate (Munirathinam and Ramaswamy 2012). The ubiquitination sites are not precisely known, although K19 and K112 could be potentially ubiquitinated (Kim et al. 2011c).

TCTP is known as IgE-dependent histamine-releasing factor (HRF) when it acts in the extracellular space during the human allergic response. The cytokine-like activity of TCTP seems to correlate with extensive posttranslational modifications that may include proteolytic cleavage, dimerization, or oxidation (Kim et al. 2013). According to the group of Lee (Kim et al. 2013), dimerization is the dominant process that activates TCTP for its extracellular cytokine-like function. TCTP contains two cysteine residues C28/C172, of which C172 seems more important for dimerization (Kim et al. 2009). This can be rationalized by the fact that residue C28 is located at the beginning of strand β_4 and its side chain is completely buried, and in contrast the C-terminal C172 is largely solvent accessible and available for self-association (Kim et al. 2009). Dimers were also observed in the C172S mutant suggesting that intermolecular C28-mediated disulfide bridge also exists (Kim et al. 2009). Dimerization as a posttranslational modification might be required for TCTP recognition by its receptor during allergy (Kim et al. 2009). To date, there is no report of the intracellular existence of such covalent dimer, suggesting that the formation of covalent dimer would be specific of the extracellular function of TCTP. Obviously, the different redox potentials in the intra- and extracellular environments might control the formation of such dimers.

2.3 Binding Properties and Structural Aspects of TCTP in Complex with Ions, Small Molecules, Carbohydrates, Peptides, and Nucleic Acids

Since its discovery, the number of TCTP ligands has continuously increased. TCTP has the ability to interact with ions, small molecules, carbohydrates, nucleic acids, and proteins for its biological functions, and several small molecules or peptides have been designed to interfere with TCTP-based cellular processes. In this part, we will introduce the different TCTP non-proteic ligands (see Table 2.1).

2.3.1 Calcium Binding

Calcium (Ca^{2+}) is one of the first molecules shown to interact with TCTP. The first evidence of calcium binding came in 1992 on the TCTP from *Trypanosoma brucei* parasite (Haghighat and Ruben 1992) and was further extended to other species such as in *Wuchereria bancrofti* (Gnanasekar et al. 2002), *Brugia malayi* (Gnanasekar et al. 2002), *Schistosoma mansoni* (Rao et al. 2002), rat (Kim et al. 2000), and human (Sanchez et al. 1997; Arcuri et al. 2004). Nevertheless, Ca^{2+}-binding is not conserved across the phyla since TCTP from ixodid ticks (Mulenga and Azad 2005) and shrimp (Bangrak et al. 2004) does not bind calcium. The functional relevance of calcium binding to TCTP is not well understood. It has been proposed that TCTP may act as a calcium scavenger in the cytosol to protect cells against Ca^{2+}-dependent apoptosis (Graidist et al. 2007). Accordingly, cells expressing TCTP mutant lacking the ability to bind calcium become more sensitive to thapsigargin-triggered apoptosis (Graidist

Table 2.1 The list of non-proteic molecules interacting with TCTP is shown in this table. The species in which the interactions have been observed and some functional insights into the interaction are also reported

Binding partner	Species	Function and distribution	References
Calcium (Ca^{2+})	*Trypanosoma brucei*, *Schistosoma mansoni*, rat, human	Regulation of Ca^{2+} homeostasis and Ca^{2+}-induced apoptosis	Feng et al. (2007), Haghighat and Ruben (1992), Rao et al. (2002), Kim et al. (2000), Sanchez et al. (1997), Arcuri et al. (2004), Mulenga and Azad (2005), Bangrak et al. (2004), Graidist et al. (2007), Lucas et al. (2014), Xu et al. (1999)
Sertraline/thioridazine/levomepromazine/buclizine	Human	Antihistaminic and antihistaminic-related small molecules.	Amson et al. (2012), Tuynder et al. (2004), Zhang (2014), Seo and Efferth (2016)
Peptides	Human	WGQWPYHC with specific cytotoxicity against tumor cells.	Kadioglu and Efferth (2016)
	Human	WYVYPSM and WEFPGWM against the covalent dimeric TCTP.	Kim et al. (2011d)
Artemisinin and analogues	*Plasmodium falciparum*, human	TCTP is targeted and covalently modified by artemisinin and analogues, possibly at multiple sites from F12 to Y22 in PfTCTP.	Eichhorn et al. (2013), Krishna et al. (2004), Bhisutthibhan and Meshnick (2001), Bhisutthibhan et al. (1999), Bhisutthibhan et al. (1998), Zhou et al. (2016), Li et al. (2016a)
Heme	Human	Heme binds TCTP and promotes its dimerization.	Lucas et al. (2014)
DNA (Sf1 promoter of *oct4*)[a]	*Xenopus* oocyte, human	Activates transcription of oct4 and nanog.	Koziol et al. (2007). See also Cheng et al. (2012)
mRNA[a]	HeLa cells	TCTP belongs to mRNA interactome	Castello et al. (2012)
Chitin, lipopolysaccharide, peptidoglycans, Bb (*Bacillus bombyseptieus*), Sm (*Serratia marcescens*)	*Bombyx mori*	TCTP as a novel opsonic molecule. Induces the production of antimicrobial peptide.	Wang et al. (2013)

[a]The interactions of TCTP with DNA and RNA were observed in oocyte extracts and in HeLa cells, respectively, and have not been confirmed yet in vitro

et al. 2007). Following this idea, the anti-apoptotic role of TCTP could be due to multiple mechanisms including the direct interaction with anti-apoptoptic proteins (including Bcl-xL and Mcl-1) to control their activity but also by preventing the Ca^{2+}-induced permeabilization of the mitochondrial membrane and the resulting release of pro-apoptotic molecules. The sequestering effect for calcium led recently to the hypothesis of a "buffer-like" role for TCTP to regulate cellular homeostasis by avoiding the unwanted excess of soluble ligands (Lucas et al. 2014). The interplay between TCTP and calcium is reinforced by the observation that calcium regulates TCTP at the transcriptional and posttranscriptional levels (Xu et al. 1999).

The binding of calcium was studied by different techniques that gave somehow inconsistent conclusions. In one study (Graidist et al. 2007), two rather high-affinity (~10 μM range) and one lower-affinity binding modes were detected. In contrast, recent NMR (Feng et al. 2007) and fluorescence (Lucas et al. 2014) studies detected low-affinity binding modes (mM range) but not the high-affinity binding modes, although both techniques are sensitive over an extreme wide range of affinity. The apparent discrepancy could be due to different purification protocols or binding conditions. From a structural perspective, the double mutant (E58A/E60A) looses the ability to bind calcium with high affinity (Graidist et al. 2007), indicating that these residues that are located in the long loop are crucial for the interaction. The same study revealed that calcium binding was accompanied by a change in the secondary structure of the protein, as judged from circular dichroism (CD) (Graidist et al. 2007). In the NMR study (Feng et al. 2007), the calcium-binding site was mapped to a region of the protein involving the C-terminal extremities of helix α_3 and of strand β_9 and the loop between strand β_9 and helix α_2. The oxygens from the side chains of residues N131, Q133, and D150 were proposed to coordinate Ca^{2+}. The chemical shift and intensity changes upon calcium binding were rather limited in amplitude and localized to a few amino acid residues. This suggests that TCTP conformation and oligomeric state is well conserved upon Ca^{2+} interaction, thus corroborating another CD study (Lucas et al. 2014) in which no secondary structure nor oligomeric change was observed up to 50 mM calcium concentration. Calcium triggers monomerization of hemin-induced dimerization (Lucas et al. 2014), possibly through direct competition against hemin binding. Indeed, hemin and calcium seem to share a similar binding area on TCTP (Lucas et al. 2014). In their study, Lucas et al. also observed that the presence of calcium contributes to destabilize TCTP by reducing the urea concentration required for denaturation (Lucas et al. 2014).

2.3.2 *Antihistaminic Drugs and the Related Sertraline/Thioridazine*

Any compound leading to reduced TCTP levels in vivo may have potential antitumor activity. Accordingly, because TCTP is a histamine-releasing factor, the group of R. Amson and A. Telerman hypothesized that antihistaminic drugs inhibiting the histaminic pathway were interesting candidates in anticancer strategies (Tuynder et al. 2004). This approach was successful, and a few antihistaminic compounds such as hydroxyzine and promethazine (see Fig. 2.4) proved efficient to kill tumor cells and to decrease the level of TCTP either directly or indirectly (Tuynder et al. 2004). Even greater antitumor activity was reported for the structurally related thioridazine and sertraline (see Fig. 2.4), although they do not display antihistaminic properties (Amson et al. 2012; Tuynder et al. 2004). The drugs thioridazine and sertraline are used for their antipsychotic and antidepressive activities, respectively. The direct interactions of sertraline and thioridazine with TCTP have been confirmed by surface plasmon resonance (SPR) (Amson et al. 2012), and dissociation constant (K_d) of 47 μM and 34 μM was estimated for sertraline and thioridazine, respectively. The interaction was later confirmed for sertraline by thermal shift assays (Zhang 2014). Surprisingly, whereas ligand binding usually stabilizes proteins, sertraline has a destabilizing effect on TCTP by reducing its melting temperature. Both thioridazine and sertraline disrupt the TCTP/MDM2 interaction in vitro which provides a rationale for the restored levels of p53 in cells treated with these pharmacological compounds (Amson et al. 2012).

To date, the structural information on TCTP/ligands complexes is very limited. Two TCTP mutants (S46E and S64E) loose their ability to bind sertraline and thioridazine (Amson et al. 2012), suggesting that the residues S46 and S64 are involved in the interaction with the drugs. Accordingly, the drugs do not interfere with the interactions between MDM2 and TCTP mutants (Amson et al. 2012). As already discussed, residues S46 and S64 are located in the long inserted flexible loop. However whether these residues are directly or indirectly involved in the interaction surface with ligands remains an open question. Additional high-resolution structural information is still awaited to better characterize TCTP/ligand complexes. Interestingly, because these S->E mutants can be seen as phosphoserine mimics, it is possible that Plk1-mediated phosphorylation perturbs TCTP/ligands interaction. Drug design programs targeting TCTP should therefore take into account the potential distinct binding properties of the molecules to phosphorylated and unphosphorylated TCTP in order to inhibit the proper TCTP forms in vivo.

A recent in silico docking study (Seo and Efferth 2016) provided new insights into the molecular interaction of TCTP with 12 antihistaminic compounds. The binding of levomepromazine or buclizine (see Fig. 2.4) was confirmed in vitro by microscale thermophoresis giving dissociation constants of 57 μM and 430 μM, respectively (Seo and Efferth 2016). All tested ligands (except cetirizine) were found to dock onto TCTP at the same position, in an area of the loop encompassing the stretches T39-I48 and E60-T65 that contains both S46 and S64, thus confirming that these two serines could be binding hotspots in TCTP (Seo and Efferth 2016). However, the simulation was carried out on a single conformation of TCTP, and loop flexibility and the potential conformation change of the loop upon binding were not taken into account. Therefore, the binding site derived from this study remains to be confirmed.

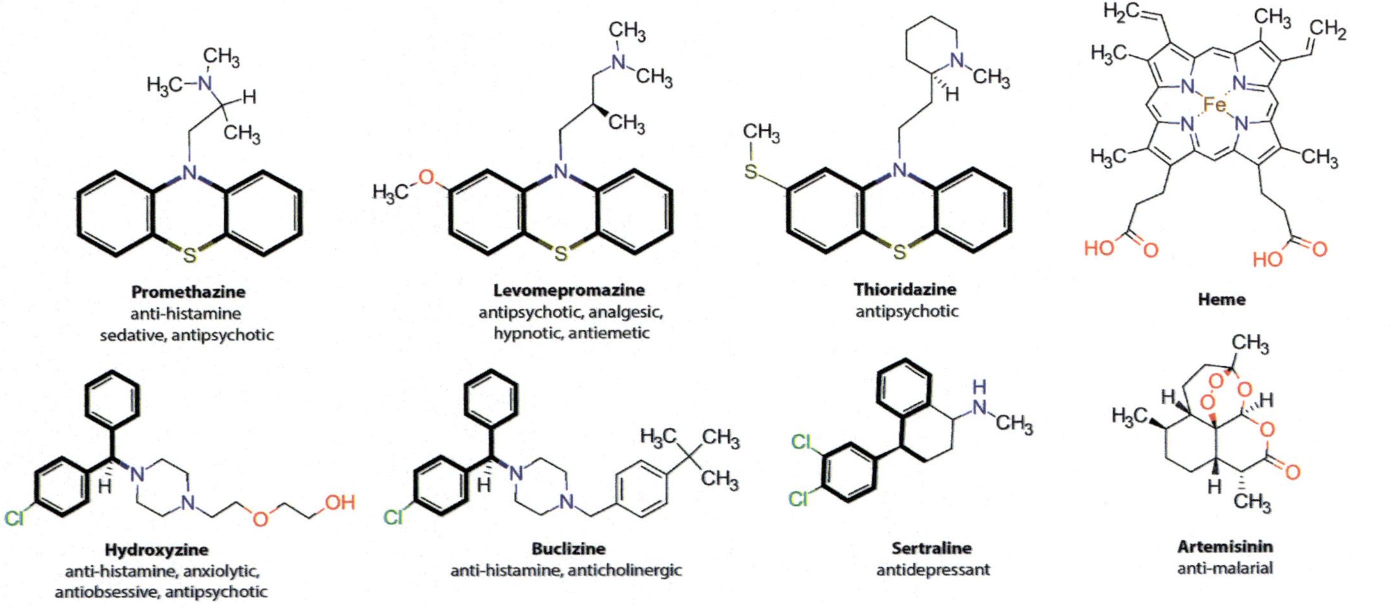

Fig. 2.4 Structures of small molecules discussed in the manuscript. When available, the biological effects of the molecules are reported. We note that several of these molecules are chiral and often found commercially as racemic mixtures

2.3.3 Peptides

TCTP is now an established pharmacological target in cancer- or allergy-related diseases and hence different groups are making efforts to develop peptide-based TCTP inhibitors to interfere with the protein–protein interactions network of TCTP (Kim et al. 2011d; Kadioglu and Efferth 2016).

In their study (Kadioglu and Efferth 2016), Kadioglu and Efferth carried out an in silico screening of peptide libraries and selected octamer peptides with predicted high affinity. The peptide sequences were rather similar with a consensus sequence WGQWPYHX, where the last residue X is the only difference between the different peptides. In spite of the small sequence difference, the docking poses of the different peptides segregated into two families differing by the binding groove. One groove is defined by the bottom of the long flexible loop and strands β_7, β_8, and β_9 on one side of the larger β-sheet whereas the other groove is located on the other side of large β-sheet and includes the C-terminus of helix α_3. Although the binding with TCTP was not confirmed in vitro, one peptide WGQWPYHC induced specific cytotoxicity against tumor cells in a TCTP-dependent manner without affecting normal cells (Kadioglu and Efferth 2016).

In another study (Kim et al. 2011d), the dimeric TCTP was targeted. The covalent dimer is thought to be the active TCTP state in inflammatory processes. With the aim to inhibit TCTP in chronic allergic diseases, three peptides dTBP1/dTBP2/dTBP3 were isolated by screening a phage-displayed 7-mer peptide library. Peptides dTBP2 (WYVYPSM) and dTBP3 (WEFPGWM) were shown to interact with TCTP and with the $TCTP_{84-108}$ peptide corresponding to the helix α_2 in TCTP. The interaction with dTBP2 was demonstrated to be specific to the dimeric versus the monomeric TCTP, and dTBP2 inhibited the cytokine-like effect of TCTP (Kim et al. 2011d). Although they have been designed for different applications and obtained by unrelated approaches, the peptides dTBP1/dTBP2 (Kim et al. 2011d) and WGQWPYHC (Kadioglu and Efferth 2016) show striking similarities: they start by a tryptophan residue and tend to contain an aromatic residue at positions 3 or 4 and a proline residue at positions 4 or 5. However, whether all these peptides share the same binding modes remain to be investigated.

2.3.4 Heme, Artemisinine, and Analogs

P. falciparum TCTP (*Pf*TCTP) is found to be one target of the antimalarial drug artemisinin (Krishna et al. 2004; Bhisutthibhan and Meshnick 2001) and forms complexes with artemisinin and its metabolites. Covalent but also non-covalent and reversible complexes have been reported (Eichhorn et al. 2013; Bhisutthibhan et al. 1998, 1999). The artemisinin-mediated alkylation of TCTP is facilitated by the presence of hemin in particular in a reducing environment (Bhisutthibhan et al. 1998, 1999; Zhou et al. 2016). To date, the exact residues of *Pf*TCTP involved in alkylation are not identified but could be mapped within three peptidic fragments of *Pf*TCTP (Eichhorn et al. 2013). A more recent study showed that multiple amino acid at the N-terminus can be modified by a reactive artemisinin analog and that F12 and C19 are key residues for the interaction (Li et al. 2016a). The reaction is thought to occur through the naturally rare endoperoxide bridge (1,2,4-trioxane structure) that becomes activated by ferrous iron, such as heme, to generate free radicals. The direct binding of heme with human TCTP has been also demonstrated (Lucas et al. 2014) and involves the dyad H76–H77. It was proposed that hemin and calcium shares a common binding pattern on TCTP (Lucas et al. 2014) and, accordingly, competes with each other. Upon complex formation with heme, TCTP forms dimers, which can be easily disrupted by calcium (Lucas et al. 2014). Therefore, ligand binding is prone to conduct to oligomers of TCTP. In *P. falciparum*, heme/TCTP interaction could be important for the fate of artemisinin in the parasite. Indeed, TCTP is associated with the parasite food vacuoles that are rich in hemin, as a product of degradation of hemoglobin by the intra-erythrocytic parasite (Slomianny 1990; Abu Bakar et al. 2010; Klonis et al. 2011). However, it is yet not fully demonstrated if such mechanism can explain the anti-malarial mode of action of artemisinin.

The heme-assisted artemisinin-alkylation of TCTP could potentially affect the various TCTP-related functions. Artemisinin can be effective in cancer (Crespo-Ortiz and Wei 2012; Krishna et al. 2008), and it has been proposed that artemisinin could adopt a similar mode of action in human cells as in parasites. Interestingly, dihydroartemisinin, a metabolite of artemisinin, binds human TCTP in vitro (K_d of 38 μM) and reduces TCTP half-like in a proteasome-dependent manner by increasing its ubiquitination (Fujita et al. 2008). Furthermore, an artemisinin analog targets human TCTP in HeLa cancer cells (Zhou et al. 2016), suggesting that artemisinin might covalently interact with TCTP from different organisms. *Pf*TCTP shares 35% sequence identity with human TCTP, and the structures of the two proteins are very similar (Eichhorn et al. 2013). Therefore, a deeper characterization of the interaction of artemisinin with TCTP and its derivatives would be helpful for a better understanding of its antimalarial activity, which is still largely unknown (O'Neill et al. 2010), but also of the role of TCTP in cancer biology.

2.3.5 Nucleic Acids

TCTP has been isolated from a search for proteins binding to the mouse *oct4* promoter region using radioactively labeled DNA incubated in *Xenopus* oocyte extract. The direct interaction between TCTP and the steroidogenic factor-1 (Sf1) site of *oct4* promoter was demonstrated in vivo from two independent studies carried out in *Xenopus* (Koziol et al. 2007) and in mouse pluripotent cells (Cheng et al. 2012). The first 60 amino acids of TCTP appear to be sufficient for Sf1 binding (Cheng et al. 2012). Three studies assessed the function of TCTP as a transcription factor with diverse outputs (Koziol et al. 2007; Cheng et al. 2012; Johansson and

Simonsson 2010). In one study carried out in *Xenopus* (Koziol et al. 2007), the transcription of a subset of genes including *oct4* was activated by TCTP. In their study, Johansson et al. (Johansson and Simonsson 2010) did not observe a change in *oct4* transcription upon shRNA knockdown of TCTP but observed that TCTP interacts with Oct4 protein in vivo. They proposed a mechanism in which TCTP controls *oct4* transcription by perturbing the self-regulatory transcriptional properties of the Oct4 transcription factor. The third study (Cheng et al. 2012) confirmed the binding of TCTP to the Sf1 site of *oct4* promoter in vivo, but demonstrated that DNA binding of TCTP negatively regulated the expression of Oct4 in mouse pluripotent cells. They proposed that different epigenetic modifications in amphibian oocytes and mammalian cells could explain the conflicting results. We retain from these works that TCTP has also the ability to act as a transcription factor, although the direct interaction between TCTP and DNA has to be confirmed in vitro.

TCTP has been captured in a systematic approach targeting RNA-binding proteins in HeLa cells (Castello et al. 2012). In this study, a "zero-distance" strategy was used to select direct contacts between proteins and RNA and to avoid protein–protein crosslinks. This work therefore suggests that TCTP has also the ability to directly bind RNA *in cellulo* (Castello et al. 2012).

2.3.6 **Bombyx mori** *TCTP as a Binding Platform for Saccharides*

In the silkworm *Bombyx mori*, *Bm*TCTP is produced in intestinal epithelial cells and is released into the hemolymph and gut lumen in response to oral microbial infection (Wang et al. 2013). A study exploring the interaction of *Bm*TCTP with a range of pathogen-associated molecular patterns (PAMP) revealed the broad binding spectrum of *Bm*TCTP (Wang et al. 2013). *Bm*TCTP interacts with chitin, a polymer formed of *N*-acetylglucosamine, and mixtures of *E. coli* lipopolysaccharides (LPS) or of *B. subtilis* peptidoglycans (PG). The binding of TCTP with highly negatively charged bacterial wall molecules was proposed to involve the lysine residues at the surface of TCTP (Wang et al. 2013). *Bm*TCTP also tends to bind bacteria such as *Bacillus bombyseptieus* or *Serratia marcescens*. In response to PAMP, *Bm*TCTP induces the production of antimicrobial peptides through the ERK pathway (Wang et al. 2013). Therefore, *Bm*TCTP contributes to the insect intestinal immunity by acting as opsonin to enhance phagocytosis. To the best of our knowledge, no mammalian TCTP has been reported to date to bind saccharides (other than nucleic acids). Considering that peptide cell-penetration often involves recognition of cell-surface carbohydrates, this study on *Bm*TCTP could inspire future research for a better characterization of the mechanism of TCTP cell penetration.

2.4 Structural Aspects of TCTP in Complex with Proteins

2.4.1 TCTP Directly Interacts with Dozens of Proteins

Over the years, the number of proteins that interact in vivo with TCTP has progressively increased and several dozens of TCTP partners have been identified and further confirmed in vitro, by pull-down assays for example. One review by Amson et al. (2013) reported the extensive list of partners known in 2013. These partners were classified according to their functions as anti-apoptotic, GTPases, p53 axis, cytoskeleton/mitotic machinery, DNA processing and repair, and RNA/ribosome/protein biogenesis. This protein repertoire has continuously expanded to include, for example, proteins such as 14-3-3 (Le et al. 2016), Apaf-1 (Jung et al. 2014), HSPA9 (Li et al. 2016b), YBX1 (Li et al. 2016b), HSP27 (Katsogiannou et al. 2014), peroxiredoxin-1 (Chattopadhyay et al. 2016), ATG16 complex (Chen et al. 2014), nucleolin (Johansson et al. 2010a), or IgE/IgG (Kashiwakura et al. 2012).

Despite the ever-accumulating evidence of the functional importance of TCTP and of its interaction with partners, the amount of structural information regarding protein–protein interaction (PPI) is yet rather limited. Each discovery of novel TCTP-related PPI is often associated to attempts to decipher the molecular basis of the PPI through peptide fragments approaches. In such strategies, peptides derived from the native proteins are designed, and the analysis of the preservation of peptide–peptide contacts leads to the identification of the protein region(s) important for the interaction under scrutiny. The TCTP-related PPI analyzed using peptide fragments were summed up in a review in 2012 (Kawakami et al. 2012). Because several novel interactions have been identified and characterized meanwhile, we propose an updated table of interactions in Table 2.2. Partners identified by coimmunoprecipitation or two-hybrid techniques may be indirect by the implication of a third partner. It is, therefore, crucial to confirm the direct interaction in vitro between recombinant proteins. In this table, two types of interaction were selected amongst the long list of known TCTP partners. On the one hand, we listed interactions confirmed in vitro, whether the biological impact is known or not. On the other hand, we chose interactions with clear biological impacts although the involvement of a third partner is not ruled out yet. The second type was included to foster future in vitro study to confirm biologically relevant interactions.

Table 2.2 clearly illustrates the versatility of TCTP to bind proteins of distinct cellular functions but also biochemical functions (enzymes, DNA/RNA/protein-binding proteins, scaffold proteins, . . .). The consequences of TCTP binding range from direct enzyme activation or inhibition, protein stabilization by promoting or preventing ubiquitination, protein stabilization in response to heat shock, facilitating or hindering the recruitment of other partners, and the control of phosphorylation of the partner. To play all these functions, TCTP evolved to interact with a large interactome and despite its relatively small size, it proposes different binding

Table 2.2 The list of proteic partners identified for TCTP is shown in this table. We have reported here partners for which the interaction has been confirmed in vitro using recombinant proteins. Other interactions waiting for in vitro confirmation have also been included (proteins indicated by a star) when functional information of particular interest has been reported. When available, the amino acids and domains of TCTP and partners involved in the interaction are indicated. nd means “not determined”. When point mutants were shown to disrupt the intermolecular interaction, the single amino acid is reported. In the case of eEF1Bδ, we listed the amino acids that are present at interface of the complex.

	Binding partner	Species	TCTP	Partner	Function and distribution	References
Cell cycle mitosis	α-tubulin	Mouse	70–130	nd	Stabilizes microtubule.	Gachet et al. (1999)
	β-tubulin	Mouse	70–130	nd	Colocalizes with microtubules in the G1, S, G2, and early M phases of cell cycle. Regulates spindle microtubule dynamics and cell shape.	Gachet et al. (1999)
	Centrosome and Microtubules*	*Xenopus laevis*, mammalian cells	nd	nd		Bazile et al. (2009), Jaglarz et al. (2012); Jeon et al. (2016). See also Li et al. (2016b)
	Actin	Rabbit	75–97	nd	TCTP binds preferentially to G-actin than to F-actin.	Tsarova et al. (2010). See also Li et al. (2016b)
	CHFR (Checkpoint protein with FHA and RING finger domain)*	Human	nd	nd	CHFR interacts and colocalizes with TCTP to the mitotic spindle.	Burgess et al. (2008)
	Plk1 (polo-like kinase 1)	Mouse/ human	107–172	Polo box	Plk1 phosphorylates TCTP on S46 and S64, which is required for proper mitosis.	Yarm (2002). See also Johnson et al. (2008)
	Nucleophosmine* (Mpm1)	Mouse	nd	nd	The Mpm1/TCTP complex is a potential biomarker for mitotic ES cells.	Johansson et al. (2010b)
	Nucleolin*	Mouse	nd	nd	Interaction with phosphorylated nucleolin observed during mitosis.	Johansson et al. (2010a)
	Rheb (Ras homolog enriched in brain)	Human/ drosophila	E12, K90, E138	Y35, K45	Preferential binds to nucleotide-free Rheb. Stimulates the GTP/GDP exchange on Rheb to control mTORC1-dependent cell growth and proliferation.	Dong et al. (2009), Hsu et al. (2007). See also Choi and Hsu (2007). Two other studies (Rehmann et al. 2008; Wang et al. 2008) seriously questioned this interaction and function.

	14-3-3 (14-3-3ε or 14-3-3ζ isoforms)	Drosophila	nd	nd	Required for TCTP–Rheb interaction.	Le et al. (2016)
Survival/ apoptosis	p53	Human	70–119	101–300 (DNA-binding domain)	Destabilizes p53 and blocks p53-induced transcriptional activation of Bax.	Rho et al. (2011)
		Human	1–70 + 121–172 + Y4 + E168	103–292 (DNA-binding domain)		Chen et al. (2011)
		Human	nd	99–293 + 294–393		Amson et al. (2012)
	NUMB	Human	1–68	20–259	Competes with NUMB for binding to MDM2.	Amson et al. (2012)
	MDM2 (murine double minute 2)	Human	1–68	134–333+ 302–435	TCTP inhibits MDM2 auto-ubiquitination and promotes MDM2-mediated ubiquitination.	Amson et al. (2012)
		Human	80–133	44–65 + M62		Funston et al. (2012)
	Bcl-xL (B-cell lymphoma-extra large)	Mouse	1–40, I20, R21, E22, D25	1–188, L90, D95	Potentiates the anti-apoptotic activity of Bcl-xL.	Yang et al. (2005)
		Human	14–29, R21	BH3-binding domain		Thebault et al. (2016)
	Mcl-1 (Induced Myeloid Leukemia Cell Differentiation Protein)	Human	5–172, R21	nd	Mcl-1 stabilizes TCTP and TCTP inhibits ubiquitination of Mcl-1.	Zhang et al. (2002)
		Mouse/ human	14–94	K257		Liu et al. (2005)
	Apaf-1 (Apoptotic peptidase activating factor 1)	Human	nd	1–97 (CARD domain)	Inhibits the etoposide-induced cell death.	Jung et al. (2014)
	TSC-22 (Transforming growth factor-beta stimulated clone-22)	Human	1–69	53–110	Prevents TSC-22-mediated apoptosis via the destabilization of TSC-22.	Lee et al. (2008)

(continued)

Table 2.2 (continued)

	Binding partner	Species	TCTP	Partner	Function and distribution	References
Protein synthesis	eEF1A (eukaryotic translation elongation factor 1 alpha)	Human	nd	nd	Preferentially binds GDP form of eEF1A and specifically antagonizes the eEF1Bδ-mediated guanine nucleotide exchange reaction. Partially colocalize around the nucleus.	Cans et al. (2003). See also Leclercq et al. (2011)
	eEF1Bδ	Human	nd	153–281		Cans et al. (2003). See also Langdon et al. (2004)
		Human	F83, K90, I92, K93, D94, M96, K97, K100, M115, T116, A118, A119, I122, M140, P142, D143	155–189 (CAR domain)		Wu et al. (2015) high resolution structural model
	40S ribosomal subunit	Yeast	nd	nd	Copurified with ribosomal complex. Required for efficient translation.	Fleischer et al. (2006)
DNA damage repair	ATM (ataxia telangiectasia mutated)	Drosophila	61–120 + 121–172 + E12	1006–1215	Enhances the binding affinity of dATM to its substrate to promote efficient DNA repair.	Hong and Choi (2013) and related (Zhang et al. 2012)
	Ku70* (aka XRCC6), Ku80*, γH2A.X*	Human	nd	nd	DNA double-strand break sensing and repair.	Zhang et al. (2012), Li et al. (2016b)

Others	TSAP6 (tumor suppressor-activated pathway 6)	Human	nd	nd	TSAP6 augments TCTP secretion through exosome.	Amzallag et al. (2004)
	Na,K-ATPase	Rat/human	102–172	Cytoplasmic domain CD3	Suppresses Na, K-ATPase activity.	Jung et al. (2004)
	Sorting nexin 6 (SNX6)*	Rat/human	nd	nd	Suppresses the TCTP-mediated inhibition of Na, K-ATPase.	Yoon et al. (2006)
	TCTP (self-association)	Rat	126–172	na	nd	Yoon et al. (2000). See also Lucas et al. (2014)
	Vitamin D3 receptor	Human	71–132	197–427	Oxidative stress leads to colocalization in the nucleus.	Rid et al. (2010)
	Fortilin-binding protein 1 (FBP1)	Shrimp	37–63	44–51 + 77–88	Protection against the widespread White Spot Syndrome Virus infection.	Tonganunt et al. (2008), Panrat et al. (2012)
	Tumor suppressor VHL (von Hippel-Lindau)	Human	nd	β domain	TCTP binds to the β domain of VHL through competition with HIF1α, which promotes VHL degradation by the ubiquitin–proteasome system and HIF1α stability.	Chen et al. (2013)
	IgE, IgG (Immunoglobulin E and G)	Mouse/human	1–19, 107–135	Fab	Stimulate IgE-bound mast cells. $K_D = \sim 1$ μM	Kashiwakura et al. (2012). See also Kawakami et al. (2014)
	Peroxiredoxin-1 (PRDX1)	Human	nd	nd	Potentiates the peroxidase activity of Peroxiredoxin-1 by regulating its phosphorylation. K_D of 125 nM.	Chattopadhyay et al. (2016). See also Li et al. (2016b)

(continued)

Table 2.2 (continued)

Binding partner	Species	TCTP	Partner	Function and distribution	References
SWI/SNF (SWItch/ Sucrose Non-Fermentable) and its homologue in drosophila Brahma	Drosophila	E12	304–747 (HSA and BRK domains)	Negatively modulates Brahma activity in transcription.	Hong and Choi (2016), Telerman et al. (2006)
YBX1 (Y-box-binding protein 1)	Human	42–83	1–129	nd	Li et al. (2016b)
Marek's disease virus (MDV)-specific protein R-LORF12	Chicken	nd	nd	nd	Niikura et al. (2004)
Oct4 (octamer-binding transcription factor 4)	Mouse	nd	nd	Controls genes involved in embryonic development.	Johansson and Simonsson (2010)
ATG16 (Autophagy protein 16)*	Human, pig, mouse	nd	nd	Regulation of autophagy	Chen et al. (2014)
TCF-4 (Transcription factor 4)*	Human	nd	nd	Enhances β-catenin/TCF-4 transcription activity.	Gu et al. (2014)
HSP27 (Heat-Shock Protein 27)*	Human	nd	nd	Protects TCTP from the ubiquitin-proteasome degradation.	Baylot et al. (2012) and the related (Zhang 2014)

modes. Table 2.2 suggests that almost all TCTP regions are potentially involved in the direct interaction with partners. For example, when mutated, residues Y4, E12, I20, R21, E22, D25, E138, or E168 abrogate binding to a range of partners. All these residues cover a wide surface on TCTP, reinforcing the idea that TCTP does not expose a unique interface for interaction. This parallels the many binding modes observed for the proteins from the β-tent family (see Fig. 2.3).

2.4.2 Structural Information on Native Complexes

The strategy consisting in deleting large portions of protein is extremely efficient when it comes to isolate interacting domains from multidomain proteins. This approach is also useful to identify short peptide fragments from independent folding units such as protein globular domains, in particular when these fragments folds as helices at the interface of protein–protein complex. Nevertheless, short fragments might not properly fold or keep the same 3D conformation as in the native protein. In such situation, the peptide fragment approach is prone to give false negative results. Oppositely, the disruption of the 3D native fold in short fragments is prone to facilitate nonnative interaction, leading to potential false positive results. These limitations may explain some discrepancies observed in the dissection of TCTP interactions as reported in Table 2.2, such as with the p53 or MDM2 partners. For these reasons, proper interaction analysis are better carried out with native proteins, preferentially with full length proteins or at least by preserving folding units, followed by point mutations. To date, our understanding of TCTP interactions using native proteins is limited to the TCTP/eEF1Bδ complex, for which a high-resolution structure have been obtained from a mixed approach based on classical NMR-based structure determination followed by molecular docking driven by experimental NMR data (Wu et al. 2015) (see Fig. 2.5a). This structure was validated by extensive site-directed mutagenesis and highlighted the role of the helical hairpin (Site I) and of a surface patch (Site II) formed by the stretch connecting helices α_1 and α_2 and including strand β_7 (around F83) and the loop β_8–β_9 (containing M140 and P142) in the interaction with the CAR domain from eEF1Bδ (Fig. 2.5a). In the complex, the negatively charged N-terminus of CAR adopts an extended conformation that wraps around the positively charged helical hairpin, including residues K90, I92, K93, M96, K97, K100, M115, T116, A118, A119, and I122. At its C-terminus, CAR adopts an α-helical conformation that docks on a surface overlapping sites I and II through hydrophobic contacts with F83, M140, P142 and electrostatic contacts with D143 and D94. Electrostatic and hydrophobic interactions both contribute to stabilize the complex.

This work provides the most convincing study of the key role of the helical hairpin in TCTP PPI. It is most likely that this structure element is also involved in other PPIs as suggested from other studies (Amson et al. 2012; Kashiwakura et al. 2012; Rid et al. 2010; Gachet et al. 1999; Rho et al. 2011; Funston et al. 2012), with strong functional impacts such as in apoptosis (Susini et al. 2008). Another hint

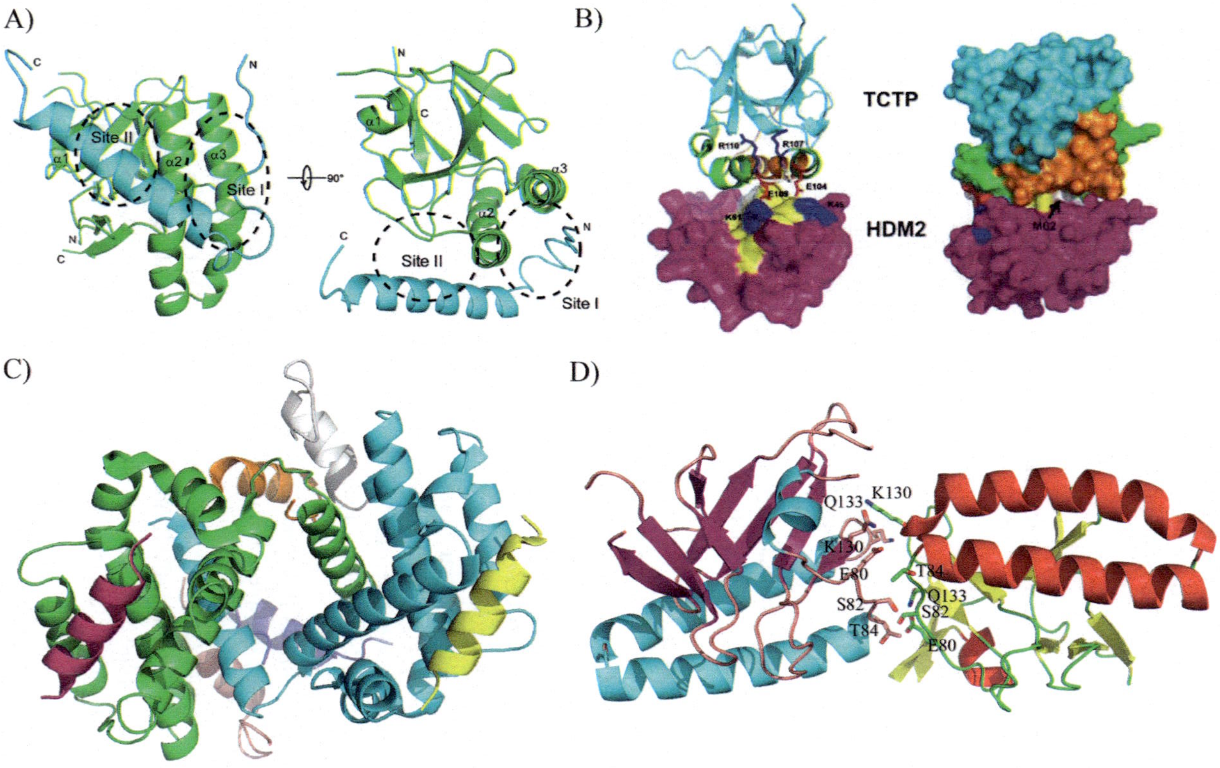

Fig. 2.5 High-resolution structures or structural models of TCTP in interaction with protein partners. (**a**) NMR-based structure of TCTP in interaction with the CAR domain of eEF1Bδ (Wu et al. 2015). The two proteins are shown as *ribbons* with TCTP and eEF1Bδ in *green* and *cyan*, respectively. Two orthogonal views are

about the key role of the helical hairpin is provided by a structural model of the complex formed by TCTP and the N-terminal domain of MDM2 (Fig. 2.5b). The molecular interface of this complex has been well characterized on MDM2 side by competition with with a well-known MDM2 binder (nutlin-3) or by using MDM2 mutants (M62A). However, the interface on TCTP still awaits validation with high resolution structural data from native protein and/or from point mutations. This could also resolve the somehow inconsistent results obtained from independent studies on TCTP/MDM2 interaction (Amson et al. 2012; Funston et al. 2012).

The TCTP/Bcl-xL complex is the only TCTP complex for which a crystal structure has been obtained (Fig. 2.5c). The TCTP/Bcl-xL complex with full-length proteins could be purified but only a complex of Bcl-xL with a peptide derived from the TCTP BH3-like domain was crystallized. This structure illustrates how the TCTP BH3-like peptide binds to the BH3-binding groove of Bcl-xL (Thebault et al. 2016). The crystal contained a swapped dimeric Bcl-xL, which appears to be an hallmark of this protein. Six TCTP peptides were observed, although only two of them were considered as significant. This work (Thebault et al. 2016) provided sound basis to demonstrate the existence of a functional BH3-like element in TCTP, but at the same time, raises novel questions. The BH3-like domain (residues 16 to 27) folds as strand β_3 in the native TCTP structure and undergoes a severe conformational rearrangement in the complex with Bcl-xL. Because residues 16 to 27 play a key role in stabilizing TCTP larger β-sheet, it is difficult to predict the impact of the global conformation change of TCTP in the complex: is strand β_3 the only element being affected? does TCTP completely unfold and remain unfolded at the exception of the helical BH3 region? does TCTP remold into another stable conformation unrelated to the fold of the unbound protein? does other region than the BH3 region of TCTP interact with Bcl-xL? Because TCTP binds at the Bax-binding groove of Bcl-xL, one would expect that TCTP competes with Bax to inhibit the Bax-mediated anti-apoptotic activity of Bcl-xL. Paradoxically, TCTP appears to potentiate the anti-apoptotic activity of Bcl-xL (Thebault et al. 2016). The apparently counter-intuitive structural mechanism by which TCTP activates Bcl-xL remains largely unknown. Clearly, additional structural, thermodynamic, and kinetic studies on full-length proteins are required to assess the plasticity of TCTP and to understand its role to control apoptosis.

Fig. 2.5 (continued) shown. Picture taken with permission from Wu et al. (2015). (**b**) Energy-minimized structural model of the complex between TCTP and the N-terminal domain of human MDM2 obtained from molecular docking based on peptide deletion analysis and MDM2 M62A mutant (Funston et al. 2012). Picture taken with permission from Funston et al. (2012). (**c**) Structure of the full-length Bcl-xL in complex with the BH3-like domain of TCTP (residues 16–27) [PDB code 4Z9V (Thebault et al. 2016)]. The proteins are shown as *ribbons*. Bcl-xL (*cyan* and *green*) crystallized as a swap dimer and six BH3-like peptides were observed (*magenta*, *yellow*, *orange*, *grey*, *salmon*, *marine*). Only the *magenta* and *yellow* peptides were considered meaningful, while the others were considered as crystallization artifacts (Thebault et al. 2016). (**d**) Potential TCTP dimer structure. The chains A and D from the 1YZ1 crystal structure of TCTP are represented as *ribbons*. The two monomers have different color codes for their secondary structures. The amino acid residues at the dimer interface are labeled

2.4.3 TCTP Tends to Self-associate

TCTP tends to self-associate. For example, non-covalent dimers/oligomers of rat TCTP have been detected from yeast two-hybrid system (Yoon et al. 2000). The deletion of the region 126–172 resulted in loss of self-interaction of TCTP in vivo (Yoon et al. 2000). The authors concluded that this region was involved in oligomer formation. The amino acids 126–172 encompass the last four β-strands of TCTP that largely contribute to create the large β-sheet at the core of TCTP structure. Hence, its deletion is prone to severely impact the proper folding of TCTP and its oligomerization properties. It is therefore highly possible that the peptide missing residues 126–172 is not able to oligomerize because of the global unfolding of the protein, which questions the involvement of these residues in oligomerization of the native TCTP. As noted before, heme tends to favor TCTP dimer in vitro with potential role in cellular homeostasis (Lucas et al. 2014). The details of the dimer interface have not been investigated yet. In the analysis of TCTP structures so far available, we have noticed that human wild-type and E12V mutant TCTP crystallize with four molecules in the unit cell although they were obtained in different space groups. Interestingly, in the two crystals, an intermolecular interface and a relative protein orientation were clearly conserved within a pair of molecules. This is illustrated by the contacts between chains A and D from the human wild-type structure (Fig. 2.5d). Because this self-association mode is observed in different crystal packings, it is possible that the interactions at this interface of these two molecules are strong enough to exist in solution. We propose that the complex observed between chains A and D could represent the structure of TCTP dimer in solution. The proposed interface is formed by the hydrophilic and charged residues such as E80, S82, T84, Q130, and K133 as highlighted in Fig. 2.5d.

2.5 Conclusions

In this chapter, we have presented the structural features of TCTP with a focus on its posttranslational modifications and interaction network. Despite its relative small size and globular nature, the TCTP structure is extremely versatile and is able to interact with ions, small molecules, carbohydrates, nucleic acids, and proteins. Although high-resolution structures and more precise delineation of complex interfaces are still required, it seems that most TCTP surface patches are potential binding hotspots, which might be the hallmark of proteins from the β-tent family.

To date, our knowledge of the structural property of TCTP in interaction is still very limited. Although the structure of TCTP is apparently highly stable, there are some lines of evidence that TCTP is prone to major rearrangement upon interaction. On the one hand, TCTP interacts with Bcl-xL by the BH3-like domain that is partially buried in the unbound TCTP, suggesting that TCTP undergoes a severe conformational change in the complex. This could explain the difficulty to form the

TCTP/Bcl-xL complex (Thebault et al. 2016). In this regard, TCTP also interacts with the Bcl-xL-related Mcl-1, although it is not clear yet if the interaction is also mediated by the BH3-like domain. The formation of TCTP/Mcl-1 complex is also rather difficult and is greatly facilitated by the truncation of the first ten residues (Liu et al. 2005). It is likely that the removal of the N-terminal residues prevents the proper folding of TCTP. Thus, by alleviating the kinetically unfavorable unfolding barrier, this truncated TCTP form probably already exposes interacting residues, possibly in the BH3-like region, thus making the formation of the complex easier. Such a truncated form was also proved to be more active to trigger the IgE response (Kim et al. 2009). Shortened TCTP has not yet been observed in vivo but a TCTP isoform lacking the first 34 amino acids is reported by UniProt (P13693-2), which could have variable binding properties compared to the canonical TCTP isoform. On the other hand, TCTP penetrates cells by using a protein transduction domain that folds as a β-strand. It is possible that TCTP also undergoes a significant fold rearrangement during cell entry to facilitate recognition and internalization. Future researches are necessary to confirm these hypotheses. Importantly, it will be crucial to identify the molecular triggers, to assess the extent of the conformational change, and to assess the functional consequences of the rearrangement on the partners. We speculate that TCTP plasticity greatly contributes to the versatility of its effects on partners and to its multifunctional nature.

One intriguing feature of TCTP is the 30–33 amino acid long flexible loop. This region is very well conserved throughout the phylum, both in length and amino acid composition and contains the TCTP1 signature at its center. Compared to structured regions, flexible regions are in general less under evolution constraints of keeping structurally important amino acids and are therefore prone to vary in length and composition. The conservation of the loop in TCTP therefore suggests that other forces drive its conservation during evolution. In particular, one may wonder if the loop directly interacts with partners, if it controls the access to other binding hotspots on TCTP, or if it contributes to the proposed conformational rearrangement. For this, it is crucial to characterize complexes between native proteins, with a focus on the dynamics of this loop. The loop might be involved in other regulatory events such as the phosphorylation of serines S46, S53, or S64. Being part of TCTP1 signature, S53 is strictly conserved and its phosphorylation could regulate biological functions shared throughout the phylum. In contrast, S46 and S64 are found only in mammalian TCTP and most likely regulate mammalian-specific functions. The impact of phosphorylation on the dynamics of the loop and, beyond, on the structure of the protein will also provide insights into the role of the loop.

Although our knowledge of TCTP functions has greatly expanded over the last years, much remains to be done to characterize the biochemical and structural features of TCTP. No doubt that such gain in knowledge will contribute to decipher the multiple functions of TCTP in physiological and pathophysiological processes.

References

Abu Bakar N, Klonis N, Hanssen E, Chan C, Tilley L (2010) Digestive-vacuole genesis and endocytic processes in the early intraerythrocytic stages of Plasmodium falciparum. J Cell Sci 123(Pt 3):441–450

Amson R, Pece S, Lespagnol A, Vyas R, Mazzarol G, Tosoni D, Colaluca I, Viale G, Rodrigues-Ferreira S, Wynendaele J, Chaloin O, Hoebeke J, Marine JC, Di Fiore PP, Telerman A (2012) Reciprocal repression between P53 and TCTP. Nat Med 18(1):91–99

Amson R, Pece S, Marine JC, Di Fiore PP, Telerman A (2013) TPT1/TCTP-regulated pathways in phenotypic reprogramming. Trends Cell Biol 23(1):37–46

Amzallag N, Passer BJ, Allanic D, Segura E, Thery C, Goud B, Amson R, Telerman A (2004) TSAP6 facilitates the secretion of translationally controlled tumor protein/histamine-releasing factor via a nonclassical pathway. J Biol Chem 279(44):46104–46112

Arcuri F, Papa S, Carducci A, Romagnoli R, Liberatori S, Riparbelli MG, Sanchez JC, Tosi P, del Vecchio MT (2004) Translationally controlled tumor protein (TCTP) in the human prostate and prostate cancer cells: expression, distribution, and calcium binding activity. Prostate 60(2): 130–140

Bae HD, Lee K (2013) On employing a translationally controlled tumor protein-derived protein transduction domain analog for transmucosal delivery of drugs. J Control Release 170(3): 358–364

Bangrak P, Graidist P, Chotigeat W, Phongdara A (2004) Molecular cloning and expression of a mammalian homologue of a translationally controlled tumor protein (TCTP) gene from Penaeus monodon shrimp. J Biotechnol 108(3):219–226

Baylot V, Katsogiannou M, Andrieu C, Taieb D, Acunzo J, Giusiano S, Fazli L, Gleave M, Garrido C, Rocchi P (2012) Targeting TCTP as a new therapeutic strategy in castration-resistant prostate cancer. Mol Ther 20(12):2244–2256

Bazile F, Pascal A, Arnal I, Le Clainche C, Chesnel F, Kubiak JZ (2009) Complex relationship between TCTP, microtubules and actin microfilaments regulates cell shape in normal and cancer cells. Carcinogenesis 30(4):555–565

Bhisutthibhan J, Meshnick SR (2001) Immunoprecipitation of [(3)H]dihydroartemisinin translationally controlled tumor protein (TCTP) adducts from Plasmodium falciparum-infected erythrocytes by using anti-TCTP antibodies. Antimicrob Agents Chemother 45(8):2397–2399

Bhisutthibhan J, Pan XQ, Hossler PA, Walker DJ, Yowell CA, Carlton J, Dame JB, Meshnick SR (1998) The Plasmodium falciparum translationally controlled tumor protein homolog and its reaction with the antimalarial drug artemisinin. J Biol Chem 273(26):16192–16198

Bhisutthibhan J, Philbert MA, Fujioka H, Aikawa M, Meshnick SR (1999) The Plasmodium falciparum translationally controlled tumor protein: subcellular localization and calcium binding. Eur J Cell Biol 78(9):665–670

Burgess A, Labbe JC, Vigneron S, Bonneaud N, Strub JM, Van Dorsselaer A, Lorca T, Castro A (2008) Chfr interacts and colocalizes with TCTP to the mitotic spindle. Oncogene 27(42): 5554–5566

Cans C, Passer BJ, Shalak V, Nancy-Portebois V, Crible V, Amzallag N, Allanic D, Tufino R, Argentini M, Moras D, Fiucci G, Goud B, Mirande M, Amson R, Telerman A (2003) Translationally controlled tumor protein acts as a guanine nucleotide dissociation inhibitor on the translation elongation factor eEF1A. Proc Natl Acad Sci USA 100(24):13892–13897

Castello A, Fischer B, Eichelbaum K, Horos R, Beckmann BM, Strein C, Davey NE, Humphreys DT, Preiss T, Steinmetz LM, Krijgsveld J, Hentze MW (2012) Insights into RNA biology from an atlas of mammalian mRNA-binding proteins. Cell 149(6):1393–1406

Chattopadhyay A, Pinkaew D, Doan HQ, Jacob RB, Verma SK, Friedman H, Peterson AC, Kuyumcu-Martinez MN, McDougal OM, Fujise K (2016) Fortilin potentiates the peroxidase activity of Peroxiredoxin-1 and protects against alcohol-induced liver damage in mice. Sci Rep 6:18701

Chen Y, Fujita T, Zhang D, Doan H, Pinkaew D, Liu Z, Wu J, Koide Y, Chiu A, Lin CC, Chang JY, Ruan KH, Fujise K (2011) Physical and functional antagonism between tumor suppressor protein p53 and fortilin, an anti-apoptotic protein. J Biol Chem 286(37):32575–32585

Chen K, Chen S, Huang C, Cheng H, Zhou R (2013) TCTP increases stability of hypoxia-inducible factor 1alpha by interaction with and degradation of the tumour suppressor VHL. Biol Cell 105(5):208–218

Chen K, Huang C, Yuan J, Cheng H, Zhou R (2014) Long-term artificial selection reveals a role of TCTP in autophagy in mammalian cells. Mol Biol Evol 31(8):2194–2211

Cheng X, Li J, Deng J, Li Z, Meng S, Wang H (2012) Translationally controlled tumor protein (TCTP) downregulates Oct4 expression in mouse pluripotent cells. BMB Rep 45(1):20–25

Choi KW, Hsu YC (2007) To cease or to proliferate: new insights into TCTP function from a Drosophila study. Cell Adhes Migr 1(3):129–130

Crespo-Ortiz MP, Wei MQ (2012) Antitumor activity of artemisinin and its derivatives: from a well-known antimalarial agent to a potential anticancer drug. J Biomed Biotechnol 2012: 247597

Cucchi U, Gianellini LM, De Ponti A, Sola F, Alzani R, Patton V, Pezzoni A, Troiani S, Saccardo MB, Rizzi S, Giorgini ML, Cappella P, Beria I, Valsasina B (2010) Phosphorylation of TCTP as a marker for polo-like kinase-1 activity in vivo. Anticancer Res 30(12):4973–4985

Cui S, Eisenacher K, Kirchhofer A, Brzozka K, Lammens A, Lammens K, Fujita T, Conzelmann KK, Krug A, Hopfner KP (2008) The C-terminal regulatory domain is the RNA 5′-triphosphate sensor of RIG-I. Mol Cell 29(2):169–179

Dephoure N, Zhou C, Villen J, Beausoleil SA, Bakalarski CE, Elledge SJ, Gygi SP (2008) A quantitative atlas of mitotic phosphorylation. Proc Natl Acad Sci USA 105(31):10762–10767

Dinkel H, Van Roey K, Michael S, Kumar M, Uyar B, Altenberg B, Milchevskaya V, Schneider M, Kuhn H, Behrendt A, Dahl SL, Damerell V, Diebel S, Kalman S, Klein S, Knudsen AC, Mader C, Merrill S, Staudt A, Thiel V, Welti L, Davey NE, Diella F, Gibson TJ (2016) ELM 2016–data update and new functionality of the eukaryotic linear motif resource. Nucleic Acids Res 44(D1):D294–D300

Dong X, Yang B, Li Y, Zhong C, Ding J (2009) Molecular basis of the acceleration of the GDP-GTP exchange of human ras homolog enriched in brain by human translationally controlled tumor protein. J Biol Chem 284(35):23754–23764

Eichhorn T, Winter D, Buchele B, Dirdjaja N, Frank M, Lehmann WD, Mertens R, Krauth-Siegel RL, Simmet T, Granzin J, Efferth T (2013) Molecular interaction of artemisinin with translationally controlled tumor protein (TCTP) of Plasmodium falciparum. Biochem Pharmacol 85(1):38–45

Feng Y, Liu D, Yao H, Wang J (2007) Solution structure and mapping of a very weak calcium-binding site of human translationally controlled tumor protein by NMR. Arch Biochem Biophys 467(1):48–57

Fischer ES, Bohm K, Lydeard JR, Yang H, Stadler MB, Cavadini S, Nagel J, Serluca F, Acker V, Lingaraju GM, Tichkule RB, Schebesta M, Forrester WC, Schirle M, Hassiepen U, Ottl J, Hild M, Beckwith RE, Harper JW, Jenkins JL, Thoma NH (2014) Structure of the DDB1-CRBN E3 ubiquitin ligase in complex with thalidomide. Nature 512(7512):49–53

Fleischer TC, Weaver CM, McAfee KJ, Jennings JL, Link AJ (2006) Systematic identification and functional screens of uncharacterized proteins associated with eukaryotic ribosomal complexes. Genes Dev 20(10):1294–1307

Fujita T, Felix K, Pinkaew D, Hutadilok-Towatana N, Liu Z, Fujise K (2008) Human fortilin is a molecular target of dihydroartemisinin. FEBS Lett 582(7):1055–1060

Funston G, Goh W, Wei SJ, Tng QS, Brown C, Jiah Tong L, Verma C, Lane D, Ghadessy F (2012) Binding of translationally controlled tumour protein to the N-terminal domain of HDM2 is inhibited by nutlin-3. PLoS One 7(8):e42642

Gachet Y, Tournier S, Lee M, Lazaris-Karatzas A, Poulton T, Bommer UA (1999) The growth-related, translationally controlled protein P23 has properties of a tubulin binding protein and associates transiently with microtubules during the cell cycle. J Cell Sci 112(Pt 8):1257–1271

Gnanasekar M, Rao KV, Chen L, Narayanan RB, Geetha M, Scott AL, Ramaswamy K, Kaliraj P (2002) Molecular characterization of a calcium binding translationally controlled tumor protein homologue from the filarial parasites Brugia malayi and Wuchereria bancrofti. Mol Biochem Parasitol 121(1):107–118

Graidist P, Yazawa M, Tonganunt M, Nakatomi A, Lin CC, Chang JY, Phongdara A, Fujise K (2007) Fortilin binds Ca2+ and blocks Ca2+-dependent apoptosis in vivo. Biochem J 408(2): 181–191

Gu X, Yao L, Ma G, Cui L, Li Y, Liang W, Zhao B, Li K (2014) TCTP promotes glioma cell proliferation in vitro and in vivo via enhanced beta-catenin/TCF-4 transcription. Neuro-oncology 16(2):217–227

Haghighat NG, Ruben L (1992) Purification of novel calcium binding proteins from Trypanosoma brucei: properties of 22-, 24- and 38-kilodalton proteins. Mol Biochem Parasitol 51(1):99–110

Hinojosa-Moya J, Xoconostle-Cazares B, Piedra-Ibarra E, Mendez-Tenorio A, Lucas WJ, Ruiz-Medrano R (2008) Phylogenetic and structural analysis of translationally controlled tumor proteins. J Mol Evol 66(5):472–483

Hong ST, Choi KW (2013) TCTP directly regulates ATM activity to control genome stability and organ development in Drosophila melanogaster. Nat Commun 4:2986

Hong ST, Choi KW (2016) Antagonistic roles of Drosophila Tctp and Brahma in chromatin remodelling and stabilizing repeated sequences. Nat Commun 7:12988

Hsu YC, Chern JJ, Cai Y, Liu M, Choi KW (2007) Drosophila TCTP is essential for growth and proliferation through regulation of dRheb GTPase. Nature 445(7129):785–788

Iovine B, Iannella ML, Bevilacqua MA (2011) Damage-specific DNA binding protein 1 (DDB1): a protein with a wide range of functions. Int J Biochem Cell Biol 43(12):1664–1667

Ito T, Ando H, Suzuki T, Ogura T, Hotta K, Imamura Y, Yamaguchi Y, Handa H (2010) Identification of a primary target of thalidomide teratogenicity. Science 327(5971):1345–1350

Itzen A, Pylypenko O, Goody RS, Alexandrov K, Rak A (2006) Nucleotide exchange via local protein unfolding--structure of Rab8 in complex with MSS4. EMBO J 25(7):1445–1455

Jaglarz MK, Bazile F, Laskowska K, Polanski Z, Chesnel F, Borsuk E, Kloc M, Kubiak JZ (2012) Association of TCTP with centrosome and microtubules. Biochem Res Int 2012:541906

Jeon HJ, You SY, Park YS, Chang JW, Kim JS, Oh JS (2016) TCTP regulates spindle microtubule dynamics by stabilizing polar microtubules during mouse oocyte meiosis. Biochim Biophys Acta 1863(4):630–637

Johansson H, Simonsson S (2010) Core transcription factors, Oct4, Sox2 and Nanog, individually form complexes with nucleophosmin (Npm1) to control embryonic stem (ES) cell fate determination. Aging 2(11):815–822

Johansson H, Svensson F, Runnberg R, Simonsson T, Simonsson S (2010a) Phosphorylated nucleolin interacts with translationally controlled tumor protein during mitosis and with Oct4 during interphase in ES cells. PLoS One 5(10):e13678

Johansson H, Vizlin-Hodzic D, Simonsson T, Simonsson S (2010b) Translationally controlled tumor protein interacts with nucleophosmin during mitosis in ES cells. Cell Cycle 9(11): 2160–2169

Johnson TM, Antrobus R, Johnson LN (2008) Plk1 activation by Ste20-like kinase (Slk) phosphorylation and polo-box phosphopeptide binding assayed with the substrate translationally controlled tumor protein (TCTP). Biochemistry 47(12):3688–3696

Jung J, Kim M, Kim MJ, Kim J, Moon J, Lim JS, Kim M, Lee K (2004) Translationally controlled tumor protein interacts with the third cytoplasmic domain of Na,K-ATPase alpha subunit and inhibits the pump activity in HeLa cells. J Biol Chem 279(48):49868–49875

Jung J, Kim HY, Maeng J, Kim M, Shin DH, Lee K (2014) Interaction of translationally controlled tumor protein with Apaf-1 is involved in the development of chemoresistance in HeLa cells. BMC Cancer 14:165

Kadioglu O, Efferth T (2016) Peptide aptamer identified by molecular docking targeting translationally controlled tumor protein in leukemia cells. Invest New Drugs 34(4):515–521

Kashiwakura JC, Ando T, Matsumoto K, Kimura M, Kitaura J, Matho MH, Zajonc DM, Ozeki T, Ra C, MacDonald SM, Siraganian RP, Broide DH, Kawakami Y, Kawakami T (2012) Histamine-releasing factor has a proinflammatory role in mouse models of asthma and allergy. J Clin Invest 122(1):218–228

Katsogiannou M, Andrieu C, Baylot V, Baudot A, Dusetti NJ, Gayet O, Finetti P, Garrido C, Birnbaum D, Bertucci F, Brun C, Rocchi P (2014) The functional landscape of Hsp27 reveals new cellular processes such as DNA repair and alternative splicing and proposes novel anticancer targets. Mol Cell Proteomics 13(12):3585–3601

Kawakami T, Ando T, Kawakami Y (2012) HRF-interacting molecules. Open Allergy J 5(41–46)

Kawakami T, Kashiwakura J, Kawakami Y (2014) Histamine-releasing factor and immunoglobulins in asthma and allergy. Allergy Asthma Immunol Res 6(1):6–12

Kim M, Jung Y, Lee K, Kim C (2000) Identification of the calcium binding sites in translationally controlled tumor protein. Arch Pharm Res 23(6):633–636

Kim M, Min HJ, Won HY, Park H, Lee JC, Park HW, Chung J, Hwang ES, Lee K (2009) Dimerization of translationally controlled tumor protein is essential for its cytokine-like activity. PLoS One 4(7):e6464

Kim M, Kim M, Kim HY, Kim S, Jung J, Maeng J, Chang J, Lee K (2011a) A protein transduction domain located at the NH2-terminus of human translationally controlled tumor protein for delivery of active molecules to cells. Biomaterials 32(1):222–230

Kim HY, Kim S, Youn H, Chung JK, Shin DH, Lee K (2011b) The cell penetrating ability of the proapoptotic peptide, KLAKLAKKLAKLAK fused to the N-terminal protein transduction domain of translationally controlled tumor protein, MIIYRDLISH. Biomaterials 32(22):5262–5268

Kim W, Bennett EJ, Huttlin EL, Guo A, Li J, Possemato A, Sowa ME, Rad R, Rush J, Comb MJ, Harper JW, Gygi SP (2011c) Systematic and quantitative assessment of the ubiquitin-modified proteome. Mol Cell 44(2):325–340

Kim M, Chung J, Lee C, Jung J, Kwon Y, Lee K (2011d) A peptide binding to dimerized translationally controlled tumor protein modulates allergic reactions. J Mol Med 89(6):603–610

Kim M, Maeng J, Lee K (2013) Dimerization of TCTP and its clinical implications for allergy. Biochimie 95(4):659–666

Kim HY, Kim S, Pyun HJ, Maeng J, Lee K (2015) Cellular uptake mechanism of TCTP-PTD in human lung carcinoma cells. Mol Pharm 12(1):194–203

Klonis N, Crespo-Ortiz MP, Bottova I, Abu-Bakar N, Kenny S, Rosenthal PJ, Tilley L (2011) Artemisinin activity against Plasmodium falciparum requires hemoglobin uptake and digestion. Proc Natl Acad Sci U S A 108(28):11405–11410

Koziol MJ, Garrett N, Gurdon JB (2007) Tpt1 activates transcription of oct4 and nanog in transplanted somatic nuclei. Curr Biol 17(9):801–807

Krishna S, Uhlemann AC, Haynes RK (2004) Artemisinins: mechanisms of action and potential for resistance. Drug Resist Updates 7(4–5):233–244

Krishna S, Bustamante L, Haynes R, Staines H (2008) Artemisinins: their growing importance in medicine. Trends Pharmacol Sci 29(10):520–527

Kronke J, Fink EC, Hollenbach PW, MacBeth KJ, Hurst SN, Udeshi ND, Chamberlain PP, Mani DR, Man HW, Gandhi AK, Svinkina T, Schneider RK, McConkey M, Jaras M, Griffiths E, Wetzler M, Bullinger L, Cathers BE, Carr SA, Chopra R, Ebert BL (2015) Lenalidomide induces ubiquitination and degradation of CK1alpha in del(5q) MDS. Nature 523(7559): 183–188

Langdon JM, Vonakis BM, MacDonald SM (2004) Identification of the interaction between the human recombinant histamine releasing factor/translationally controlled tumor protein and elongation factor-1 delta (also known as elongation factor-1B beta). Biochim Biophys Acta 1688(3):232–236

Lange OF, Rossi P, Sgourakis NG, Song Y, Lee HW, Aramini JM, Ertekin A, Xiao R, Acton TB, Montelione GT, Baker D (2012) Determination of solution structures of proteins up to 40 kDa using CS-Rosetta with sparse NMR data from deuterated samples. Proc Natl Acad Sci U S A 109(27):10873–10878

Le TP, Vuong LT, Kim AR, Hsu YC, Choi KW (2016) 14-3-3 proteins regulate Tctp-Rheb interaction for organ growth in Drosophila. Nat Commun 7:11501

Leclercq TM, Moretti PA, Pitson SM (2011) Guanine nucleotides regulate sphingosine kinase 1 activation by eukaryotic elongation factor 1A and provide a mechanism for eEF1A-associated oncogenesis. Oncogene 30(3):372–378

Lee JH, Rho SB, Park SY, Chun T (2008) Interaction between fortilin and transforming growth factor-beta stimulated clone-22 (TSC-22) prevents apoptosis via the destabilization of TSC-22. FEBS Lett 582(8):1210–1218

Lee J, Kim S, Shin DH, Kim HJ, Lee K (2011) Neuroprotective effect of Cu,Zn-superoxide dismutase fused to a TCTP-derived protein transduction domain. Eur J Pharmacol 666(1–3): 87–92

Leung DW, Amarasinghe GK (2012) Structural insights into RNA recognition and activation of RIG-I-like receptors. Curr Opin Struct Biol 22(3):297–303

Leung DW, Amarasinghe GK (2016) When your cap matters: structural insights into self vs non-self recognition of 5′ RNA by immunomodulatory host proteins. Curr Opin Struct Biol 36:133–141

Li W, Zhou Y, Tang G, Xiao Y (2016a) Characterization of the artemisinin binding site for translationally controlled tumor protein (TCTP) by bioorthogonal click chemistry. Bioconjug Chem 27(12):2828–2833

Li S, Chen M, Xiong Q, Zhang J, Cui Z, Ge F (2016b) Characterization of the translationally controlled tumor protein (TCTP) interactome reveals novel binding partners in human cancer cells. J Proteome Res 15:3741–3751

Liu H, Peng HW, Cheng YS, Yuan HS, Yang-Yen HF (2005) Stabilization and enhancement of the antiapoptotic activity of mcl-1 by TCTP. Mol Cell Biol 25(8):3117–3126

Lowther WT, Weissbach H, Etienne F, Brot N, Matthews BW (2002) The mirrored methionine sulfoxide reductases of Neisseria gonorrhoeae pilB. Nat Struct Biol 9(5):348–352

Lu C, Xu H, Ranjith-Kumar CT, Brooks MT, Hou TY, Hu F, Herr AB, Strong RK, Kao CC, Li P (2010) The structural basis of 5′ triphosphate double-stranded RNA recognition by RIG-I C-terminal domain. Structure 18(8):1032–1043

Lucas AT, Fu X, Liu J, Brannon MK, Yang J, Capelluto DG, Finkielstein CV (2014) Ligand binding reveals a role for heme in translationally-controlled tumor protein dimerization. PLoS One 9(11):e112823

Lucibello M, Adanti S, Antelmi E, Dezi D, Ciafre S, Carcangiu ML, Zonfrillo M, Nicotera G, Sica L, De Braud F, Pierimarchi P (2015) Phospho-TCTP as a therapeutic target of Dihydroartemisinin for aggressive breast cancer cells. Oncotarget 6(7):5275–5291

Lupas AN, Zhu H, Korycinski M (2015) The thalidomide-binding domain of cereblon defines the CULT domain family and is a new member of the beta-tent fold. PLoS Comput Biol 11(1): e1004023

Maeng J, Kim M, Lee H, Lee K (2015) Insulin induces phosphorylation of serine residues of translationally controlled tumor protein in 293T cells. Int J Mol Sci 16(4):7565–7576

Mulenga A, Azad AF (2005) The molecular and biological analysis of ixodid ticks histamine release factors. Exp Appl Acarol 37(3–4):215–229

Munirathinam G, Ramaswamy K (2012) Sumoylation of human translationally controlled tumor protein is important for its nuclear transport. Biochem Res Int 2012:831940

Niikura M, Liu HC, Dodgson JB, Cheng HH (2004) A comprehensive screen for chicken proteins that interact with proteins unique to virulent strains of Marek's disease virus. Poult Sci 83(7): 1117–1123

O'Neill PM, Barton VE, Ward SA (2010) The molecular mechanism of action of artemisinin–the debate continues. Molecules 15(3):1705–1721

Panrat T, Sinthujaroen P, Nupan B, Wanna W, Tammi MT, Phongdara A (2012) Characterization of a novel binding protein for Fortilin/TCTP–component of a defense mechanism against viral infection in Penaeus monodon. PLoS One 7(3):e33291

Petzold G, Fischer ES, Thoma NH (2016) Structural basis of lenalidomide-induced CK1alpha degradation by the CRL4(CRBN) ubiquitin ligase. Nature 532(7597):127–130

Ranaivoson FM, Neiers F, Kauffmann B, Boschi-Muller S, Branlant G, Favier F (2009) Methionine sulfoxide reductase B displays a high level of flexibility. J Mol Biol 394(1):83–93

Rao KV, Chen L, Gnanasekar M, Ramaswamy K (2002) Cloning and characterization of a calcium-binding, histamine-releasing protein from Schistosoma mansoni. J Biol Chem 277 (34):31207–31213

Rehmann H, Bruning M, Berghaus C, Schwarten M, Kohler K, Stocker H, Stoll R, Zwartkruis FJ, Wittinghofer A (2008) Biochemical characterisation of TCTP questions its function as a guanine nucleotide exchange factor for Rheb. FEBS Lett 582(20):3005–3010

Rho SB, Lee JH, Park MS, Byun HJ, Kang S, Seo SS, Kim JY, Park SY (2011) Anti-apoptotic protein TCTP controls the stability of the tumor suppressor p53. FEBS Lett 585(1):29–35

Rid R, Onder K, Trost A, Bauer J, Hintner H, Ritter M, Jakab M, Costa I, Reischl W, Richter K, MacDonald S, Jendrach M, Bereiter-Hahn J, Breitenbach M (2010) H2O2-dependent translocation of TCTP into the nucleus enables its interaction with VDR in human keratinocytes: TCTP as a further module in calcitriol signalling. J Steroid Biochem Mol Biol 118(1–2):29–40

Robert X, Gouet P (2014) Deciphering key features in protein structures with the new ENDscript server. Nucleic Acids Res 42(Web Server issue):W320–W324

Sanchez JC, Schaller D, Ravier F, Golaz O, Jaccoud S, Belet M, Wilkins MR, James R, Deshusses J, Hochstrasser D (1997) Translationally controlled tumor protein: a protein identified in several nontumoral cells including erythrocytes. Electrophoresis 18(1):150–155

Seo EJ, Efferth T (2016) Interaction of antihistaminic drugs with human translationally controlled tumor protein (TCTP) as novel approach for differentiation therapy. Oncotarget 7(13): 16818–16839

Slomianny C (1990) Three-dimensional reconstruction of the feeding process of the malaria parasite. Blood Cells 16(2–3):369–378

Stellfox ME, Nardi IK, Knippler CM, Foltz DR (2016) Differential binding partners of the Mis18alpha/beta YIPPEE domains regulate Mis18 complex recruitment to centromeres. Cell Rep 15(10):2127–2135

Subramanian L, Medina-Pritchard B, Barton R, Spiller F, Kulasegaran-Shylini R, Radaviciute G, Allshire RC, Arockia Jeyaprakash A (2016) Centromere localization and function of Mis18 requires Yippee-like domain-mediated oligomerization. EMBO Rep 17(4):496–507

Susini L, Besse S, Duflaut D, Lespagnol A, Beekman C, Fiucci G, Atkinson AR, Busso D, Poussin P, Marine JC, Martinou JC, Cavarelli J, Moras D, Amson R, Telerman A (2008) TCTP protects from apoptotic cell death by antagonizing bax function. Cell Death Differ 15(8): 1211–1220

Telerman A, Amson R, Cans C, Nancy-Portebois V, Passer BJ (2006) US Patent 20060140970 (06/29/2006)

Teshima S, Rokutan K, Nikawa T, & Kishi K (1998) Macrophage colony-stimulating factor stimulates synthesis and secretion of a mouse homolog of a human IgE-dependent histamine-releasing factor by macrophages in vitro and in vivo. J Immunol 161(11):6356–6366

Thaw P, Baxter NJ, Hounslow AM, Price C, Waltho JP, Craven CJ (2001) Structure of TCTP reveals unexpected relationship with guanine nucleotide-free chaperones. Nat Struct Biol 8(8): 701–704

Thebault S, Agez M, Chi X, Stojko J, Cura V, Telerman SB, Maillet L, Gautier F, Billas-Massobrio I, Birck C, Troffer-Charlier N, Karafin T, Honore J, Senff-Ribeiro A, Montessuit S, Johnson CM, Juin P, Cianferani S, Martinou JC, Andrews DW, Amson R, Telerman A, Cavarelli J (2016) TCTP contains a BH3-like domain, which instead of inhibiting, activates Bcl-xL. Sci Rep 6:19725

Tonganunt M, Nupan B, Saengsakda M, Suklour S, Wanna W, Senapin S, Chotigeat W, Phongdara A (2008) The role of Pm-fortilin in protecting shrimp from white spot syndrome virus (WSSV) infection. Fish Shellfish Immunol 25(5):633–637

Tsarova K, Yarmola EG, Bubb MR (2010) Identification of a cofilin-like actin-binding site on translationally controlled tumor protein (TCTP). FEBS Lett 584(23):4756–4760

Tuynder M, Fiucci G, Prieur S, Lespagnol A, Geant A, Beaucourt S, Duflaut D, Besse S, Susini L, Cavarelli J, Moras D, Amson R, Telerman A (2004) Translationally controlled tumor protein is a target of tumor reversion. Proc Natl Acad Sci U S A 101(43):15364–15369

Vedadi M, Lew J, Artz J, Amani M, Zhao Y, Dong A, Wasney GA, Gao M, Hills T, Brokx S, Qiu W, Sharma S, Diassiti A, Alam Z, Melone M, Mulichak A, Wernimont A, Bray J, Loppnau P, Plotnikova O, Newberry K, Sundararajan E, Houston S, Walker J, Tempel W, Bochkarev A, Kozieradzki I, Edwards A, Arrowsmith C, Roos D, Kain K, Hui R (2007) Genome-scale protein expression and structural biology of Plasmodium falciparum and related Apicomplexan organisms. Mol Biochem Parasitol 151(1):100–110

Wang X, Fonseca BD, Tang H, Liu R, Elia A, Clemens MJ, Bommer UA, Proud CG (2008) Re-evaluating the roles of proposed modulators of mammalian target of rapamycin complex 1 (mTORC1) signaling. J Biol Chem 283(45):30482–30492

Wang F, Hu C, Hua X, Song L, Xia Q (2013) Translationally controlled tumor protein, a dual functional protein involved in the immune response of the silkworm, Bombyx mori. PLoS One 8(7):e69284

Wu H, Gong W, Yao X, Wang J, Perrett S, Feng Y (2015) Evolutionarily conserved binding of translationally controlled tumor protein to eukaryotic elongation factor 1B. J Biol Chem 290(14):8694–8710

Xiao B, Chen D, Luo S, Hao W, Jing F, Liu T, Wang S, Geng Y, Li L, Xu W, Zhang Y, Liao X, Zuo D, Wu Y, Li M, Ma Q (2016) Extracellular translationally controlled tumor protein promotes colorectal cancer invasion and metastasis through Cdc42/JNK/MMP9 signaling. Oncotarget 7(31):50057–50073

Xu A, Bellamy AR, Taylor JA (1999) Expression of translationally controlled tumour protein is regulated by calcium at both the transcriptional and post-transcriptional level. Biochem J 342:683–689

Xu Q, Hou YX, Langlais P, Erickson P, Zhu J, Shi CX, Luo M, Zhu Y, Xu Y, Mandarino LJ, Stewart K, Chang XB (2016) Expression of the cereblon binding protein argonaute 2 plays an important role for multiple myeloma cell growth and survival. BMC Cancer 16:297

Yang Y, Yang F, Xiong Z, Yan Y, Wang X, Nishino M, Mirkovic D, Nguyen J, Wang H, Yang XF (2005) An N-terminal region of translationally controlled tumor protein is required for its antiapoptotic activity. Oncogene 24(30):4778–4788

Yarm FR (2002) Plk phosphorylation regulates the microtubule-stabilizing protein TCTP. Mol Cell Biol 22(17):6209–6221

Yoon T, Jung J, Kim M, Lee KM, Choi EC, Lee K (2000) Identification of the self-interaction of rat TCTP/IgE-dependent histamine-releasing factor using yeast two-hybrid system. Arch Biochem Biophys 384(2):379–382

Yoon T, Kim M, Lee K (2006) Inhibition of Na,K-ATPase-suppressive activity of translationally controlled tumor protein by sorting nexin 6. FEBS Lett 580(14):3558–3564

Zhang X (2014) Etude de complexes protéine-protéine impliquant la chaperone de bas poids moléculaire HSP27: Implications dans le cancer de la prostate. PhD Thesis, Aix-Marseille Université

Zhang D, Li F, Weidner D, Mnjoyan ZH, Fujise K (2002) Physical and functional interaction between myeloid cell leukemia 1 protein (MCL1) and Fortilin. The potential role of MCL1 as a fortilin chaperone. J Biol Chem 277(40):37430–37438

Zhang J, de Toledo SM, Pandey BN, Guo G, Pain D, Li H, Azzam EI (2012) Role of the translationally controlled tumor protein in DNA damage sensing and repair. Proc Natl Acad Sci U S A 109(16):E926–E933

Zhou Y, Li W, Xiao Y (2016) Profiling of multiple targets of artemisinin activated by hemin in cancer cell proteome. ACS Chem Biol 11(4):882–888

Chapter 3
Structure-Function Relationship of TCTP

Beatriz Xoconostle-Cázares and Roberto Ruiz-Medrano

Abstract The translationally controlled tumor protein (TCTP) is a small, multifunctional protein found in most, if not all, eukaryotic lineages, involved in a myriad of key regulatory processes. Among these, the control of proliferation and inhibition of cell death, as well as differentiation, are the most important, and it is probable that other responses are derived from the ability of TCTP to influence them in both unicellular and multicellular organisms. In the latter, an additional function for TCTP stems from its capacity to be secreted via a nonclassical pathway and function in a non-cell autonomous (paracrine) manner, thus affecting the responses of neighboring or distant cells to developmental or environmental stimuli (as in the case of serum TCTP/histamine-releasing factor in mammals and phloem TCTP in *Arabidopsis*). The additional ability to traverse membranes without a requirement for transmembrane receptors adds to its functional flexibility. The long-distance transport of TCTP mRNA and protein in plants via the vascular system supports the notion that an important aspect of TCTP function is its ability to influence the response of neighboring and distant cells to endogenous and exogenous signals in a supracellular manner. The predicted tridimensional structure of TCTPs indicates a high degree of conservation, more than its amino acid sequence similarity could suggest. However, subtle differences in structure could lead to different activities, as evidenced by TCTPs secreted by *Plasmodium* spp. Similar structural variations in animal and plant TCTPs, likely the result of convergent evolution, could lead to deviations from the canonical function of this group of proteins, which could have an impact from a biomedical and agricultural perspectives.

R. Ruiz-Medrano (✉)
Department of Biotechnology and Bioengineering, Center for Research and Advanced Studies of the National Polytechnic Institute, Avenida IPN 2508, Colonia San Pedro Zacatenco, México City 07360, México
e-mail: rmedrano@cinvestav.mx

A. Telerman, R. Amson (eds.), *TCTP/tpt1 - Remodeling Signaling from Stem Cell to Disease*, Results and Problems in Cell Differentiation 64,
DOI 10.1007/978-3-319-67591-6_3

3.1 Introduction

There are multiple mechanisms through which cells regulate proliferation and differentiation. Some genes are specific to certain taxa, but the function of others is quite conserved among phylogenetically distant groups. These genes hint to a core mechanism controlling such fundamental processes; although, it is expected that unicellular organisms should have a simpler machinery underlying cell division and differentiation, recent evidence indicates that, at least in some eukaryotic lineages, co-option of components of a preexisting machinery, such as the retinoblastoma pathway, gave rise to multicellularity (Hanschen et al. 2016). Similarly, the evolution of multicellularity in animals required the emergence of novel genes but also the co-option of preexisting ones (Suga et al. 2013). Furthermore, basic gene modules have been adapted for novel functions during evolution, to which modified versions of these genes have been added, as in the evolution of the plant vascular system (Martínez-Navarro et al. 2013).

In eukaryotes, the translationally controlled tumor protein, or TCTP, is a central regulator of cell division, differentiation, and a host of other essential processes, being present in most extant eukaryote taxa analyzed to date. The messenger RNA for TCTP was initially identified in an erythroleukemia cell line and, as its name implies, accumulated to high levels without the corresponding accumulation of its encoded protein (Bommer and Thiele 2004; Amson et al. 2013). Translational regulation may occur with TCTP mRNAs from different taxa, but this has not been determined experimentally in most cases. High levels of accumulation of this transcript have been observed in several species (mostly animals), which is proportionally correlated with proliferation. Moreover, the repression of this gene is linked to tumor reversion (Tuynder et al. 2004). Indeed, increased accumulation of TCTP has been observed in different types of cancer, correlating with poor prognosis (Chan et al. 2012; Xiao et al. 2016). However, multitude of studies indicate that TCTP is involved in several other phenomena (Amson et al. 2013) and thus supports the notion that these proteins function as adapters within core modules involved in regulating a variety of processes. These include cell proliferation and differentiation, as well as others that may be termed idiosyncratic, since these are taxon, developmental, or tissue specific. For example, some regulate development of reproductive structures in plants, as well as plant regeneration and lateral roots. Still others may be involved in certain types of molecular mimicry in some parasites. Thus, the study of these proteins may yield significant insight on a wealth of phenomena, some of which could appear unrelated but are connected by the involvement of TCTP, as will be mentioned below.

3.2 Conserved Functions of TCTP Across Kingdoms

Much work has been done recently to elucidate the function of TCTP. As mentioned before, the mRNA of a human TCTP was first isolated from an erythroleukemia line. Such abundant RNA was not associated to any protein, and therefore it was thought to be regulated at the translational level. Abundant transcripts for TCTP were isolated from other organisms, mostly from developing tissues, which suggested that TCTP is involved in promoting development and/or cell proliferation. Several lines of evidence indicate that the primary function of TCTP is the regulation of proliferation. Indeed, human TCTP binding to microtubules is regulated by the polo kinase (Plk), which is essential for the control of polar spindle dynamics; overexpression of a TCTP mutant lacking the Plk phosphorylation site resulted in multinucleate cells (Yarm 2002). Thus, phosphorylated TCTP binds less efficiently to and stabilizes microtubules, leading to cell division arrest. Furthermore, phosphorylated mouse TCTP (and probably several other TCTPs from diverse organisms) promotes depolymerization of polar, but not kinetochore microtubules during oocyte meiosis (Jeon et al. 2016). Therefore, the primary function of all TCTPs may be promoting cytokinesis, although whether this occurs mostly in germline or in all somatic cells (which seems more likely) remains to be determined. However, it has been suggested that TCTP, based on structural similarities with the Mss family of chaperones (for they which have also been termed Mss4), is a guanine nucleotide exchange factor (GEF), which are involved in many processes, such as the polymerization of microtubules (Thaw et al. 2001). Furthermore, it has been demonstrated that TCTP inhibits the dissociation of GDP from the eukaryotic translation initiation factor eEF1A (Cans et al. 2003), implicating it in the regulation of protein synthesis. This binding is evolutionarily conserved, since the human TCTP can bind eEF1A from fission yeast and chlorophytes in vitro (Wu et al. 2015). The notion of TCTP as a second messenger involved in growth regulation and responses to external stimuli is supported by the fact that it is a calcium-binding protein, and this has been confirmed in several systems (Bommer and Thiele 2004).

As a general regulator of growth in eukaryotes, it has been assumed that TCTP is also at the helm of pathways involved in such general processes, for example, the target of rapamycin (TOR) pathway. This is central in the control of cell growth in response to nutrients as well as to internal signals in extant eukaryotes (Albert and Hall 2015; Dobrenel et al. 2016). One of the inducers of TOR complex activity is the small G-protein Ras enriched in brain (Rheb) through their interaction in the lysosomal membrane (Heard et al. 2014). This implies that the regulation of this pathway necessitates a guanine nucleotide exchange factor (GEF); indeed, *Drosophila* TCTP binds Rheb, and such interaction is necessary for the regulation of growth and cell size (Hsu et al. 2007), although this has been disputed (Wang et al. 2008). Support that at least in *Drosophila* TCTP functions as a bona fide GEF comes from the fact that Rheb and TCTP interaction controls organ growth, which is in turn regulated by members of the large family of 14-3-3 general transcription factors (Le et al. 2016). It remains to be established how general is this interaction, but it must be considered that analysis of genome databases yielded no Rheb homologs in plants or apicomplexans, such as *Plasmodium*. Insulin is another general regulator of growth in animals and other eukaryotes, which acts via, among several others, the TOR pathway (Albert and Hall 2015). TCTP appears to mediate insulin signaling, since this induces phosphorylation of the former in an embryonic kidney cell line, although not in HeLa cells (Maeng et al. 2015).

The complementation of TCTP mutants between evolutionarily distant species supports the notion that some functions of TCTP are conserved among kingdoms, as evidenced by wild-type phenotype rescue by expression of an *Arabidopsis* TCTP in a *Drosophila* mutant, and vice versa (Brioudes et al. 2010). Null mutations in TCTP are lethal during early embryonic development in *Drosophila*, underscoring its essential role in viability; hypomorphic mutations basically lead to a decrease in cell number and size, which in turn affects organ size (Hsu et al. 2007). Null alleles of *Arabidopsis* TCTP are also lethal during early stages of development, again supporting the hypothesis that the function of TCTP is essential for survival and it is required throughout the whole life cycle of an organism, but more importantly during phases that require sustained cell division (Brioudes et al. 2010; Toscano-Morales et al. 2015).

TCTP is known to inhibit programmed cell death during development in some animal models as well as in plants, in response to pathogen infection (Susini et al. 2008; Hoepflinger et al. 2013).

It is possible that these taxon-specific functions are derived from the aforementioned activities. Therefore, TCTP activity in the function or development of specialized structures in a wide variety of taxa (such as axons or pollen tubes) could have originated from their ability to interact with G proteins or with the outer mitochondrial membrane, as well as by stabilizing or destabilizing pro- or antiapoptotic factors.

3.3 Taxon-Specific Functions of TCTP

A *TCTP* mRNA was first isolated from a human erythroleukemia cell line, and the gene and its expression pattern analyzed in rabbit (Bommer and Thiele 2004). It was realized first that at least in some species, this gene was subject to translational regulation.

An interesting feature of TCTP is that its mRNA is capable of inducing protein kinase R. This kinase is induced by double-stranded RNA (dsRNA) and phosphorylates the translation elongation factor eEIF2, resulting in the inhibition of protein synthesis (Bommer et al. 2002). This indicates that (a) some TCTP mRNAs may have a considerable secondary structure, suggesting that these may be subjected to translational regulation; (b) the mRNA itself has a function other than being translated; and (c) that it may participate in the interferon pathway. Human TCTP has also been found to induce histamine release from basophils (and hence its other term histamine-releasing factor or HRF), and thus it is involved in the onset of inflammation and allergic processes. Indeed, a dimerized form of TCTP is the active factor responsible for triggering inflammation and allergies (Kim et al. 2009), but it is not known whether this dimerized form is found only in pathological conditions or can be found carrying out normal functions in healthy individuals. This underscores an important point regarding TCTP localization. It is evident that TCTP/HRF is probably expressed in most cell types, but in the case of blood mononuclear cells, it is not clear whether there are two separate pools or, alternatively, the majority is secreted. Additionally, the possibility that other cell types secrete this protein is supported by the fact that a factor required for its export is expressed in tumor cells (and, incidentally, suggests a cell-to-cell signaling function during normal and abnormal proliferation) (Amzallag et al. 2004). In mice, TCTP is expressed during early stages of pancreas differentiation, where it is required for islet β-cell proliferation, and, in later stages, during adaptation to insulin resistance (Tsai et al. 2014).

Additional roles, at least at the biochemical level, indicate that TCTP binds to microtubules, which is regulated by phosphorylation. Such activity may regulate a host of functions, such as cytokinesis, vesicle trafficking, or organelle movement within the cell (Amson et al. 2013). All observations carried out in human cell lines point to an essential role in cell proliferation as well as Ca^{2+}-dependent inhibition of apoptosis (Amson et al. 2013). TCTP also appears to be a moonlighting protein regarding its cytokine function, although it may be related to its ability to interact with receptors in the surface of certain cell types; however, as will be mentioned below, it appears that this protein does not require transmembrane receptors. Earlier work in other species (such as *Oryctolagus*) indicated that the TCTP gene responds to heavy metal accumulation; similar results have been reported in earthworm. This suggests that TCTP has also a role in the response to abiotic stress, but its significance in this regard is not completely clear (Bommer and Thiele 2004; Amson et al. 2013). Importantly, a decrease in TCTP levels correlates with tumor reversion in breast cancer, which supports a role of this protein in inducing cell proliferation (Tuynder et al. 2004). Other roles of TCTP in vertebrates implicating the apoptotic machinery and more specifically its interaction with the P53 pathway and Bcl-xL have already been described extensively (Liu et al. 2005; Thébault et al. 2016). Its function in *Drosophila* has been reviewed in detail in this work by Kwang W. Choi.

3.3.1 Fungi

TCTP is found in the genome of extant fungi, although its function in these taxa has seldom been explored with detail. However, while TCTP inhibits apoptosis through interaction with P53 in human and mouse cell lines, in *Saccharomyces cerevisiae* (yeast) reports have shown to promote it by interacting with the external mitochondrial membrane in response to oxidative stress (Rinnerthaler et al. 2006). Furthermore, it is capable of interacting with stress granules where it modulates proteasome activity, which could indirectly regulate cell death (Rinnerthaler et al. 2013). It has been proposed that the function of the yeast TCTP, and possibly in other organisms, is to protect proteins in granules after heat shock. Evidently this must be a highly regulated process, but how are pro- and antiapoptotic activities of TCTP balanced, if they do occur in yeast, as well as how general is this phenomenon, remains to be elucidated. An example of taxon-specific functions of TCTP can be found in *Aspergillus nidulans*, a polymorphic fungus, in which this protein is involved in branching as well as in sexual differentiation (Oh et al. 2013). Its localization is either nuclear or cytoplasmic, depending on the growth phase, similar to what has been found in human cell lines, yeast, or *Arabidopsis* (Ma and Zhu 2012; Rinnerthaler et al. 2013; Toscano-Morales et al. 2015). Interestingly, a human pathogenic fungus that causes eumycetoma, *Madurella mycetomatis*, secretes a TCTP homolog, which could be involved in the development of tumors as a result of infection with this pathogen (van de Sande et al. 2006). Similarly, mycelia of the human pathogen *Paracoccidioides* spp. secrete TCTP, which could have a role in infection and/or colonization of its host (Weber et al. 2012). The human opportunistic pathogen *Candida albicans* also secretes TCTP via exosomes; thus, secretion of this protein by pathogenic fungi could be a general strategy to suppress the host's defense response (Vargas et al. 2015). However, the role of TCTP may be more general, since a free-living fungus, yeast, also secretes a TCTP form through exosomes (Oliveira et al. 2010).

3.3.2 Plants

TCTP-like mRNAs were detected in plants early on. Indeed, a TCTP mRNA was found to increase its levels after dark treatment of *Pharbitis*, a plant in which flower induction is dependent on day length, suggesting a role in circadian rhythms (Sage-Ono et al. 1998). Another remarkable feature of this and other emerging plant sequences was the similarity to animal TCTP sequences. While to date the structure of a plant TCTP has not yet been resolved, given the sequence conservation and structural similarity between TCTPs from evolutionarily diverse organisms such as fission yeast, human, and *Plasmodium*, it can be confidently assumed that the structure of plant TCTPs is similar. This notion is supported by the fact that the *Arabidopsis* and *Drosophila* functions are, at least in part, exchangeable. Indeed, the expression of the *Drosophila* TCTP can rescue an *Arabidopsis* TCTP mutant and vice versa (Brioudes et al. 2010). There are not many studies on plant TCTP function, although in *Arabidopsis* its knockout causes early lethality; silenced plants also harbor defective pollen; furthermore, it controls cell cycle duration (Berkowitz et al. 2008; Brioudes et al. 2010). It could be assumed that TCTP genes could have overlapping or completely redundant functions in organisms harboring more than one copy; however, this may not always be the case, as will be detailed below. For instance, *Arabidopsis* itself harbors two TCTP genes, termed *AtTCTP1* and *AtTCTP2* (accession numbers At3g16640 and At3g05540, respectively). The data regarding TCTP function corresponds to *AtTCTP1*, while *AtTCTP2* was initially considered a pseudogene (Berkowitz et al. 2008). However, work in our group demonstrated that the latter is not a pseudogene, and the loss of function is also lethal, although at later stages in development (Toscano-Morales et al. 2015). Thus, the wild-type *AtTCTP1* allele cannot compensate for the loss of the *AtTCTP*2 allele and vice versa, suggesting that these genes may have nonoverlapping functions. Developmental arrest in *AtTCTP1* null mutants occurs at early embryonic stages, while in *AtTCTP2* this occurs during the early rosette stage (Toscano-Morales et al. 2015). Additionally, while *AtTCTP2* expressed in *Agrobacterium rhizogenes* is capable of regenerating whole tobacco plants, *AtTCTP1* is not (Toscano-Morales et al. 2015). Furthermore, *AtTCTP2* mRNA is more difficult to detect than the *AtTCTP1* mRNA, not only because of lower expression levels but probably also because of more intricate secondary structure. All this evidence reinforces the notion that these genes may have at least partially different functions.

Despite their little overlap in activity, AtTCTP1 and AtTCTP2 differ mostly in a 13-amino-acid deletion in AtTCTP2 relative to AtTCTP1 in positions 35 to 47, plus a few substitutions distributed along the entire sequence. Insertion of this sequence in AtTCTP2 results in a marked decrease in its ability to induce plant regeneration, while the deletion of this sequence in AtTCTP1 leads to an in increase in its ability to induce regeneration (Toscano-Morales et al. 2015). Furthermore, AtTCTP2 localizes to nuclei in roots and AtTCTP1 in cytoplasm in this same plant organ, but the localization of the corresponding modified proteins is exchanged. Also, the predicted structure of the modified AtTCTP2 resembles more than that of AtTCTP1 and vice versa. The predicted structural modification is located in the putative G-protein-binding pocket, a modification predicted to be harbored by proteins from diverse taxa (see below). These results also highlight the ability of TCTP to shuttle between nucleus and cytoplasm, conceivably in response to signaling molecules, a phenomenon also observed in TCTPs from animals. Indeed, the TCTP localization pattern oscillates between nucleus and cytoplasm in a time-dependent manner in a human cell line (Ma and Zhu 2012). The mechanism for TCTP import to and export from the nucleus is not known, given the probable absence of a clear nuclear localization signal in TCTP. Although a recent report in which a role for TCTP in potyvirus infection in plants predicts that the tomato TCTP harbors a nuclear export signal (NES) (Bruckner et al. 2017) requires to be experimentally tested, it should be noted that this putative NES does not coincide with the region that blocks AtTCTP2 entry into the nucleus (i.e., the 13 amino acids present in AtTCTP1 but not in AtTCTP2). Recently, a role in regulating cell death in response to pathogens has been observed for TCTP in tobacco, as well as in the response to ethylene; it is well established that both phenomena are related in plants (Hoepflinger et al. 2013). Interestingly, overexpression of TCTP in tobacco decreased cell death induced in leaf disks by different effectors, among them BAX (a protein that induces cell death in mammals and other organisms). This supports the notion that protection against cell death is mediated in animals and plants via similar mechanisms involving TCTP. Furthermore, this activity likely requires calcium binding. TCTP in the plant-parasitic nematode *Meloidogyne enterolobii* has been recruited to block programmed cell death in response to pathogens, thus effectively favoring colonization of its host; thus, it is likely that plant and TCTPs from other taxa display an antiapoptotic activity (Zhuo et al. 2016). Another interesting feature of some TCTPs is that they have been found in the phloem translocation stream of certain species (Lin et al. 2009; Rodríguez-Medina et al. 2011). This suggests that in vascular plants, TCTP may function in a non-cell autonomous manner, although the nature of such function is still a matter of speculation. Conversely, this protein could be involved in the reorganization of its cytoskeletal system; given the enucleate nature of the phloem sieve tube, its longevity, and its role in long-distance transport of nutrients and chemical signals, it is evident that it requires structural reinforcement, an activity that could be partially provided by this protein, although this is also speculative. The extracellular proteome of different species was searched for TCTP, with no matches found. Thus, in contrast to certain vertebrate TCTPs, the corresponding plant proteins are probably not secreted in normal conditions. However, AtTCTP1 has been observed in cell wall-rich fractions in *Arabidopsis*, implying that it could be secreted in some cell types (Jamet et al. 2008). Additionally, the tomato TCTP has been found in the xylem proteome in response to *Fusarium* effectors (Gawehns et al. 2015). More recently, AtTCTP1 has been found in the pollen tube secretome; mutants in this gene show defects in fertilization, indicating an important role in this regard (Hafidh et al. 2016). Thus, it has been suggested that TCTP has an important role in pollen tube guidance, although its precise function in this case is not clear. Interestingly, in this same work, proteins that are secreted via a nonclassical pathway were also detected in the pollen secretome, i.e., proteins lacking a discernible signal peptide, suggesting that this secretion pathway is more conspicuous than previously thought. Additionally, nonclassical secretion pathways in plants are not known, although TCTP export could occur via exosomes, which have been observed in soybean protoplasts (Tanchak and Fowke, 1987).

In the few cases analyzed in plants, TCTP displays dual intracellular localization, while AtTCTP1 localizes to nuclei in mesophyll and accumulates mostly in cytoplasm in roots; the opposite has been observed in AtTCTP2 (Toscano-Morales et al. 2015). How general is this dual localization in plants and other organisms remains to be determined.

On the other hand, it is necessary to study whether the function of TCTP in other plant species, particularly nonvascular plants and chlorophytes, is similar. It must be mentioned that in the five chlorophyte genomes sequenced to date, a TCTP gene homolog is found in only one of this, *Coccomyxa subellipsoidea*, although erroneous annotation or incomplete coverage (i.e., if these sequences were found close to centromeres) could explain this observation (Gutiérrez-Galeano et al. 2014).

3.3.3 Blood-Borne and Other Vertebrate Parasites

In a previous work, we scoured for TCTP sequences in extant taxa genomic databases, and, as expected, most harbored at least one gene (Hinojosa-Moya et al. 2008). Interestingly, it had been reported that several evolutionarily distant organisms secrete a form of TCTP; these organisms have in common that they either colonize or feed on blood from vertebrates. These include apicomplexa, arthropods, and filarial parasites (Chmelar et al. 2008; Hewitson et al. 2008; Silverman et al. 2008; Weir et al. 2010). Secretomes of other parasites, such as *Trypanosoma* spp., also include TCTP homologs, although in most cases their function is far from clear (Bayer-Santos et al. 2013). In the particular case of *Plasmodium falciparum*, it has been observed that in some cases a very high concentration of TCTP is found in sera of infected individuals (approximately 7 μg/ml). *E. coli*-expressed *P. falciparum* TCTP has been shown to induce histamine release from mast cells, thus possessing a cytokine activity (although its efficacy is much lower than the purified human homolog, likely because of expression efficiency in this heterologous system, and/or missing possible posttranslational modifications; MacDonald et al. 2001). Thus, its function, at least in vitro, appears to be similar to its human counterpart. Additionally, as implied by its ability to induce secretion of histamine when applied exogenously, it also functions in a non-cell autonomous manner. Since the host TCTP induces inflammation, as a response to microbial infection, it would not be clear the role of the *Plasmodium* TCTP in the pathogenic process if this were its sole function. We have previously analyzed the reported structure of *P. knowlesi* and *P. falciparum* and compared them with the structure of human and *S. pombe* TCTPs (Hinojosa-Moya et al. 2008). It must be considered that the structure of protein in solution, particularly in its physiological milieu, is quite dynamic, and thus in its crystalized form may represent only a possible conformation. Nonetheless, the aforementioned analysis suggested that, while the predicted structure of TCTPs from diverse clades is strikingly similar, the *Plasmodium* protein showed a structural alteration relative to *S. pombe*, in that an alpha helix was present in the former, instead of a beta sheet close to the N end. Interestingly, the predicted secondary structure of other members of the *Plasmodium* genus suggested the presence of this helix in the same position. This region is part of the potential G-protein-binding pocket that includes the conserved glutamic acid in position 12 (in *Drosophila*) (Hsu et al. 2007). Thus, while the evidence that the ability of TCTP to bind G proteins may be lacking, it is evident that this region is essential for this protein to sustain cell proliferation. In the case of *Plasmodium*, its target is unlikely to be Rheb, since this gene ortholog has not been found in the extant genomes of this genus (Ruiz-Medrano and Xoconostle-Cázares, unpublished observations).

The extra alpha helix in the *P. knowlesi* and *P. falciparum* TCTPs causes a distortion in its potential G-protein-binding pocket (Hinojosa-Moya et al. 2008). This alteration in the putative G-protein-binding pocket results in this displaying a more "open" conformation in the parasite TCTP relative to the *S. pombe* and human proteins. We hypothesized that this structural alteration could lead to functional differences with the host TCTP. This could be relevant, considering the high amount that *Plasmodium* secretes into the host's bloodstream. We therefore tested this hypothesis by comparing the effect of recombinant human and *P. falciparum* TCTPs (HsTCTP and PfTCTP, respectively) on mouse spleen B cell proliferation. As mentioned earlier, human TCTP induced the proliferation of these cells (Kang et al. 2001). Interestingly, PfTCTP induced B cell proliferation at much lower levels than those of HsTCTP (Calderón-Pérez et al. 2014). Furthermore, when these proteins were incubated with isolated B cells, PfTCTP was incorporated much more efficiently than HsTCTP; additionally, the intracellular distribution of fluorescently labeled proteins appeared different, although more work is needed in this direction. In all, these results suggest that PfTCTP and HsTCTP have different activities, or at least these do not overlap completely. It remains to be determined whether there is an intracellular pool of PfTCTP, and if so, what is its function in the parasite life cycle, although some data indicates that there is indeed intracellular *Plasmodium* TCTP that localizes to cytoplasm and food vacuolar membrane (Bhisutthibhan et al. 1999). *Plasmodium* TCTP has received attention because it has been proposed as a target of the antimalarial drug artemisinin; more recently, it has been shown that artemisinin binds to various proteins, including TCTP; such interactions are activated by the heme group from the host's ingested blood (Wang et al. 2015). Importantly, most of the target proteins are required for parasite survival. That *Plasmodium* TCTP could be a therapeutic target for treatment of malaria is supported by the fact that immunization of mice with this protein leads to increased resistance (Taylor et al. 2015). It is tempting to speculate that a fragment encompassing the region that is structurally different from the host TCTP could be used for more effective immunization. Furthermore, a similar structural variation of certain vertebrate parasites TCTPs has been predicted.

TCTP is almost ubiquitous in eukaryotes, and some hematophagous arthropods and certain parasites (blood-borne ones as well), secrete a TCTP homolog. It is possible that these have a role in facilitating infection or suppressing the host immune response, although this has not been established with detail. To explore whether the aforementioned structural variation of the PfTCTP was also observed with these proteins, we have obtained predictive structures of TCTPs from representative human parasites, using the Protein Homology/analogY Recognition Engine V 2.0 (http://www.sbg.bio.ic.ac.uk/phyre2; Kelley et al. 2015). These are shown in Figs. 3.1, 3.2 and 3.3. To assess the precision of this analysis, the human TCTP isoform 2 structure was also predicted (as mentioned before, the structure of this protein has been solved; Fig. 3.1a). The structural similarity between all these proteins is striking; however, subtle variations can be observed. *P. berghei*, *Trypanosoma cruzi*, and *Toxoplasma gondii* TCTPs display a potential G-protein-binding pocket in a more open conformation relative to the human TCTP isoform 2 (Fig. 3.1b–d). In addition, the clamp that is central to this domain is perpendicular to the central alpha helices (1 and 2) in parasite TCTPs, but not in the human TCTP. Furthermore, the predicted *Cryptosporidium hominis*, *Theileria annulata*, and *Schistosoma mansoni* TCTPs show similar structural variations relative to human isoform 2; however, *Ixodes scapularis* TCTP resembles more the latter, harboring a

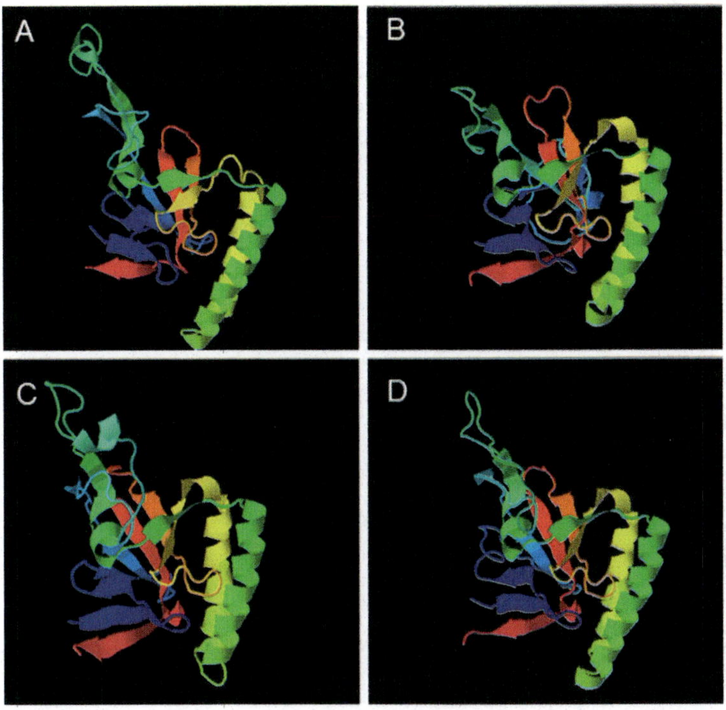

Fig. 3.1 Predictive structures of TCTP obtained with the Protein Homology/analogY Recognition Engine V 2.0 (http://www.sbg.bio.ic.ac.uk/phyre2; Kelley et al. 2015) from the following organisms: (**a**) *Homo sapiens* TCTP (isoform 2, GenBank accession number NP_003286.1); (**b**) *Plasmodium berghei* TCTP (GenBank accession number CXI61325.1); (**c**) *Trypanosoma cruzi* TCTP (GenBank accession number XP_806523.1); and (**d**) *Toxoplasma gondii* TCTP (GenBank accession number CEL78687.1)

clamp that is not completely perpendicular to the aforementioned alpha helices (Fig. 3.2a–d). Different structural isoforms of TCTP may be present in some organisms, including human. Indeed, in TCTP isoform 1 the clamp is perpendicular to the main helices, suggesting additional or different functions of this isoform relative to the more thoroughly studied isoform 2 (Fig. 3.3a). Human TCTP isoform 3 shows a similar structural variation, with a disordered lower part of the clamp (Fig. 3.3b). It must be mentioned that isoform 1 is 197 aa in length, isoform 2 is 172 aa, and isoform 3 is much shorter, 136 aa, further supporting different activities for these proteins. However, there is some confusion regarding the function of these proteins, since in several studies they are mentioned interchangeably. On the other hand, *Leishmania* TCTP structure lacks the lower part of the clamp and appears less organized overall (Fig. 3.3c). TCTP from *Giardia lamblia* (causal agent of giardiasis) may have a similar structure to parasite TCTPs, although it is not known whether it is secreted (Fig. 3.3d).

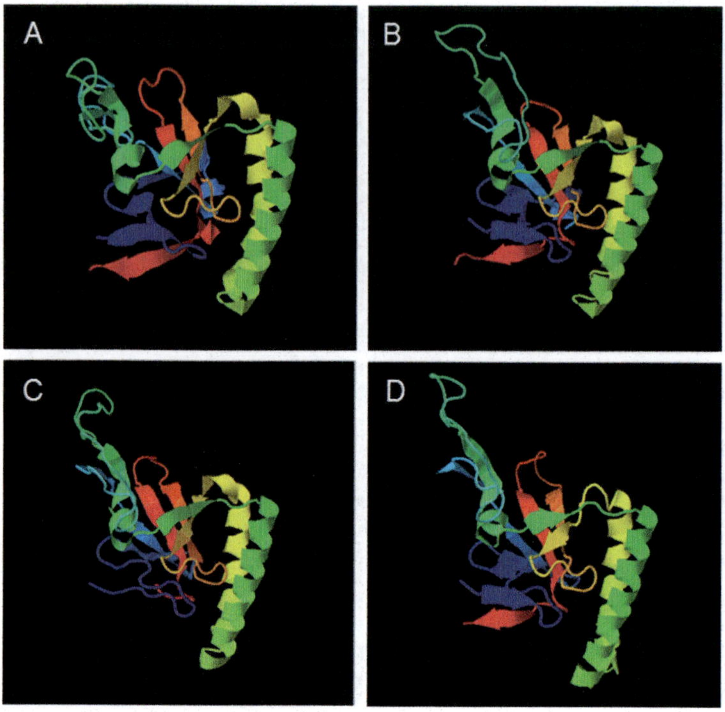

Fig. 3.2 Predictive structures of TCTP from the following organisms: (**a**) *Cryptosporidium hominis* TCTP (GenBank accession number XP_668673.1); (**b**) *Theileria annulata* TCTP; (GenBank accession number CAI73774.1); (**c**) *Schistosoma mansoni* TCTP (Swiss-Prot accession number: Q95WA2.1); and (**d**) *Ixodes scapularis* TCTP (GenBank accession number AAY66972.1)

These predictions suggest that in certain vertebrate parasites, a secreted form of TCTP may have a function that is different from the host TCTP; this could help to downregulate the immune response by mimicking and interfering with the host TCTP activity. Finally, this knowledge could be used to devise strategies for protection against blood-borne or blood-feeding parasites. Thus, it will be important to determine experimentally the tridimensional structure of these proteins. Recently, a role for the *P. berghei* TCTP/HRF has been established; this protein suppresses the secretion of interleukin-6, which is known to inhibit liver infection by the parasite (Mathieu et al. 2015). Furthermore, host infection with parasites in which its TCTP/HRF gene had been deleted protects against subsequent infection, which underscores the importance of TCTP for the infection process in malaria and, possibly, other diseases (Demarta-Gatsi et al. 2016).

There is evidence suggesting that the phylogeny of plant TCTPs cannot be resolved completely, in contrast to animals, in which the phylogeny derived from TCTP is in agreement with that obtained using other sequences (Hinojosa-Moya et al. 2008; Gutiérrez-Galeano et al. 2014). The reasons for this are probably

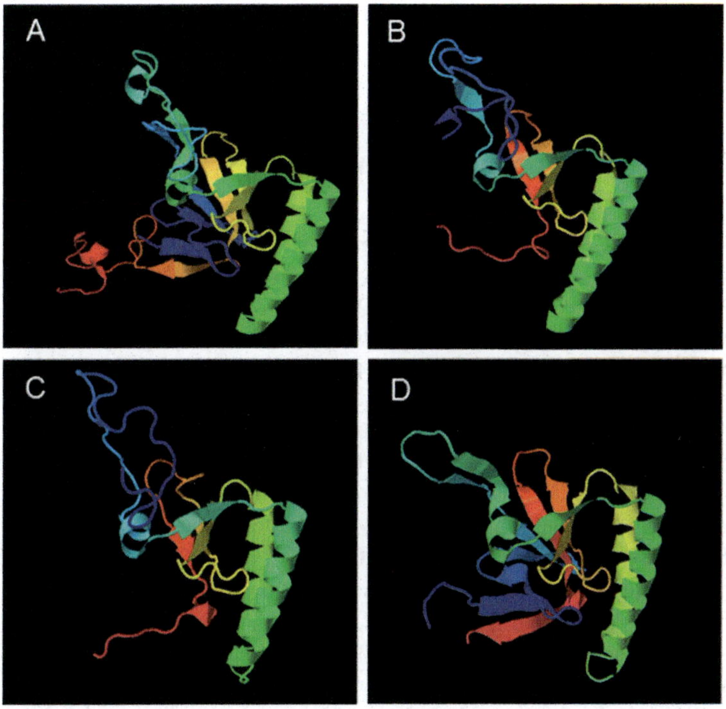

Fig. 3.3 Predictive structures of TCTP from the following organisms: (**a**) *Homo sapiens* TCTP (isoform 1, GenBank accession number NP_001273201.1); (**b**) *Homo sapiens* TCTP (isoform 3, GenBank accession number NP_001273202.1); (**c**) *Leishmania major* TCTP (GenBank accession number XP_001683667.1); and (**d**) *Giardia lamblia* TCTP (GenBank accession number EDO80348.1)

manifold and could include unequal substitution rates between clades and horizontal gene transfer. However, the predicted structure of some plant TCTPs suggests that these proteins can be classified in two large groups, one that is similar to *Arabidopsis* AtTCTP1 and the other to CmTCTP and AtTCTP2. Interestingly, AtTCTP1 predicted structure resembles that of *Schizosaccharomyces pombe* TCTP (SpTCTP), while CmTCTP/AtTCTP2 is more similar to the corresponding proteins from *P. falciparum* and *P. knowlesi* (PfTCTP and PkTCTP, respectively; Gutiérrez-Galeano et al. 2014). If these predictions based on its structure are correct, a more general classification of these proteins could be envisaged. Additionally, a broader function for members of each of these two groups could be proposed; indeed, it is possible that the proteins belonging to the SpTCTP/AtTCTP1 clade have a more direct role in cell division and proliferation, while members of the CmTCTP/AtTCTP2/PfTCTP group possibly function in diverse roles more related to differentiation. In the case of PfTCTP, this role would be more precisely to compete against the endogenous TCTP, thus helping suppress the immune response of the host. Also, the predicted structure of human TCTP isoform

1 is more similar to the *P. falciparum* TCTP and as such would be expected to have similar functional properties.

3.4 Non-cell Autonomous Functions of TCTP

All extant evidence indicates that TCTP is a multifunctional protein, with activities ranging from regulation and promotion of proliferation to intercellular signaling. It could be inferred, also, that this protein localizes to different subcellular compartments under different circumstances. Indeed, this has been observed in different systems. The isoform 2 of TCTP corresponds to the histamine releasing factor (HRF); thus, it is secreted into the bloodstream (MacDonald et al. 1995). As implied by its name, this protein induces secretion of histamine, as well as interleukins, from eosinophils and basophils (Amson et al. 2013). TCTP lacks a signal peptide, so it is secreted by an alternative mechanism. It has been shown that its secretion requires the TSAP6 protein that is also involved in exosome production (Amzallag et al. 2004; Lespagnol et al. 2008; Feng 2010). Secreted TCTP also promotes liver regeneration and, on the other hand, enhances colorectal cancer invasion, which further supports a role of this protein in non-cell autonomous regulation of cell growth (Hao et al. 2016; Xiao et al. 2016).

In the case of secreted TCTP, it is reasonable to assume that the target cell should have specific mechanisms to bind, in some cases internalize, and respond to a potential signaling cascade triggered by this protein. However, in a human lung carcinoma cell line, TCTP can be internalized via endocytosis and macropinocytosis processes that are dependent of the cytoskeleton (Kim et al. 2011, 2015). Indeed, the first ten amino acids of the N-terminus of TCTP act as a protein transduction domain (PTD) that enable this protein, or unrelated cargo fused to it, to enter cells independently of a putative specific transmembrane receptor (Kim et al. 2011). Although the precise mechanism through which PTDs are capable of entering cells is not completely known, extant evidence suggests that this process requires interaction of basic amino acids of the PTD with the membrane surface, after which endocytosis ensues, and thus it does not require interaction with a specific receptor. Moreover, lipid rafts and caveolin, but not clathrin, appear to be involved in the internalization of TCTP (Kim et al. 2011, 2015). How general is this process is not known, but given the high structural conservation of this protein family, it is likely that most, if not all, TCTPs harbor a structurally related PTD. On the other hand, with the exception of unicellular parasites that secrete TCTP into its human host bloodstream, it is not clear whether TCTP is also secreted in unicellular eukaryotes, and if so, what would be the function of such secretion. Extracellular TCTP in humans can undergo dimerization, which results from proteolytic processing by an extracellular protease (and which eliminates its N-end); this dimerized form, in contrast to in vitro synthesized TCTP, acts as a cytokine and inducing histamine secretion from mast cells and basophils and therefore has an important role in allergies; indeed, this dimerized form of TCTP is found in sera of individuals affected by allergies (Kim et al. 2009).

Interestingly, this dimerized form is bound by immunoglobulin E (IgE), which is in turn recognized by an IgE receptor, suggesting that only the dimerized form of TCTP requires a receptor, in this case, to trigger histamine release (Kashiwakura et al. 2012). While it has not been established that this dimerization occurs in other conditions or is involved in additional processes other than allergies in humans (and, conceivably, in vertebrates), it has been suggested that the *Brugia malayi* TCTP undergoes a similar modification, since this protein is also secreted into the serum of its host, suggesting that this form is also functionally relevant (Gnanasekar et al. 2002).

As mentioned in the present work, the secretion of TCTP, and thus its likely non-cell autonomous function, has been observed in several different systems, pointing to an important role in intercellular signaling. This function ranges from a cytokine-like one in the immune response to molecular mimicry by different parasitic protozoa. More recently, TCTP has been implicated in the regulation of axon development of the visual system of *Xenopus laevis* (Roque et al. 2016). An extracellular form TCTP is required to control this process, and this seems to be related to its ability to interact with the antiapoptotic Mcl1 protein. Thus, TCTP maintains viability of axons during development. The importance of TCTP in central nervous system function throughout different developmental stages is illustrated by the fact that lower levels of this protein are found in individuals with Down syndrome or afflicted with Alzheimer's disease (Kim et al. 2001). A fascinating analogy could be suggested with the pollen tube secretome in plants; it is possible that this protein is required for pollen tube survival during growth, rather than for its guidance (Hafidh et al. 2016). As mentioned before, the pollen tube secretome includes several other proteins that lack a signal peptide, indicating that these are secreted via unconventional pathways, possibly via exosomes. These vesicles have been observed in plants and are thought to have a role in defense signaling (Tanchak and Fowke 1987; An et al. 2006; Hafidh et al. 2016).

An important aspect to consider is that TCTP has an essential role in regulating cell proliferation and viability through its interaction with multiple factors, such as microtubules and antiapoptotic proteins. These roles have been found mostly in multicellular organisms, but it is not clear whether the same applies to unicellular eukaryotes, more importantly in blood-borne parasites described in the previous sections. Additionally, it is not known whether free-living unicellular or colonial eukaryotes require secretion of TCTP for their viability. It will be of interest to determine if the extracellular functions of TCTP are essential for survival; extant data suggests that this is the case (see, e.g., Roque et al. 2016). *Plasmodium* TCTP has been localized intracellularly in different developmental stages in the parasite (Bhisutthibhan et al. 1999); the factors that regulate the "partition" of this protein between the cytoplasm, food vacuoles, and the extracellular space are not known; intriguingly, no TCTP has been observed in the nucleus of *Plasmodium,* although more work is needed in this direction. Additionally, TCTP from model multicellular organisms (fundamentally human, *Drosophila,* and *Arabidopsis*) is expressed in most tissues, but it remains to be determined if this protein has the potential to be secreted from all cell types. Since TSAP6, which is required for

TCTP export, is expressed in several different organs and tissues, including skin, cardiac muscle, lungs peripheral blood, and central nervous system (see https://www.ebi.ac.uk/gxa/home), it is then likely that TCTP is secreted in all these tissues. Secretion of TCTP occurs in normal as well as in pathological conditions, as evidenced by the role of the dysregulation of this protein in endothelial cells under serum starvation and in prostate and colorectal cancers (Sirois et al. 2011; Xiao et al. 2016). The mechanism through which TCTP could act in a paracrine manner (in normal and pathologic conditions) is likely through downregulation of apoptosis in cells that bind and internalize the extracellular form of this protein; a decrease in TCTP increases apoptosis, while extracellular TCTP induces an increase in colony formation in a metastatic cell line (Sirois et al. 2011). TCTP induces migration and invasion of colon cancer cells via activation of the CDC42 small GTPase, which is a central regulator of the cell cycle, and the mitogen activated protein kinase (MAPK) JNK; this same mechanism could be involved in liver cancer and metastasis (Chan et al. 2012; Xiao et al. 2016).

Monomeric TCTP appears not to require a transmembrane receptor to induce the aforementioned effects. Given the presence of a PTD, TCTP is probably internalized through caveolar endocytosis (lipid rafts) and released from the endoplasmic reticulum into the cytoplasm, where it conceivably interacts with target proteins (Kim et al. 2015).

TCTP has been found in the phloem sap proteome of several plant species. Given that these proteins reside in the enucleate sieve tubes (ST), which deliver fixed carbon from photosynthetic to heterotrophic tissues and where there is probably no protein synthesis (Lucas et al. 2013), it is possible that TCTP functions in the maintenance of the ST system and/or long-distance (and hence non-cell autonomous) signaling. The long-distance transport of TCTP mRNA and protein suggests that this is indeed the case (Toscano-Morales et al. 2014). An intriguing possibility is that TCTP that resides in the ST system could regulate the differentiation of the sieve elements (SE), which are the individual cells that form the STs. Indeed, since plant TCTPs inhibit programmed cell death induced by pathogen infection (similar to the animal counterparts; Hoepflinger et al. 2013), these proteins could arrest the process before plasma membrane disruption, which in xylem leads to cell death, so that SEs retain a functional plasma membrane. This evidently requires experimental testing; if true, ectopic expression of TCTP in the cells that give rise to xylem should inhibit mature xylem formation. Intriguingly, *TCTP* mRNA has been found also in the phloem long-distance translocation stream of lupin, and *AtTCTP1* is one of the more conspicuous mobile RNAs found both in heterografts between different *Arabidopsis* ecotypes, as well as between *Arabidopsis* and a parasitic plant, which illustrates its efficient long-distance mobility (Rodríguez-Medina et al. 2011; Kim et al. 2014; Thieme et al. 2015). *AtTCTP2* mRNA also moves long distance from a transgenic to a non-transgenic tobacco graft, where it possibly induces the formation of aerial roots (Toscano-Morales et al. 2014). The function of the long-distance transport of *TCTP* mRNA and protein is not known; the only instance of a possible function is precisely the induction of aerial roots, which are formed in response to stress (Toscano-Morales et al. 2014). Mobile TCTP mRNA could be translated in systemic tissues or serve another function, such as a source of phosphate and nitrogen. The function of this protein in nonvascular plants, as well as in chlorophytes, has not been reported. In the latter case, the apparent absence of this gene in all chlorophyte genomes, except for one species, *Coccomyxa subellipsoidea*, needs to be evaluated (whether this results from incomplete coverage of these genomes, genuine absence of this gene, or very low similarity to extant TCTP sequences). In the case of nonvascular plants, a non-cell autonomous function has not been observed.

The secretion of TCTP by fungi, either free-living or pathogenic, suggests that this non-cell autonomous function is pivotal to maintain homeostasis; the function in this case is far from clear, but, on a speculative note, it constitutes an extracellular signal required for survival in hostile environments; further, it is possible that this secreted form is endocyted by neighboring cells, where conceivably an antiapoptotic program would be triggered.

It has become increasingly clear that TCTP bears key functions in both cell autonomous and non-cell autonomous manner in eukaryotes, ranging from vertebrates to plants and fungal pathogens.

3.5 Perspectives

Growing evidence indicates that TCTP is a central regulator of proliferation and cell survival through inhibition of apoptosis both in animals and plants. Also, a role in differentiation as well as in the modulation of the defense response in diverse taxa has been found. The knowledge of the precise mechanisms through which this protein (and conceivably its mRNA in some instances) regulates a plethora of seemingly disparate processes will be essential to understand those that are of interest from biomedical and agricultural perspectives. Thus, recent work indicates that it could constitute a therapeutic target in metastasis in different types of cancer, infectious diseases, and neurodegenerative disorders, as well as a tool for increasing plant productivity. Several questions remain to be answered regarding the diverse roles of this polyfacetic protein, such as whether the function of secreted and internalized TCTP is the same as non-secreted protein, the mechanisms regulating the secretion of this protein in different organisms (and, in particular, in several blood-borne parasites and fungal pathogens), and the non-cell autonomous function of TCTP in animals and plants. To understand the function of this protein, it will also be essential to determine whether the microtubule destabilization function and the inhibition of apoptosis are different aspects of the same activity of TCTP. It is evident that an integrated approach is required to help answering such questions, including those involving cell and molecular biology as well as massive analysis techniques. Finally, the therapeutic potential of this protein in diseases caused by the aforementioned pathogens, in cancer, and in developmental maladies (such as those pertaining to the central nervous system) cannot be overstated. From an evolutionary viewpoint, the multiple roles of this protein are a fascinating example of structural conservation despite a high-sequence divergence in plants and, if confirmed, a case of convergent evolution of molecular mimicry in unrelated taxa that share a common strategy to downregulate the immune response to certain human parasites.

Acknowledgments Work in our laboratory described here was supported by CONACYT grants Nos. 109885 (to BX-C) and 156162 (to RR-M) and SENASICA-2014, 2015, and 2016 (to BX-C). We apologize to colleagues whose results were not included in this work due to space constraints.

References

Albert V, Hall MN (2015) mTOR signaling in cellular and organismal energetics. Curr Opin Cell Biol 33:55–66. https://doi.org/10.1016/j.ceb.2014.12.001

Amson R, Pece S, Marine JC, Di Fiore PP, Telerman A (2013) TPT1/TCTP-regulated pathways in phenotypic reprogramming. Trends Cell Biol 23:37–46. https://doi.org/10.1016/j.tcb.2012.10.002

Amzallag N, Passer BJ, Allanic D, Segura E, Théry C, Goud B, Amson R, Telerman A (2004) TSAP6 facilitates the secretion of translationally controlled tumor protein/histamine-releasing factor via a nonclassical pathway. J Biol Chem 279(44):46104–46112

An Q, Ehlers K, Kogel KH, van Bel AJ, Hückelhoven R (2006) Multivesicular compartments proliferate in susceptible and resistant MLA12-barley leaves in response to infection by the biotrophic powdery mildew fungus. New Phytol 172:563–576

Bayer-Santos E, Aguilar-Bonavides C, Rodrigues SP, Cordero EM, Marques AF, Varela-Ramirez-A, Choi H, Yoshida N, da Silveira JF, Almeida IC (2013) Proteomic analysis of Trypanosoma cruzi secretome: characterization of two populations of extracellular vesicles and soluble proteins. J Proteome Res 12:883–897. https://doi.org/10.1021/pr300947g

Berkowitz O, Jost R, Pollmann S, Masle J (2008) Characterization of TCTP, the translationally controlled tumor protein, from Arabidopsis thaliana. Plant Cell Online 20(12):3430–3447

Bhisutthibhan J, Philbert MA, Fujioka H, Aikawa M, Meshnick SR (1999) The Plasmodium falciparum translationally controlled tumor protein: subcellular localization and calcium binding. Eur J Cell Biol 78:665–670

Bommer UA, Borovjagin AV, Greagg MA, Jeffrey IW, Russell P, Laing KG, Lee M, Clemens MJ (2002) The mRNA of the translationally controlled tumor protein P23/TCTP is a highly structured RNA, which activates the dsRNA-dependent protein kinase PKR. RNA *8*:478–496

Bommer UA, Thiele BJ (2004) The translationally controlled tumour protein (TCTP). Int J Biochem Cell Biol 36:379–385

Brioudes F, Thierry AM, Chambrier P, Mollereau B, Bendahmane M (2010) Translationally controlled tumor protein is a conserved mitotic growth integrator in animals and plants. Proc Natl Acad Sci U S A 107:16384–16389. https://doi.org/10.1073/pnas.1007926107

Bruckner FP, Xavier ADS, Cascardo RDS, Otoni WC, Zerbini FM, Alfenas-Zerbini P (2017) Translationally controlled tumour protein (TCTP) from tomato and is necessary for successful infection by a potyvirus. Mol Plant Pathol 18(5):672–683

Calderón-Pérez B, Xoconostle-Cázares B, Lira-Carmona R, Hernández-Rivas R, Ortega-López J, Ruiz-Medrano R (2014) The Plasmodium falciparum translationally controlled tumor protein (TCTP) is incorporated more efficiently into B cells than its human homologue. PLoS One 9: e85514. https://doi.org/10.1371/journal.pone.0085514

Cans C, Passer BJ, Shalak V, Nancy-Portebois V, Crible V, Amzallag N, Allanic D, Tufino R, Argentini M, Moras D, Fiucci G, Goud B, Mirande M, Amson R, Telerman A (2003) Translationally controlled tumor protein acts as a guanine nucleotide dissociation inhibitor on the translation elongation factor eEF1A. Proc Natl Acad Sci U S A 100:13892–13897

Chan TH, Chen L, Liu M, Hu L, Zheng BJ, Poon VK, Huang P, Yuan YF, Huang JD, Yang J, Tsao GS, Guan XY (2012) Translationally controlled tumor protein induces mitotic defects and chromosome missegregation in hepatocellular carcinoma development. Hepatology 55:491–505

Chmelar J, Anderson JM, Mu J, Jochim RC, Valenzuela JG, Kopecký J (2008) Insight into the sialome of the castor bean tick, Ixodes ricinus. BMC Genomics 9:233. https://doi.org/10.1186/1471-2164-9-233

Demarta-Gatsi C, Smith L, Thiberge S, Peronet R, Commere PH, Matondo M, Apetoh L, Bruhns P, Ménard R, Mécheri S (2016) Protection against malaria in mice is induced by blood stage-arresting histamine-releasing factor (HRF)-deficient parasites. J Exp Med 213(8):1419–1428

Dobrenel T, Caldana C, Hanson J, Robaglia C, Vincentz M, Veit B, Meyer C (2016) TOR signaling and nutrient sensing. Annu Rev Plant Biol 67:261–285. https://doi.org/10.1146/annurev-arplant-043014-114648

Feng Z (2010) p53 regulation of the IGF-1/AKT/mTOR pathways and the endosomal compartment. Cold Spring Harb Perspect Biol 2(2):a001057

Gawehns F, Ma L, Bruning O, Houterman PM, Boeren S, Cornelissen BJ, Rep M, Takken FL (2015) The effector repertoire of Fusarium oxysporum determines the tomato xylem proteome composition following infection. Front Plant Sci 6:967. https://doi.org/10.3389/fpls.2015.00967

Gnanasekar M, Rao KV, Chen L, Narayanan RB, Geetha M, Scott AL, Ramaswamy K, Kaliraj P (2002) Molecular characterization of a calcium binding translationally controlled tumor protein homologue from the filarial parasites Brugia malayi and Wuchereria bancrofti. Mol Biochem Parasitol 121:107–118

Gutiérrez-Galeano DF, Toscano-Morales R, Calderón-Pérez B, Xoconostle-Cázares B, Ruiz-Medrano R (2014) Structural divergence of plant TCTPs. Front Plant Sci 5:361. https://doi.org/10.3389/fpls.2014.00361

Hafidh S, Potěšil D, Fíla J, Čapková V, Zdráhal Z, Honys D (2016) Quantitative proteomics of the tobacco pollen tube secretome identifies novel pollen tube guidance proteins important for fertilization. Genome Biol 17:81. https://doi.org/10.1186/s13059-016-0928-x

Hanschen ER, Marriage TN, Ferris PJ, Hamaji T, Toyoda A, Fujiyama A, Neme R, Noguchi H, Minakuchi Y, Suzuki M et al (2016) The Gonium pectorale genome demonstrates co-option of cell cycle regulation during the evolution of multicellularity. Nat Commun 7:11370. https://doi.org/10.1038/ncomms11370

Hao S, Yu Q, Yin S, He J, He D, Wang C (2016) Serum translationally controlled tumor protein is involved in rat liver regeneration after hepatectomy. Hepatol Res 46(13):1392–1401

Heard JJ, Fong V, Bathaie SZ, Tamanoi F (2014) Recent progress in the study of the Rheb family GTPases. Cell Signal 26:1950–1957. https://doi.org/10.1016/j.cellsig.2014.05.011

Hewitson JP, Harcus YM, Curwen RS, Dowle AA, Atmadja AK, Ashton PD, Wilson A, Maizels RM (2008) The secretome of the filarial parasite, Brugia malayi: proteomic profile of adult excretory-secretory products. Mol Biochem Parasitol 160:8–21. https://doi.org/10.1016/j.molbiopara.2008.02.007

Hinojosa-Moya J, Xoconostle-Cázares B, Piedra-Ibarra E, Méndez-Tenorio A, Lucas WJ, Ruiz-Medrano R (2008) Phylogenetic and structural analysis of translationally controlled tumor proteins. J Mol Evol 66:472–483. https://doi.org/10.1007/s00239-008-9099-z

Hoepflinger MC, Reitsamer J, Geretschlaeger AM, Mehlmer N, Tenhaken R (2013) The effect of translationally controlled tumour protein (TCTP) on programmed cell death in plants. BMC Plant Biol 13:135. https://doi.org/10.1186/1471-2229-13-135

Hsu YC, Chern JJ, Cai Y, Liu M, Choi KW (2007) Drosophila TCTP is essential for growth and proliferation through regulation of dRheb GTPase. Nature 445:785–788

Jamet E, Albenne C, Boudart G, Irshad M, Canut H, Pont-Lezica R (2008) Recent advances in plant cell wall proteomics. Proteomics 8:893–908. https://doi.org/10.1002/pmic.200700938

Jeon HJ, You SY, Park YS, Chang JW, Kim JS, Oh JS (2016) TCTP regulates spindle microtubule dynamics by stabilizing polar microtubules during mouse oocyte meiosis. Biochim Biophys Acta 1863:630–637. https://doi.org/10.1016/j.bbamcr.2016.01.012

Kang HS, Lee MJ, Song H, Han SH, Kim YM, Im JY, Choi I (2001) Molecular identification of IgE-dependent histamine-releasing factor as a B cell growth factor. J Immunol 166:6545–6554

Kashiwakura JC, Ando T, Matsumoto K, Kimura M, Kitaura J, Matho MH, Zajonc DM, Ozeki T, Ra C, MacDonald SM et al (2012) Histamine-releasing factor has a proinflammatory role in mouse models of asthma and allergy. J Clin Invest 122:218–228. https://doi.org/10.1172/JCI59072

Kelley LA, Mezulis S, Yates CM, Wass MN, Sternberg MJ (2015) The Phyre2 web portal for protein modeling, prediction and analysis. Nat. Protocols 10:845–858. https://doi.org/10.1038/nprot.2015.053

Kim SH, Cairns N, Fountoulakisc M, Lubec G (2001) Decreased brain histamine-releasing factor protein in patients with Down syndrome and Alzheimer's disease. Neurosci Lett 300:41–44

Kim M, Min HJ, Won HY, Park H, Lee JC, Park HW, Chung J, Hwang ES, Lee K (2009) Dimerization of translationally controlled tumor protein is essential for its cytokine-like activity. PLoS One 4:e6464

Kim M, Kim M, Kim HY, Kim S, Jung J, Maeng J, Chang J, Lee K (2011) A protein transduction domain located at the NH2-terminus of human translationally controlled tumor protein for delivery of active molecules to cells. Biomaterials 32:222–230. https://doi.org/10.1016/j.biomaterials.2010.08.077

Kim G, LeBlanc ML, Wafula EK, dePamphilis CW, Westwood JH (2014) Plant science. Genomic-scale exchange of mRNA between a parasitic plant and its hosts. Science 345:808–811. https://doi.org/10.1126/science.1253122

Kim HY, Kim S, Pyun HJ, Maeng J, Lee K (2015) Cellular uptake mechanism of TCTP-PTD in human lung carcinoma cells. Mol Pharm 12:194–203. https://doi.org/10.1021/mp500547f

Le TP, Vuong LT, Kim AR, Hsu YC, Choi KW (2016) 14-3-3 proteins regulate TCTP-Rheb interaction for organ growth in Drosophila. Nat Commun 7:11501. https://doi.org/10.1038/ncomms11501

Lespagnol A, Duflaut D, Beekman C, Blanc L, Fiucci G, Marine J-C, Vidal M, Amson R, Telerman A (2008) Exosome secretion, including the DNA damage-induced p53-dependent secretory pathway, is severely compromised in TSAP6/Steap3-null mice. Cell Death Differ 15 (11):1723–1733

Lin MK, Lee YJ, Lough TJ, Phinney BS, Lucas WJ (2009) Analysis of the pumpkin phloem proteome provides insights into angiosperm sieve tube function. Mol. Cell Proteomics 8:343–356. https://doi.org/10.1074/mcp.M800420-MCP200

Liu H, Peng HW, Cheng YS, Yuan HS, Yang-Yen HF (2005) Stabilization and enhancement of the antiapoptotic activity of mcl-1 by TCTP. Mol Cell Biol 25:3117–3126

Lucas WJ, Groover A, Lichtenberger R, Furuta K, Yadav S-R, Helariutta Y, He X-Q, Fukuda H, Kang J, Brady SM, Patrick JW, Sperry J, Yoshida A, López-Millán A-F, Grusak MA, Kachroo P (2013) The plant vascular system: evolution, development and functions. J Integr Plant Biol 55(4):294–388

Ma Y-P, Zhu W-L (2012) Cytoplasmic and nuclear localization of TCTP in normal and cancer cells. Biochem Res Int 2012:871728. https://doi.org/10.1155/2012/871728

MacDonald SM, Rafnar T, Langdon J, Lichtenstein LM (1995) Molecular identification of an IgE-dependent histamine-releasing factor. Science 269:688–690

MacDonald SM, Bhisutthibhan J, Shapiro TA, Rogerson SJ, Taylor TE, Tembo M, Langdon JM, Meshnick SR (2001) Immune mimicry in malaria: Plasmodium falciparum secretes a functional histamine-releasing factor homolog in vitro and in vivo. Proc Natl Acad Sci U S A 98:10829–10832

Maeng J, Kim M, Lee H, Lee K (2015) Insulin induces phosphorylation of serine residues of translationally controlled tumor protein in 293T cells. Int J Mol Sci 16:7565–7576. https://doi.org/10.3390/ijms16047565

Mathieu C, Demarta-Gatsi C, Porcherie A, Brega S, Thiberge S, Ronce K, Smith L, Peronet R, Amino R, Ménard R, Mécheri S (2015) Plasmodium berghei histamine-releasing factor favours liver-stage development via inhibition of IL-6 production and associates with a severe outcome of disease. Cell Microbiol 17:542–558. https://doi.org/10.1111/cmi.12382

Martínez-Navarro AC, Galván-Gordillo SV, Xoconostle-Cázares B, Ruiz-Medrano R (2013) Vascular gene expression: a hypothesis. Front Plant Sci 4:261. https://doi.org/10.3389/fpls.2013.00261

Oh YT, Ahn CS, Jeong YJ, Kim JG, Ro HS, Han KH, Lee CW, Kim JW (2013) Aspergillus nidulans translationally controlled tumor protein has a role in the balance between asexual and sexual differentiation and normal hyphal branching. FEMS Microbiol Lett 343:20–25. https://doi.org/10.1111/1574-6968.12120

Oliveira DL, Nakayasu ES, Joffe LS, Guimarães AJ, Sobreira TJ, Nosanchuk JD, Cordero RJ, Frases S, Casadevall A, Almeida IC, Nimrichter L, Rodrigues ML (2010) Characterization of yeast extracellular vesicles: evidence for the participation of different pathways of cellular traffic in vesicle biogenesis. PLoS One 5:e11113. https://doi.org/10.1371/journal.pone.0011113

Rinnerthaler M, Jarolim S, Heeren G, Palle E, Perju S, Klinger H, Bogengruber E, Madeo F, Braun RJ, Breitenbach-Koller L et al (2006) MMI1 (YKL056c, TMA19), the yeast orthologue of the translationally controlled tumor protein (TCTP) has apoptotic functions and interacts with both microtubules and mitochondria. Biochim Biophys Acta 1757:631–638

Rinnerthaler M, Lejskova R, Grousl T, Stradalova V, Heeren G, Richter K, Breitenbach-Koller L, Malinsky J, Hasek J, Breitenbach M (2013) Mmi1, the yeast homologue of mammalian TCTP, associates with stress granules in heat-shocked cells and modulates proteasome activity. PLoS One 8:e77791. https://doi.org/10.1371/journal.pone.0077791

Rodriguez-Medina C, Atkins CA, Mann AJ, Jordan ME, Smith PM (2011) Macromolecular composition of phloem exudate from white lupin (Lupinus albus L). BMC Plant Biol 11:36. https://doi.org/10.1186/1471-2229-11-36

Roque CG, Wong HH, Lin JQ, Holt CE (2016) Tumor protein Tctp regulates axon development in the embryonic visual system. Development 143:1134–1148. https://doi.org/10.1242/dev.131060

Sage-Ono K, Ono M, Harada H, Kamada H (1998) Dark-induced accumulation of mRNA for a homolog of translationally controlled tumor protein (TCTP) in Pharbitis. Plant Cell Physiol 39:357–360

Silverman JM, Chan SK, Robinson DP, Dwyer DM, Nandan D, Foster LJ, Reiner NE (2008) Proteomic analysis of the secretome of Leishmania donovani. Genome Biol 9:R35. https://doi.org/10.1186/gb-2008-9-2-r35

Sirois I, Raymond MA, Brassard N, Cailhier JF, Fedjaev M, Hamelin K, Londono I, Bendayan M, Pshezhetsky AV, Hébert MJ (2011) Caspase-3-dependent export of TCTP: a novel pathway for antiapoptotic intercellular communication. Cell Death Differ 18:549–562

Suga H, Chen Z, de Mendoza A, Sebé-Pedrós A, Brown MW, Kramer E, Carr M, Kerner P, Vervoort M, Sánchez-Pons N et al (2013) The Capsaspora genome reveals a complex unicellular prehistory of animals. Nat Commun 4:2325. https://doi.org/10.1038/ncomms3325

Susini L, Besse S, Duflaut D, Lespagnol A, Beekman C, Fiucci G, Atkinson AR, Busso D, Poussin P, Marine JC et al (2008) TCTP protects from apoptotic cell death by antagonizing bax function. Cell Death Differ 15:1211–1220. https://doi.org/10.1038/cdd.2008.18

Tanchak MA, Fowke LC (1987) The morphology of multivesicular bodies in soybean protoplasts and their role in endocytosis. Protoplasma 138:173–182

Taylor KJ, Van TT, MacDonald SM, Meshnick SR, Fernley RT, Macreadie IG, Smooker PM (2015) Immunization of mice with Plasmodium TCTP delays establishment of Plasmodium infection. Parasite Immunol 37:23–31. https://doi.org/10.1111/pim.12158

Thaw P, Baxter NJ, Hounslow AM, Price C, Waltho JP, Craven CJ (2001) Structure of TCTP reveals unexpected relationship with guanine nucleotide-free chaperones. Nat Struct Biol 8:701–704

Thébault S, Agez M, Chi X, Stojko J, Cura V, Telerman SB, Maillet L, Gautier F, Billas-Massobrio I, Birck C, Troffer-Charlier N, Karafin T, Honoré J, Senff-Ribeiro A, Montessuit S, Johnson CM, Juin P, Cianférani S, Martinou J-C, Andrews DW, Amson R, Telerman A, Cavarelli J (2016) TCTP contains a BH3-like domain, which instead of inhibiting, activates Bcl-xL. Sci Rep 6(1):19725

Thieme CJ, Rojas-Triana M, Stecyk E, Schudoma C, Zhang W, Yang L, Miñambres M, Walther D, Schulze WX, Paz-Ares J et al (2015) Endogenous Arabidopsis messenger RNAs transported to distant tissues. Nat Plants 1:15025. https://doi.org/10.1038/nplants.2015.25

Toscano-Morales R, Xoconostle-Cázares B, Martínez-Navarro AC, Ruiz-Medrano R (2014) Long distance movement of an Arabidopsis Translationally Controlled Tumor Protein (AtTCTP2) mRNA and protein in tobacco. Front Plant Sci 5:705. https://doi.org/10.3389/fpls.2014.00705

Toscano-Morales R, Xoconostle-Cázares B, Cabrera-Ponce JL, Hinojosa-Moya J, Ruiz-Salas JL, Galván-Gordillo SV, Guevara-González RG, Ruiz-Medrano R (2015) AtTCTP2, an Arabidopsis thaliana homolog of translationally controlled tumor protein, enhances in vitro plant regeneration. Front Plant Sci 6:468. https://doi.org/10.3389/fpls.2015.00468

Tsai MJ, Yang-Yen HF, Chiang MK, Wang MJ, Wu SS, Chen SH (2014) TCTP is essential for β-cell proliferation and mass expansion during development and β-cell adaptation in response to insulin resistance. Endocrinology 155:392–404. https://doi.org/10.1210/en.2013-1663

Tuynder M, Fiucci G, Prieur S, Lespagnol A, Géant A, Beaucourt S, Duflaut D, Besse S, Susini L, Cavarelli J et al (2004) Translationally controlled tumor protein is a target of tumor reversion. Proc Natl Acad Sci U S A 101:15364–15369

van de Sande WW, Janse DJ, Hira V, Goedhart H, van der Zee R, Ahmed AO, Ott A, Verbrugh H, van Belkum A (2006) Translationally controlled tumor protein from Madurella mycetomatis, a marker for tumorous mycetoma progression. J Immunol 177:1997–2005

Vargas G, Rocha JD, Oliveira DL, Albuquerque PC, Frases S, Santos SS, Nosanchuk JD, Gomes AM, Medeiros LC, Miranda K et al (2015) Compositional and immunobiological analyses of extracellular vesicles released by Candida albicans. Cell Microbiol 17:389–407. https://doi.org/10.1111/cmi.12374

Wang X, Fonseca BD, Tang H, Liu R, Elia A, Clemens MJ, Bommer UA, Proud CG (2008) Re-evaluating the roles of proposed modulators of mammalian target of rapamycin complex 1 (mTORC1) signaling. J Biol Chem 283:30482–30492. https://doi.org/10.1074/jbc.M803348200

Wang J, Zhang CJ, Chia WN, Loh CC, Li Z, Lee YM, He Y, Yuan LX, Lim TK, Liu M et al (2015) Haem-activated promiscuous targeting of artemisinin in Plasmodium falciparum. Nat Commun 6:10111. https://doi.org/10.1038/ncomms10111

Weber SS, Parente AF, Borges CL, Parente JA, Bailão AM, de Almeida Soares CM (2012) Analysis of the secretomes of Paracoccidioides mycelia and yeast cells. PLoS One 7:e52470. https://doi.org/10.1371/journal.pone.0052470

Weir W, Karagenç T, Baird M, Tait A, Shiels BR (2010) Evolution and diversity of secretome genes in the apicomplexan parasite Theileria annulata. BMC Genomics 11:42. https://doi.org/10.1186/1471-2164-11-42

Wu H, Gong W, Yao X, Wang J, Perrett S, Feng Y (2015) Evolutionarily conserved binding of translationally controlled tumor protein to eukaryotic elongation factor 1B. J Biol Chem 290:8694–8710. https://doi.org/10.1074/jbc.M114.628594

Xiao B, Chen D, Luo S, Hao W, Jing F, Liu T, Wang S, Geng Y, Li L, Xu W et al (2016) Extracellular translationally controlled tumor protein promotes colorectal cancer invasion and metastasis through Cdc42/JNK/MMP9 signaling. Oncotarget. 10.18632/oncotarget.10315

Yarm FR (2002) Plk phosphorylation regulates the microtubule-stabilizing protein TCTP. Mol Cell Biol 22:6209–6221

Zhuo K, Chen J, Lin B, Wang J, Sun F, Hu L, Liao J (2016) A novel Meloidogyne enterolobii effector MeTCTP promotes parasitism by suppressing programmed cell death in host plants. Mol Plant Pathol. https://doi.org/10.1111/mpp.12374

Chapter 4
The Translational Controlled Tumour Protein TCTP: Biological Functions and Regulation

Ulrich-Axel Bommer

Abstract The Translational Controlled Tumour Protein TCTP (gene symbol TPT1, also called P21, P23, Q23, fortilin or histamine-releasing factor, HRF) is a highly conserved protein present in essentially all eukaryotic organisms and involved in many fundamental cell biological and disease processes. It was first discovered about 35 years ago, and it took an extended period of time for its multiple functions to be revealed, and even today we do not yet fully understand all the details. Having witnessed most of this history, in this chapter, I give a brief overview and review the current knowledge on the structure, biological functions, disease involvements and cellular regulation of this protein.

TCTP is able to interact with a large number of other proteins and is therefore involved in many core cell biological processes, predominantly in the response to cellular stresses, such as oxidative stress, heat shock, genotoxic stress, imbalance of ion metabolism as well as other conditions. Mechanistically, TCTP acts as an anti-apoptotic protein, and it is involved in DNA-damage repair and in cellular autophagy. Thus, broadly speaking, TCTP can be considered a cytoprotective protein. In addition, TCTP facilitates cell division through stabilising the mitotic spindle and cell growth through modulating growth signalling pathways and through its interaction with the proteosynthetic machinery of the cell. Due to its activities, both as an anti-apoptotic protein and in promoting cell growth and division, TCTP is also essential in the early development of both animals and plants.

Apart from its involvement in various biological processes at the cellular level, TCTP can also act as an extracellular protein and as such has been involved in modulating whole-body defence processes, namely in the mammalian immune system. Extracellular TCTP, typically in its dimerised form, is able to induce the release of cytokines and other signalling molecules from various types of immune cells. There are also several examples, where TCTP was shown to be involved in antiviral/antibacterial defence in lower animals. In plants, the protein appears to

U.-A. Bommer (✉)
School of Medicine, Graduate Medicine, University of Wollongong, Wollongong, NSW 2522, Australia
e-mail: ubommer@uow.edu.au

A. Telerman, R. Amson (eds.), *TCTP/tpt1 - Remodeling Signaling from Stem Cell to Disease*, Results and Problems in Cell Differentiation 64,
DOI 10.1007/978-3-319-67591-6_4

have a protective effect against phytotoxic stresses, such as flooding, draught, too high or low temperature, salt stress or exposure to heavy metals. The finding for the latter stress condition is corroborated by earlier reports that TCTP levels are considerably up-regulated upon exposure of earthworms to high levels of heavy metals.

Given the involvement of TCTP in many biological processes aimed at maintaining cellular or whole-body homeostasis, it is not surprising that dysregulation of TCTP levels may promote a range of disease processes, foremost cancer. Indeed a large body of evidence now supports a role of TCTP in at least the most predominant types of human cancers. Typically, this can be ascribed to both the anti-apoptotic activity of the protein and to its function in promoting cell growth and division. However, TCTP also appears to be involved in the later stages of cancer progression, such as invasion and metastasis. Hence, high TCTP levels in tumour tissues are often associated with a poor patient outcome. Due to its multiple roles in cancer progression, TCTP has been proposed as a potential target for the development of new anti-cancer strategies in recent pilot studies. Apart from its role in cancer, TCTP dysregulation has been reported to contribute to certain processes in the development of diabetes, as well as in diseases associated with the cardiovascular system.

Since cellular TCTP levels are highly regulated, e.g. in response to cell stress or to growth signalling, and because deregulation of this protein contributes to many disease processes, a detailed understanding of regulatory processes that impinge on TCTP levels is required. The last section of this chapter summarises our current knowledge on the mechanisms that may be involved in the regulation of TCTP levels. Essentially, expression of the TPT1 gene is regulated at both the transcriptional and the translational level, the latter being particularly advantageous when a rapid adjustment of cellular TCTP levels is required, for example in cell stress responses. Other regulatory mechanisms, such as protein stability regulation, may also contribute to the regulation of overall TCTP levels.

4.1 Introduction

Despite there being more than 300 scientific publications on 'TCTP' that have been published on this protein over a period of 35 years, it is still difficult to give a brief, comprehensive statement on its functional importance. Typically, papers on this protein start with a sentence such as: 'TCTP is a highly-conserved, multifunctional protein involved in important biological and disease processes.' This reflects the fact that by now, many functional aspects of this protein have been revealed, many of which are highly relevant to core biological and medical problems. The recent increasing interest in this protein is also reflected by the fact that 15 review articles on TCTP have already been published, all but two (Bommer and Thiele 2004; Telerman and Amson 2009) within the last 5 years. However, most of these articles focus on specific aspects of the TCTP literature, and they are spread over a wide range of journals. I feel therefore it is a very worthwhile undertaking by Amson and Telerman to assemble this new collection of up-to-date reviews into one single book. It should be a useful guide for the still growing 'TCTP community', which currently comprises research groups from about 25 countries.

4.1.1 The 'Translationally Controlled Tumour Protein TCTP': Names and History

The 'TCTP story' had a humble beginning with less than ten papers published during the 1980s by three research groups, all interested in protein synthesis and in the hunt for translationally controlled proteins. Initially, the protein was termed according to its approximate molecular mass as Q23 (G. Thomas, Basel; Thomas et al. 1981), P21 (G. Brawermann, Boston; Yenofsky et al. 1982) or as P23 (H. Bielka, Berlin-Buch; Bohm et al. 1989). At this time, the function of the protein was completely unknown, and its most distinguished property was the ability to be rapidly up-regulated upon growth induction of serum-starved murine cells, in a manner insensitive to inhibition by actinomycin D (Thomas et al. 1981; Bohm et al. 1989). Even the publication of the first cDNA sequences of mouse P21 (Chitpatima et al. 1988) and of its human homologue (Gross et al. 1989) did not shed any light on its possible function, as the derived amino acid sequence did not display similarity to any other known protein sequence, and since then, the protein is listed in the databases as a separate 'family'. It was in the latter publication (Gross et al. 1989) that the name 'translationally controlled tumour protein' was coined, since the cDNA sequence had been derived from a human mammary carcinoma.

The first report on a functional association of TCTP appeared in 1995. Susan MacDonald and co-workers (Baltimore) identified an extracellular function of this protein as 'histamine-releasing factor' (HRF) present in biological fluids of allergic patients (MacDonald et al. 1995). This discovery led to a series of studies exploring the extracellular signalling function of HRF/TCTP in allergic and inflammatory responses, as recently reviewed by Dr. MacDonald (2012a, b).

The first studies on intracellular functions of TCTP reported its Ca^{2+}-binding activity (Haghighat and Ruben 1992) and its microtubule association. Our study revealed that TCTP/P23 is bound to microtubules, inclusive of the mitotic spindle, in a cell-cycle-dependent manner, and that it has microtubule-stabilising activity (Gachet et al. 1999). Subsequently, a whole array of additional interaction partners of TCTP/HRF have been identified (reviewed in Kawakami et al. 2012; Bommer 2012; Acunzo et al. 2014).

Another milestone in the unravelling of TCTP's function was the demonstration of its anti-apoptotic activity by Ken Fujise's group (Houston) (Li et al. 2001), exactly 20 years after its initial discovery (Thomas et al. 1981; Yenofsky et al. 1982). Based on their findings, these authors invented the name 'fortilin' for this protein. Their discovery led to a large number of additional studies on the cytoprotective role of

TCTP/fortilin, which showed that the protein is rapidly up- (or down-)regulated in response to a wide array of cellular stresses and that it is involved in a range of cellular defence pathways (reviewed in Bommer 2012; Acunzo et al. 2014).

Despite its name, a convincing demonstration that TCTP is implicated in cancer came only relatively late. Instrumental in this was the experimental 'tumour reversion model' developed by the Telerman group (Telerman and Amson 2009; Tuynder et al. 2002, 2004). Naturally, the cancer aspect of TCTP's 'function' has recently attracted increasing attention, as documented by the about 40 papers and 6 review articles (Telerman and Amson 2009; Bommer 2012; Acunzo et al. 2014; Amson et al. 2013, 2011; Chan et al. 2012a; Efferth 2006; Koziol and Gurdon 2012) on 'TCTP and cancer' that appeared only within the last 10 years.

Further additions to the complex story of the naming of this protein are in reports on TCTP orthologues in non-vertebrate animals, lower eukaryotes and plants. Typically, in these papers, a two-letter suffix before 'TCTP' indicates the species, from which the protein sequence was derived (see Gutierrez-Galeano et al. (2014) for an example). The yeast orthologue of TCTP was named 'microtubule and mitochondria interacting protein' (Mmi1p) (Rinnerthaler et al. 2006), but in another paper, it was also identified as the 'translation-machinery-associated protein 19' (TMA19) (Fleischer et al. 2006), and yet the gene symbol in yeast is YKL056c. In contrast, the gene symbol in higher eukaryotes is TPT1, for 'tumour protein, translationally controlled-1' (Thiele et al. 2000).

4.1.2 Gene Structure and mRNA

The first structure of a TPT1 gene (for rabbit) was characterised in Bernd Thiele's group in Berlin; the gene was found to comprise about 4000 nucleotides, beginning at the transcription start site, and to consist of 6 exons and 5 introns, where the last exon comprises the entire 3′-UTR of the mRNA (Thiele et al. 1998). A very similar structure was also reported for the mouse (Fiucci et al. 2003) and the human (Andree et al. 2006) TPT1 gene. The latter has been mapped to chromosome 13, q12-->q14 (MacDonald et al. 1999), the mouse gene to chromosome 14 (Fiucci et al. 2003) and the pig TPT1 gene to chromosome 11 (Yubero et al. 2009), in a genomic context that is similar to that of the human chromosomal location.

The promoter regions of the active TPT1 gene were further investigated for the rabbit (Thiele et al. 1998), mouse (Fiucci et al. 2003) and human (Andree et al. 2006) genes. They displayed a large number of transcription factor-binding motifs, i.e. 10–15 motifs within the 450 nucleotides adjacent to the transcription start site. Of these, seven motifs were conserved between all five mammalian species inspected, and the activity of the tandem CREB transcription factor-binding site was experimentally demonstrated for the human TPT1 gene (Andree et al. 2006).

The early characterisation and mapping of the TPT1 gene was complicated by the fact that there is a considerable number of processed or unprocessed pseudogenes in mammalian genomes, which is apparently not the case in lower eukaryotes, fungi, plants and non-mammalian animals (Hinojosa-Moya et al. 2008). Several of these pseudogenes have been studied in some detail in the rabbit (Thiele et al. 2000) and in mouse (Fiucci et al. 2003), and only few were found to have the potential to produce a protein product.

Northern blot analysis of TCTP mRNAs from a range of different tissues revealed that all cells produce two different TCTP mRNA species in both rabbits (Thiele et al. 2000) and in humans (Andree et al. 2006), the longer one bearing a 320 nucleotide extension at the 3′-end, which is generated by an additional polyadenylation site. These two isoforms comprise a total of about 800 and 1200 nucleotides, respectively (Andree et al. 2006). The ratio between the two mRNA species differs considerably between different tissues, but the short mRNA is always in excess (Thiele et al. 2000). The importance for the production of these two different mRNA species has not yet been elucidated, but interestingly, in mouse tissues, there was only one TCTP mRNA isoform detected (Fiucci et al. 2003).

Mammalian TCTP mRNAs contain an open reading frame of 519 nt (172 amino acids), a 5′-UTR of about 100 nt and a 3′-UTR of about 200 nt or of 520 nt for the longer isoform (Thiele et al. 2000). A particular feature of TCTP mRNAs, indicative for translational regulation, is a 5′-terminal oligopyrimidine tract (5′-TOP): sequence 5′-CTTTTCCG . . . for the human, mouse and rabbit mRNAs (Thiele et al. 1998, 2000; Fiucci et al. 2003). The 5′-TOP motif is a hallmark for a specific group of mRNAs that are under translational control through the mTOR complex 1 (mTORC1) signalling pathway (Meyuhas and Kahan 2015; Yamashita et al. 2008). We have recently demonstrated that growth induction of TCTP expression is indeed regulated through this pathway (Bommer et al. 2015).

A second feature is the high CG-content in the 5′-UTR (about 80% in the mouse sequence), which indicates the potential of the molecule to form a high degree of secondary structure. Our lab has indeed shown that full-length TCTP mRNA of the mouse is a very structured molecule, the structure formation being dependent on the presence of the 5′- and the 3′-UTR. Consequently, the full-length mRNA molecule is poorly translated, in contrast to a truncated version devoid of the 5′- and 3′-UTRs (Bommer et al. 2002). Yet another feature is the presence of several AUUUA elements in the 3'-UTR of the TCTP mRNA (Gross et al. 1989; Thiele et al. 2000). Such elements typically target cytokine mRNAs for destabilisation; however, the AUUUA motifs present in TCTP mRNA do not comprise the complete consensus sequence for such destabilisation elements. To my knowledge, there is currently only limited experimental evidence for regulated degradation of TCTP mRNA. In contrast, the aforementioned high degree of structure formation (Bommer et al. 2002) would indicate that it is a rather stable mRNA molecule, consistent with the early observation that a large proportion of the mRNA exits in untranslated mRNP particles (Yenofsky et al. 1982).

4.1.3 Molecular Structure, Conservation and Interactions

As mentioned above, elucidation of the amino acid sequence of TCTP did not reveal anything about the functional implication of the protein. Only the first publication of the solution structure of TCTP from the fission yeast (Thaw et al. 2001) provided some structure-derived information about a possible functional association of the protein. It revealed similarity to the Mss4/Dss4 family of proteins, a group also called guanine nucleotide-free chaperones.

The amino acid sequences of TCTP are highly conserved throughout all eukaryotic taxa; several sequence comparisons have been published recently (Yubero et al. 2009; Hinojosa-Moya et al. 2008; Thayanithy 2005), and a detailed phylogenetic analysis using 93 different TCTP sequences revealed that, by and large the phylogeny of TCTP sequences is consistent with the evolutionary history, although there were some exceptions (Hinojosa-Moya et al. 2008).

The three-dimensional structures of the following TCTP proteins were solved and are deposited in the databases: The solution structures of fission yeast (Thaw et al. 2001) and human TCTP (Feng et al. 2007a), as well as the crystal structures of *Plasmodium knowlesi* (Vedadi et al. 2007) and human TCTP (Susini et al. 2008) at 2 Å resolution. A mutant version of human TCTP showing a structure very similar to the normal protein (Dong et al. 2009) can be found as well. Figure 4.1 shows the solution structure of human TCTP as an example, since in the crystal structure, the flexible loop domain is not displayed. The figure demonstrates the three major domains of the TCTP protein.

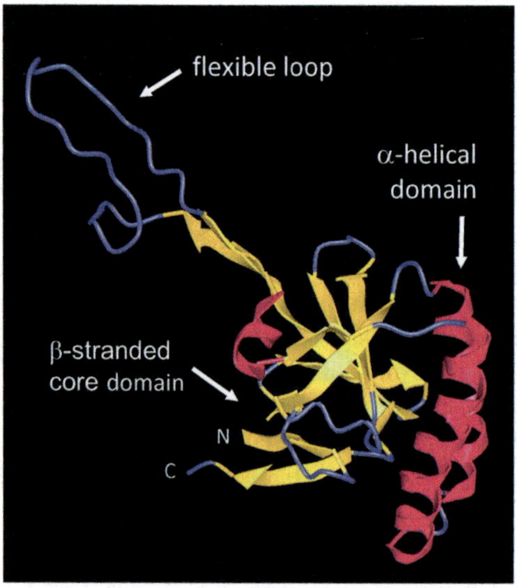

Fig. 4.1 NMR structure of human TCTP. Secondary structure elements are highlighted as follows: *magenta*: α-helices, *yellow*: β-sheets, *blue lines*: unstructured areas; major domains are indicated (from: Feng et al. 2007a, PDB ID: 2HR9)

A considerable number of 3D structures of other members of the TCTP family, in particular from various plant species, have been predicted (Gutierrez-Galeano et al. 2014; Hinojosa-Moya et al. 2008; Berkowitz et al. 2008), based on the experimentally determined 3D-structures of TCTP and on sequence comparisons. Phylogenetic comparisons showed that the predicted tertiary structure of TCTP proteins is in general highly conserved throughout evolution (Gutierrez-Galeano et al. 2014; Hinojosa-Moya et al. 2008), although there are some deviations in the TCTP structure of lower eukaryotes (Hinojosa-Moya et al. 2008) and some species of chlorophyta do not seem to have a TCTP gene at all (Gutierrez-Galeano et al. 2014). In the plant kingdom, two principal types of TCTP structures are being distinguished, with some plants having the gene for one of the two types, whereas other ones have genes for both types (Gutierrez-Galeano et al. 2014).

Since TCTP is involved in a wide range of biological processes, it is not surprising that it undergoes interactions with a large number of partner proteins. The distinct domain structure (Fig. 4.1) is likely to enable the protein to interact with many other proteins (and with itself) in a specific manner. In 2012, Toshi Kawakami published a review article summarising the TCTP/HRF interactions known at the time (Kawakami et al. 2012). In the meantime, a considerable number of additional interaction partners were identified, so it is timely now to compile an updated list, as shown here in Table 4.1. This table provides an overview on the many interaction partners and biological processes TCTP/fortilin/HRF is involved in, some of which will be discussed in the following sections. Recent interaction screens indicate that a much larger number of proteins have the potential to interact with TCTP (see e.g. Li et al. 2016), however only a limited number of those have actually been validated today.

4.2 Biological Functions of TCTP

4.2.1 Maintaining Cell Homoeostasis and Survival

4.2.1.1 Anti-apoptotic Activity: Discovery and Mechanisms

The anti-apoptotic activity of TCTP was originally discovered in Ken Fujise's laboratory, where the name 'fortilin' was coined (Li et al. 2001). Their study described the cytoprotective effect of TCTP/fortilin overexpression against etoposide-induced apoptosis in HeLa cells. Since then, the anti-apoptotic activity of TCTP has been reported on numerous occasions and in a range of different settings, which will be discussed below. Major support for the cytoprotective role of TCTP came from gene-knockout studies in mice, which demonstrated that TCTP-knockout was embryonic lethal, due to excessive apoptosis at an early embryonic state (Susini et al. 2008; Chen et al. 2007a; Koide et al. 2009). The following mechanisms have been proposed to underlie the anti-apoptotic activity of TCTP/fortilin:

Table 4.1 TCTP interaction partners

Biological function	Binding partner	Binding site in TCTP[a]	Biological importance	References
Stress response	Na-K-ATPase	C-terminus	Na-K-ATPase inhibition	Jung et al. (2004), Kim et al. (2008a)
	Sorting nexin 6 (SNX6)		Na-K-ATPase activation	Yoon et al. (2006)
	Ca^{2+}	N- and C- terminus	Ca^{2+}-scavenging	Feng et al. (2007a), Graidist et al. (2007), Kim et al. (2000)
	Vitamin-D3 receptor	α-Helical domain	Ca^{2+}-related processes	Rid et al. (2010)
	Peroxiredoxin-1		Protection from ROS	Chattopadhyay et al. (2016)
	ATM kinase and γH2A.X		DNA damage response (DDR)	Zhang et al. (2012)
	Brahma		Genome stability	Hong and Choi (2016)
	Y-box binding protein 1	N-terminus	Stress response, DNA damage response (DDR)	Li et al. (2016)
	Stress granules		Heat shock protection	Rinnerthaler et al. (2013)
	Ubp3, Cdc48 in yeast		Proteasome inhibition	Rinnerthaler et al. (2013)
	VHL Tumour suppressor		Ubiquitinylation of VHL, stabilisation of HIF-1α	Chen et al. (2013a)
Survival, apoptosis prevention	p53	α-Helical domain	Degradation of p53	Amson et al. (2012), Chen et al. (2011), Rho et al. (2011)
	MDM2/HDM2		Degradation of p53	Amson et al. (2012), Funston et al. (2012)
	Mcl-1	N-terminus	Stabilisation of Mcl-1/TCTP	Liu et al. (2005), Roque et al. (2016), Zhang et al. (2002)
	Bcl-XL	α-Helical domain	T-cell survival	Yang et al. (2005), Thebault et al. (2016)
	Apaf-1		Chemoresistance	Jung et al. (2014)
	Hsp27		Stabilisation of TCTP	Baylot et al. (2012)
	Mitochondria	α-Helical domain	Apoptosis prevention	Rinnerthaler et al. (2006), Susini et al. (2008)

Cytoskeleton, cell cycle/ division	α- and β-tubulin	α-Helical domain		Gachet et al. (1999)
	Microtubules	α-Helical domain	Microtubule stabilisation	Gachet et al. (1999), Jaglarz et al. (2012)
	Mitot./meiotic spindle		Spindle stabilisation	Gachet et al. (1999), Jeon et al. (2016)
	Chfr (checkpoint protein)		Located at spindle	Burgess et al. (2008)
	Actin filaments			Bazile et al. (2009), Tsarova et al. (2011)
	Plk-1 kinase	Flexible loop	Cell cycle progression	Yarm (2002), Cucchi et al. (2010)
	Pim-3 kinase	N-terminus	Stabilisation of Pim-3	Zhang et al. (2013)
	Cdc25C phosphatase		Cdc25C degradation	Chan et al. (2012b)
Early development	ATM kinase		Organ development	Hong and Choi (2013)
	Nucleophosmin		ES cell proliferation	Johansson et al. (2010a)
	Nucleolin		ES cell proliferation	Johansson et al. (2010b)
	Oct-4		ES cell maintenance	Johansson and Simonsson (2010)
	ATG16		Regulation of autophagy	Chen et al. (2014a)
Protein synthesis; growth regulation	eEF1A		Stabilisation of eEF1A-GDP	Cans et al. (2003)
	eEF1Bβ	α-Helical domain and core domain	Inhibition of the GEF activity of eEF1Bβ	Cans et al. (2003), Langdon et al. (2004), Wu et al. (2015)
	40S ribosomal subunit		Ribosome binding in yeast	Fleischer et al. (2006)
	Rheb GTPase	Core domain	Drosophila, TOR activation	Dong et al. (2009), Hsu et al. (2007)
	14-3-3 proteins	(TCTP and Rheb)	Drosophila, organ growth	Le et al. (2016)
Extracellular function and allergic response	TSAP6		TCTP export in exosomes	Amzallag et al. (2004)
	TCTP		Dimerisation	Kim et al. (2009a), Lucas et al. (2014)
	Specific antibodies	TCTP-dimer		Kashiwakura et al. (2012)
	Haemin	His76/His77	TCTP dimerisation	Lucas et al. (2014)

[a]Where determined, the location of the binding site in the respective TCTP domain (Fig. 4.1) is indicated

1. *Cooperation with other anti-apoptotic proteins.* Two anti-apoptotic proteins of the Bcl-2 family have been shown to interact with TCTP, i.e. Mcl-1 (Liu et al. 2005; Zhang et al. 2002) and Bcl-XL (Yang et al. 2005). In the case of Mcl-1, it was reported that TCTP stabilises Mcl-1 (Liu et al. 2005) and vice versa (Zhang et al. 2002), but another study also showed that both proteins are able to exert their anti-apoptotic activity independently of each other (Graidist et al. 2004). Recently, the interaction of TCTP with Bcl-XL was investigated in more detail (Thebault et al. 2016). This study demonstrated that TCTP has a BH3-like domain. Binding to such BH3 domains (Bcl2 homology domain 3) typically inhibits anti-apoptotic proteins; however, TCTP's BH3-like domain actually activates the anti-apoptotic activity of Bcl-XL (Thebault et al. 2016). Thus, in both cases, Mcl-1 and Bcl-XL, interaction with TCTP stimulates or maintains their activity.
2. *Preventing apoptotic mechanisms.* The pro-apoptotic protein Bax promotes the execution of apoptosis by inserting itself via an α-helical domain into the mitochondrial membrane, forming a homo- or heterodimer (with Bak) and inducing membrane permeability and cytochrome c release. A detailed study by the Telerman-group demonstrated that the α-helical domain of TCTP (Fig. 4.1) resembles that domain of Bax, which is crucial for its activity (Susini et al. 2008). They also showed that TCTP, with its α-helical domain, is able to insert itself into the mitochondrial membrane and to prevent the dimerisation and activation of Bax. Translocation of TCTP (Mml1) to the mitochondrial surface was also observed in yeast after mild oxidative stress (Rinnerthaler et al. 2006). Another apoptotic mechanism affected by TCTP is the formation of the apoptosome. A core component of this structure is the apoptotic protease activating factor (Apaf-1), whose inactivation has been implicated in carcinogenesis and the development of anticancer drug resistance (Fadeel et al. 2008). A recent report by the group of Kyunglim Lee documented that TCTP binds to Apaf-1 via its caspase recruitment domain and inhibits the activation of caspase 9 (Jung et al. 2014).
3. *Antagonism to p53.* The tumour suppressor protein p53 is a very powerful pro-apoptotic agent, so interfering with its activity is an essential strategy for anti-apoptotic players. Conversely, down-regulation of anti-apoptotic proteins forms part of the armory of p53 to promote apoptosis. Indeed down-regulation of TCTP through activation of a (temperature-sensitive) mutant protein of p53 (Tuynder et al. 2002; Bommer et al. 2010) was an initial observation in this context. The mutual antagonism between p53 and TCTP was then described by three groups in brief succession (Amson et al. 2012; Chen et al. 2011; Rho et al. 2011). The underlying mechanisms involved in this included the following: (1) TCTP stimulates degradation of p53 (Rho et al. 2011) by binding to P53-MDM2 complexes [or to HDM2 (Funston et al. 2012)] and promoting MDM2-mediated ubiquitination and degradation of P53 (Amson et al. 2012). (2) P53 as a transcription factor binds to a P53 responsive element in the promoter region of the TPT1 gene and inhibits its transcription (Amson et al. 2012) [although p53-dependent *induction* of TPT1 transcription by has also been reported (Chen et al. 2013b)]. (3) TCTP levels are also translationally regulated

through the PI3-kinase/Akt/mTORC1 signalling pathway (Bommer et al. 2015). Since this pathway is targeted by p53, through induction of its negative effectors PTEN and TSC2 (Feng et al. 2007b), this implies that p53 also interferes with TCTP synthesis at the translational level.

4.2.1.2 Involvement in Cellular Stress Responses

The cytoprotective action of TCTP has been demonstrated in a variety of cellular stress responses, such as heat shock, oxidative, genotoxic or Ca^{2+}-stress and in a range of different cell lines. These effects of TCTP also contribute to its involvement in the development of drug resistance, a frequent problem in chemotherapeutic treatment of cancer.

1. *TCTP and heat shock.* Work on TCTP proteins from parasitic organisms (*Trichinella* and Filarials) showed that TCTP behaves as a heat-shock protein with chaperone-like activity. Its synthesis is induced after heat-shock treatment (Gnanasekar et al. 2009; Mak et al. 2007). Both human and filiarial TCTP (from *Schistosoma mansoni*) bind to denatured and native proteins, protecting the latter from thermal denaturation (Gnanasekar et al. 2009). The yeast homologue of TCTP, Mml1, has been shown to associate with stress granules in heat-shocked cells and to modulate proteasome activity (Rinnerthaler et al. 2013).
2. *TCTP in oxidative stress.* Another filarial TCTP protein (from *Brugia malayi*) was reported to display anti-oxidant activity. The reduced form of BmTCTP was able to protect DNA from oxidative damage (Gnanasekar and Ramaswamy 2007). Similarly, TCTP prevented hydrogen peroxide-induced cell death in murine fibroblasts (Nagano-Ito et al. 2009). Lucibello et al. studied the behaviour and role of TCTP in a panel of cancer cell lines under oxidative stress conditions. They observed that TCTP is up-regulated in these cells under mild oxidative stress, whereas strong oxidative stress resulted in reduced TCTP levels and subsequent apoptotic cell death. Increased TCTP levels partially protected cells from oxidation-induced cell death (Lucibello et al. 2011). Similarly, overexpression of TCTP (TMA19) in yeast conferred resistance to arsenite to the cells (Takahashi et al. 2010). A mechanism, by which TCTP (fortilin) could exert this protective effect against ROS-induced cell death, was recently proposed by the Fujise laboratory (Chattopadhyay et al. 2016). They reported that TCTP protects the enzyme peroxiredoxin-1 from degradation and keeps it in an active state by blocking its deactivating phosphorylation by the protein kinase Mst1.
3. *TCTP, genotoxic stress and genome stability.* The participation of TCTP in the sensing and repair of DNA damage was first demonstrated by Zhang and colleagues in 2012 (Zhang et al. 2012). This involved the up-regulation of TCTP after γ-irradiation of cells, its complex formation with the ataxia-telangiectasia mutated (ATM) kinase and with the histone γH2A.X and colocalisation with other DNA-damage marker proteins. Lack of TCTP resulted in severe deficiency in chromosome damage repair. A more recent report on the interaction of

Drosophila TCTP confirmed the (genetic) interaction with the ATM kinase and its activation by TCTP (Hong and Choi 2013). These studies demonstrated the importance of TCTP in maintaining genome stability under genotoxic stress, which is consistent with observations from Fujise's and our own laboratory, showing that TCTP is able to protect cancer cells against the cytotoxicity exerted by DNA-damaging anticancer drugs (Graidist et al. 2004; Bommer et al. 2017). The involvement of TCTP in the development of chemoresistance in cancer chemotherapy has also been emphasised by several other papers (Jung et al. 2014; Takahashi et al. 2010; He et al. 2015; Sinha et al. 2000).

4. *Ca^{2+}-stress and maintenance of ion homeostasis*. Possibly the earliest functional association of the TCTP protein to be discovered was the Ca^{2+}-binding activity of a protein from Trypanosomes (Haghighat and Ruben 1992). Later, it was shown that TCTP levels are regulated in response to perturbations of intracellular Ca^{2+}-homeostasis (Xu et al. 1999). Although TCTP does not have a canonical Ca^{2+}-binding site (Kim et al. 2000), its Ca^{2+}-binding activity has been confirmed by several laboratories using a range of methods (Feng et al. 2007a; Graidist et al. 2007; Kim et al. 2000; Lucas et al. 2014; Arcuri et al. 2004, 2005; Sanchez et al. 1997). However, the different approaches used to map the Ca^{2+}-binding site at the TCTP protein yielded different results (Feng et al. 2007a; Graidist et al. 2007; Kim et al. 2000). Feng et al. demonstrated that it is a weak binding site (Feng et al. 2007a); they provided the NMR structure of human TCTP (Fig. 4.1) and mapped the Ca^{2+}-binding site to a conserved part of the β-stranded core domain, close to the 'hinge' region that connects it to the α-helical domain. This result was consolidated and further refined in a very recent study (Lucas et al. 2014). TCTP/fortilin has been reported to act as Ca^{2+}-scavenger, and this was proposed as a yet another mechanism of its anti-apoptotic activity (Graidist et al. 2007). Indeed, overexpression of TCTP has been shown to protect cells against Ca^{2+}-dependent apoptosis induced by thapsigargin (Graidist et al. 2007; Bommer et al. 2010), a reagent that inhibits the Ca^{2+}-pump of the ER membrane, resulting in a significant increase in cytosolic Ca^{2+}-levels. The notion of TCTP acting as a Ca^{2+}- scavenger, or more generally as a 'buffer-like' molecule for Ca^{2+}, is consistent with its high abundance and with the fact that its binding site is rather weak (Lucas et al. 2014). The importance of such a Ca^{2+}-buffer-like function for TCTP was described for trophoblast cells of the placenta, where it is involved in Ca^{2+}-handling and in its provision for the foetal blood circulation (Arcuri et al. 2005). Similarly, in human prostate epithelial cells, TCTP is the Ca^{2+}-binding protein with the highest expression levels (Arcuri et al. 2004).

Another important player in maintaining intracellular ion homeostasis that is influenced by TCTP is the membrane-bound Na^+,K^+-ATPase. Kyunglim Lee's group discovered that TCTP interacts with the third cytoplasmic domain of the alpha subunit of the Na^+,K^+-ATPase in HeLa cells and inhibits its activity (Jung et al. 2004). They also identified another protein, sorting nexin 6 (SNX6), that binds to TCTP and neutralises TCTP's negative effector activity on the Na^+,K^+-ATPase, but does not have an inhibitory activity towards the enzyme on its own

(Yoon et al. 2006). Since the Na^+,K^+-ATPase has been implicated in the pathogenesis of hypertension, the group went on to employ a transgenic mouse model that overexpresses TCTP. They showed that these mice developed systemic arterial hypertension about 6 weeks after birth, due to reduced Na^+,K^+-ATPase activity and increased intracellular calcium (Kim et al. 2008a).

5. *Other stress conditions*. Regulation of TCTP levels in response to various stress situations in plants has been studied recently. Protection against high or low temperatures, salt stress or flooding are particularly important for plant survival, and TCTP is regulated under all these conditions. For example, a recent study on maize leaves subjected to flooding stress, reported that TCTP was increased in response to flooding stress, possibly mediated by increased production of hydrogen peroxide (Chen et al. 2014b). TCTP from the rubber tree (*Hevea brasiliensis*) was found to be regulated by drought, low temperature, high salt, ethylene, hydrogen peroxide or wounding (Deng et al. 2016; Li et al. 2013). This is consistent with the observation that overexpression of TCTP in *Arabidopsis* resulted in increased drought tolerance of the plants (Kim et al. 2012a). Moreover, in the cabbage (*Brassica oleracea*), TCTP was found to be involved in maintaining tolerance to heat, cold and to salt stresses (Cao et al. 2010). A study on TCTP from cassava (*Manihot esculenta Crantz*) reported that the promoter region of the TCTP gene harbours regulatory elements responsive to sodium chloride and to pathogen infection, and that expression of this gene in bacteria conferred a protective effect against salt stress (Santa Brigida et al. 2014). Two other types of phytotoxic stresses, where TCTP appears to have a protective role, is the stress induced by an excess of aluminium in slightly acidic soil (Ermolayev et al. 2003) and the effect of mercury, as investigated in rice (Wang et al. 2012). These findings relate to one of the earliest observations, which reported an up-regulation of TCTP levels in cellular stresses. This study investigated the effect of heavy metal-contaminated soils from mining areas on earthworms and found that the presence of copper or cadmium in these soils led to an extreme up-regulation of TCTP levels in these animals (Sturzenbaum et al. 1998). Subsequently in mammalian cells, a metal-responsive element was discovered in the TPT1 promoter that is highly responsive to copper (Schmidt et al. 2007), confirming that this regulation occurs largely at the transcriptional level.

4.2.2 *Involvement in the Cell Cycle and in Early Development*

4.2.2.1 TCTP, a Microtubule Stabilising Protein at the Mitotic and the Meiotic Spindle

The property of TCTP/P23 as a tubulin-binding protein that associates with microtubules in a cell cycle-dependent manner was discovered in our laboratory in the 1990s (Gachet et al. 1999). Using deletion mutations, we characterised the tubulin-binding domain of TCTP as the part of the molecule, which was later identified as the α-helical domain (Fig. 4.1; Thaw et al. 2001). This domain is characterised by a rather basic charge, compared to the overall very acidic charge of the protein; 58% of all basic amino acids are located in this domain, which also shows similarity to the tubulin-binding domain of the canonical microtubule-associated protein MAP-1B (Gachet et al. 1999). Furthermore, we established that overexpression of TCTP/P23 results in microtubule stabilisation, and we showed that the protein binds to the mitotic spindle, but is detached from the spindle during metaphase-anaphase transition (Gachet et al. 1999). Not shown in this paper are our results on mitotic phosphorylation of TCTP/P23 (Fig. 4.2). In this experiment, we tested the TCTP phosphorylating activity in cell extracts from synchronised HeLa cells at different time points after release from the S-Phase block (Fig. 4.2a). The isoform pattern of TCTP in these cells was analysed by 2D electrophoresis in the lab of Jean-Charles Sanchez, Geneva (Fig. 4.2b). These results demonstrated that TCTP phosphorylating activity is low throughout the cell cycle, but peaks sharply (nearly ten-fold) in mitosis, which is consistent with a brief occurrence of an additional

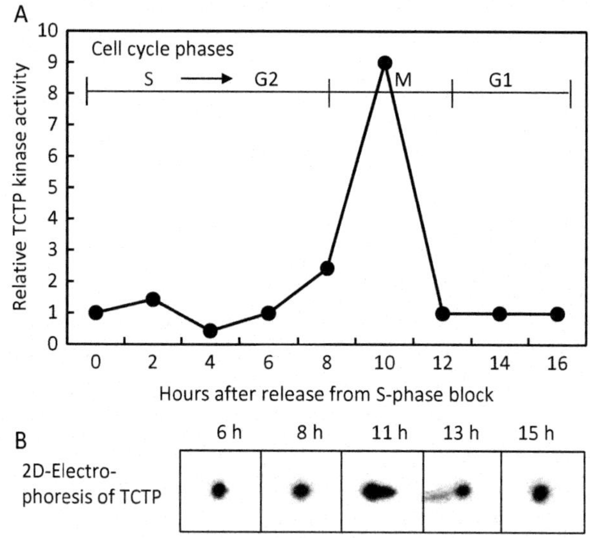

Fig. 4.2 TCTP is phosphorylated by a mitotic protein kinase activity. HeLa cells were synchronised by an S-phase block using aphidicoline. Cell extracts were prepared at the indicated time points after release from the block. The cell cycle phases indicated at the top of the graph were verified by FACS-scan analysis (not shown). (**a**) Cell extracts were tested for TCTP kinase activity by incubating of cell extracts (30 μg total protein) with 1 μg of GST–TCTP fusion protein in the presence of 5 μCi [γ-^{32}P]ATP in 50 mM MOPS buffer, pH 7.2, containing 1 mM DTT, 5 mM $MgCl_2$ and 5 mM p-NPP. The GST–TCTP fusion protein was recovered from the mixture on glutathione-agarose beads and analysed for ^{32}P incorporation by SDS electrophoresis, which was quantified using a phosphoimager. The assay conditions were optimised before, and the incorporation was shown to be specific for TCTP. (**b**) Cells were harvested at the indicated time points after release from the aphidicoline block and analysed by 2D-electrophoresis for TCTP isoforms (Y. Gachet, M. Lee, I. Demalte, J.C. Sanchez and U.A. Bommer, unpublished results)

isoform of TCTP/P23 at the same time (Fig. 4.2b). The protein kinase that catalyses the mitotic phosphorylation of TCTP was subsequently characterised by Frederic Yarm as the mitotic master kinase Plk-1 (Yarm 2002). He also identified the phosphorylation sites for Plk-1 as the serine residues 46 and 64 in the flexible loop of the TCTP protein. Expression of a mutant protein bearing serine to alanine mutations in these sites resulted in a dramatic increase in the number of multinucleated cells, indicating that Plk-1-dependent phosphorylation of TCTP is mediating the detachment of TCTP from the spindle, which in turn is essential for the orderly progression through mitosis. In later studies, TCTP phosphorylation has even been exploited as a marker for Plk-1 activity (Cucchi et al. 2010) and as a potential prognostic marker in testing the efficacy of anticancer drugs (Lucibello et al. 2015).

Localisation of TCTP to the mitotic spindle, particularly the spindle poles, was also reported by other laboratories (Jaglarz et al. 2012; Burgess et al. 2008; Bazile et al. 2009). A very recent paper even demonstrated the importance of TCTP for the *meiotic* spindle in mouse oocytes (Jeon et al. 2016). These authors found that TCTP is predominantly bound to the spindle poles and contributes to the stability of pole microtubules, but not of kinetochore microtubules. TCTP is phosphorylated during meiosis [as also in seen in bovine oocytes (Tani et al. 2007)], and overexpression of a non-phosphorylatable mutant of TCTP led to disturbances of meiotic maturation (Jeon et al. 2016).

A detailed study on plant (*Arabidopsis thaliana; At*) TCTP established that the protein acts as mitotic growth integrator in both plants and animals by controlling the duration of the cell cycle. In this function, the *Arabidopsis* and *Drosophila* proteins are exchangeable in their respective systems (Brioudes et al. 2010). The microtubule-binding activity of AtTCTP has been investigated in yet another paper (Kim et al. 2012a); the binding to actin filaments (Bazile et al. 2009) and identification of an actin-binding site (Tsarova et al. 2011) were also reported for TCTP.

The importance of TCTP in cell cycle progression has been specifically studied in cancer cells. For example, TCTP promoted cell cycle progression in pancreatic cancer cells through stabilisation of the protein kinase Pim-3 (Zhang et al. 2013). In hepatocellular carcinoma (HCC), the oncogene CHD1L was found to drive the overexpression of TCTP, resulting in an increased number of mitotic defects (Chan et al. 2012b). TCTP in turn promoted the degradation of Cdc25C during mitosis, leading to a faster mitotic exit and miss-segregation of chromosomes and consequently to chromosomal instability.

4.2.2.2 Roles for TCTP in Early Development

1. *TCTP in the reproductive system.* The first reports on the role for TCTP in the mammalian reproductive system appeared about 15 years ago. Specifically, the developmental expression of TCTP in the rat and human testes were studied by the group of Sanchez and Hochstrasser in Geneva (Guillaume et al. 2001), and the importance of TCTP as a Ca^{2+}-binding protein in the prostate (Arcuri et al.

2004) and the placenta (Arcuri et al. 2005) were demonstrated by Arcuri et al. A more recent study showed that TCTP is essential for the implantation of embryos in the uterus of mice (Li et al. 2011). An important role for TCTP in egg production in nematodes, such as *C. elegans*, was demonstrated by TCTP knock-down, which severely reduced the number of eggs produced by these worms (Meyvis et al. 2009), consistent with the role of TCTP for meiotic maturation demonstrated in mouse oocytes (Jeon et al. 2016).

2. *TCTP in early development*. The most impressive evidence for the importance of TCTP in early development was provided by TPT1-gene-knockouts in mice, which resulted in embryonic lethality (Susini et al. 2008; Chen et al. 2007a; Koide et al. 2009). The explanation typically given for this effect was 'excessive apoptosis at an early embryonic state' (Susini et al. 2008; Chen et al. 2007a), whereas another report reasons that the lack of TCTP results in an overactivity of the BMP4 (bone morphogenetic protein 4) pathway, which is normally inhibited by TCTP (Koide et al. 2009). The authors of this paper also show that in Xenopus embryos TCTP/fortilin is particularly important for the formation of neural tissue, even in the brain. This is consistent with a very recent report showing that TCTP regulates axon development in the embryonic visual system (Roque et al. 2016).

   The importance of TCTP in early development was also demonstrated in *Drosophila*, where TCTP-knockdown experiments established a role for TCTP in the regulation of cell size and number, organ growth (Hsu et al. 2007) and development (Hong and Choi 2013). Interaction partners of TCTP in this context are the small GTPase Rheb, an upstream regulator of mTOR (mechanistic target of rapamycin), and 14-3-3 proteins (Le et al. 2016). Knockout of TCTP in plants (*Arabidopsis thaliana*) resulted in a male gametophytic phenotype with impaired pollen tube growth. Moreover, TCTP knock-down resulted in severe developmental aberrations, such as slow vegetative growth, reduced leaf expansion and lateral root formation and impaired root hair development (Berkowitz et al. 2008). A recent paper reports on the importance of TCTP (Rp41) for the nodulation and root hair formation in *Robinia pseudoacacia* as another example of TCTP's involvement in plant developmental processes (Chou et al. 2016).

3. *TCTP and pluripotency in somatic cell nuclei and ES cells*. In 2007, several papers documented the involvement of TCTP in early development and the establishment of pluripotency after transplantation of somatic nuclei. Chen et al. monitored the expression of TCTP in the eggs of the cephalochordate *Amphioxus* after fertilisation and found that it is expressed in zygotes and the early, but not the late, cleavage stages (Chen et al. 2007b). Koziol and colleagues reported that TCTP, the gene product of Tpt1, acts as a transcription factor that activates the transcription of oct4 and nanog in transplanted somatic nuclei (Koziol et al. 2007). Oct4 and nanog are transcription factors that are critical for reprogramming of somatic nuclei, when transplanted into oocytes or eggs. Tani et al. observed that phosphorylated TCTP facilitates the first step of somatic cell reprogramming in bovine oocytes (Tani et al. 2007), complementing the study by Koziol. In contrast, another paper reported 5 years later that TCTP

inhibited the Oct4 transcription and also decreased the pluripotency of murine embryonic stem cells (Cheng et al. 2012). In 2010, Johansson and colleagues investigated the interactions of TCTP/tpt1 in mouse embryonic stem cells (ES cells) in more detail. They reported that TCTP interacts with the nucleolar protein nucleolin (Ncl) (Johansson et al. 2010b) and with nucleophosmin (Npm1) (Johansson et al. 2010a), and both these interactions peaked at mitosis, but are independent of TCTP phosphorylation by Plk1. In a third paper, the group also documents the direct interaction of both nucleoplasmin and TCTP with Oct4, independently of each other (Johansson and Simonsson 2010). From these interactions in ES cells, the authors deduce that TCTP has a role in ES cell proliferation and maintenance (Table 4.1).

4.2.3 *TCTP in Cell Growth Regulation, Protein Synthesis and Degradation*

4.2.3.1 TCTP and Cell Growth Regulation

One of the earliest observations on TCTP is the translational induction of its synthesis upon growth stimulation of mammalian cells (Thomas et al. 1981; Bohm et al. 1989, 1991; Thomas and Thomas 1986). Typically, we observed a four-fold increase of TCTP levels after serum-stimulation of mammalian cells (Bommer et al. 2002, 2015). The involvement of TCTP in cellular growth regulation is documented in various settings. One example, already mentioned above, described the role for TCTP in the regulation of cell size and number, as well as organ growth (Hsu et al. 2007) and development (Hong and Choi 2013) in *Drosophila*. In plants, it was shown that knockout of TCTP resulted in a phenotype with a slow vegetative growth and impaired pollen tube growth (Berkowitz et al. 2008). Another recent paper showed that in the Tobacco plant, TCTP is able to interact with the ethylene receptor histidine kinase-1 and to enhance plant growth through promotion of cell proliferation (Tao et al. 2015).

The regenerating rat liver was frequently used as an experimental model for studying processes of cell growth regulation in rodents. In 2008, Zhu et al. observed that in rat liver, TCTP mRNA levels are transiently up-regulated in the time period up to 12 h after partial hepatectomy, which indicates that the protein is required for tissue growth during liver regeneration (Zhu et al. 2008). This is consistent with a recent report showing that both the expression of intracellular TCTP and the release of TCTP protein into serum were significantly increased in rats after partial hepatectomy, and that enhanced TCTP levels promoted hepatocyte proliferation (Hao et al. 2016).

The influence of TCTP on cellular signalling processes that are involved in the regulation of growth and survival was studied by the group of Kyunglim Lee. They reported that overexpression of TCTP in HeLa cells resulted in tyrosine phosphorylation of the epidermal growth factor receptor and in activation of both the Ras/Raf/

ERK and the PI3K/Akt pathways (Kim et al. 2009b). In a more recent paper, they provided a link to the activity of TCTP in binding to and inhibiting the Na,K-ATPase, which they had observed earlier (Jung et al. 2004). The authors found that TCTP in binding to the third cytoplasmic domain of the Na,K-ATPase results in release and activation of the protein kinase Src and consequently in the activation of the PI3K/Akt and the Ras/Raf/ERK pathways, as well as of additional signalling pathways (Jung et al. 2011). A more recent paper from this group also described a novel activity of (recombinant) TCTP in enhancing the neurotransmitter release from the neurosecretory pheochromocytoma (PC12) cells (Seo et al. 2016).

The signalling pathway most intimately linked to cell growth regulation is the PI3K/Akt-mTORC1 (mechanistic target of rapamycin complex 1) pathway. It is part of a complex signalling network that regulates several anabolic processes, inclusive of protein synthesis (Laplante and Sabatini 2012; Zoncu et al. 2011). In 2007, Hsu et al. published a paper describing TCTP in *Drosophila* as a direct activator of this pathway, acting as a guanine nucleotide exchange factor (GEF) for the small GTPase Rheb, upstream of mTORC1 (Hsu et al. 2007). Consistent with this, reducing *Drosophila* TCTP levels reduced cell size, cell number and organ size, similar to Rheb mutant phenotypes. Subsequently, this group also reported that the 14-3-3 proteins regulate the interaction between TCTP and Rheb, thus playing an important role in regulating organ growth in *Drosophila* (Le et al. 2016). The TCTP–Rheb interaction was further confirmed for the human proteins by molecular modelling studies (Dong et al. 2009). The first description of a NMR structure of TCTP had revealed a similarity of TCTP to the MSS4/DSS4 proteins, which bind to the Rab family of small GTPases (Thaw et al. 2001). Thus, the idea that TCTP acts as a GEF for Rheb appeared attractive, but it remains controversial, since later studies failed to support the results described for Drosophila: We found that reducing TCTP levels did not reproducibly affect mTORC1 signalling in human cells, and we were unable to detect a stable interaction between TCTP and Rheb (Wang et al. 2008). Similarly, Rehmann and colleagues did not detect GEF activity of TCTP for Rheb or any interaction between TCTP and Rheb (Rehmann et al. 2008).

Last but not least, the frequent reports on TCTP overexpression in cancer cells and human tumours (reviewed in Bommer 2012; Acunzo et al. 2014; Chan et al. 2012a; Koziol and Gurdon 2012) and on its down-regulation in the tumour reversion model (Telerman and Amson 2009; Amson et al. 2013) provide additional compelling evidence that TCTP is a growth-promoting protein. These aspects are discussed below, in Sect. 4.3.1.

4.2.3.2 The Involvement of TCTP in Protein Synthesis

Cell and organ growth is dependent on the up-regulation of anabolic pathways, in particular protein synthesis. The core regulatory hub for this regulation is the mTORC1 pathway, which regulates protein synthesis through several mechanisms. For example, mTORC1 enhances the proteosynthetic capacity of the cell through stimulating ribosome synthesis, resulting in an increase in ribosome numbers. This is achieved through regulation of rRNA synthesis and the selective translational activation of a subset of mRNAs, the 'TOP-mRNAs' whose joint feature is the presence of the 5′-terminal-oligopyrimidine tract (5′-TOP). This group of mRNAs largely comprises those mRNAs coding for components of the translational apparatus, in particular ribosomal proteins and translation factors (Meyuhas and Kahan 2015; Meyuhas 2000). The observations that the mRNA of TCTP also bears a 5′-TOP (Yamashita et al. 2008; Meyuhas 2000) and is regulated through the mTORC1 pathway (Bommer et al. 2015) would suggest that TCTP also participates in the activities (or regulation) of the translational machinery. In fact, as early as 2003, the Telerman group identified translation elongation factor eEF1A and its guanine nucleotide exchange factor eEF1Bβ as TCTP-interacting partners (Cans et al. 2003). TCTP stabilised the GDP form of eEF1A, and impaired the GDP exchange reaction promoted by eEF1Bβ, thereby acting as an inhibitor of the GEF function, in contrast to the activation of the GEF activity of TCTP for Rheb, mentioned above. The interaction of TCTP with eEF1Bβ was later confirmed by Langdon et al. (2004), and a more recent structural analysis of this interaction demonstrated that it represents the most conserved interaction of TCTP, indicating that this might be a primary function of the protein (Wu et al. 2015). A functional screen of proteins associated with ribosomal complexes in yeast identified TCTP as a translation-machinery-associated (TMA) protein, TMA19 (Fleischer et al. 2006). Analysis of yeast mutant strains, deleted in TMA19, revealed that such strains have a reduced rate of protein synthesis and alterations in polysome profiles, lending further support to the notion that TCTP is involved in protein synthesis.

4.2.3.3 TCTP in the Regulation of Protein Degradation

There are a few instances, where TCTP was also reported to be involved in the regulation of protein degradation. Typically, these were examples of degradation of *specific* proteins or participation in autophagy, which either serve to accomplish cell-cycle-dependent processes or otherwise to maintain cellular homeostasis. The involvement of TCTP in the regulation of autophagy in mammalian cells and the underlying mechanism was studied in a recent paper. The authors presented data showing that TCTP interacts with the ATG16L1 complex, which is directly engaged in the autophagy pathway (Chen et al. 2014a). They demonstrated that TCTP positively regulates autophagy through the AMP-activated protein kinase (AMPK) pathway, which also involves mTORC1. This is in contrast to a more recent study, which showed that TCTP inhibits the process of autophagy (see 'Note Added in Proof').

The same group investigated the interaction of TCTP with the tumour suppressor protein 'von Hippel–Lindau protein' (VHL), which functions as an E3 ubiquitin ligase and is involved the degradation of *hypoxia-inducible factor HIF1*. They demonstrated that TCTP binds specifically to the VHL protein and promotes its ubiquitinylation and subsequent degradation, in this way stabilising the HIF1 protein (Chen et al. 2013a). Another important example of TCTP being involved in regulation of protein levels via modulation of protein degradation is the *tumour suppressor protein p53*. That TCTP overexpression promotes P53 degradation has been demonstrated independently by three groups (Amson et al. 2012; Rho et al. 2011; Funston et al. 2012). It does this by binding to P53-MDM2-containing complexes and inhibiting MDM2 auto-ubiquitination, thereby promoting MDM2-mediated ubiquitination and degradation of P53 (Amson et al. 2012; Funston et al. 2012).

An interesting point is that TCTP was shown to stabilise the proto-oncogenic *protein kinase Pim-3*, which is involved in promoting cell cycle progression and the development of pancreatic cancer (Zhang et al. 2013). In this case, TCTP prevents the degradation of Pim-3 via the ubiquitin–proteasome pathway. Yet another example of a protein that is involved in cell-cycle regulation, and whose levels are modulated by TCTP, is the cell-cycle-dependent *phosphatase Cdc25C*. The ubiquitin–proteasome-dependent degradation of Cdc25C is an important step in mitotic progression. It was shown that in hepatocellular cancer (HCC) this step is promoted by overexpression of TCTP, leading to a faster mitotic exit and consequently to chromosome miss-segregation (Chan et al. 2012b). On the other hand, TCTP itself has been found to be subject to proteasomal degradation after the first embryonic mitosis in *Xenopus laevis* (Kubiak et al. 2008).

The involvement of TCTP with the proteasomal machinery in more general terms has also been demonstrated in two other studies. Rinnerthaler and colleagues showed that in yeast, under heat stress conditions, the TCTP homologue Mmi1 binds to components of the proteasomal complex and to cytoplasmic stress granules. These proteasomal components, to which TCTP colocalises, are typically involved in protecting protein substrates from degradation (Rinnerthaler et al. 2013). Another proteomics study on colon cancer cells identified 27 proteins, whose levels were altered after TCTP knockdown. In particular, components of ubiquitin–proteasome system and proteins involved in the cytoskeleton were affected under this condition (Ma et al. 2010).

4.2.4 Extracellular Functions of TCTP

4.2.4.1 TCTP as Extracellular 'Signaling Molecule' in Immune Reactions

Since other chapters of this book will certainly consider the extracellular functions of TCTP in more detail, this aspect of TCTP function will be covered only briefly at this point. In 1995, Susan Macdonald's group first described the presence of an 'IgE-dependent histamine-releasing factor (HRF)' in biological fluids of allergic patients, which was also produced by lymphocytes of atopic children (MacDonald et al. 1995). The molecular characterisation revealed that this protein is identical to TCTP (or p21/p23). Since then, a plethora of papers was published relating to this extracellular function of TCTP/HRF, which I will briefly summarise under the following points. [For a more detailed account, the reader is referred to three recent review articles on this matter (MacDonald 2012a, b; Maeng et al. 2012).]

1. *Release of local signalling molecules from immune cells, triggered by extracellular TCTP/HRF*. The focus of the initial publication, which identified TCTP as HRF, was on the IgE-dependent of histamine release in biological fluids from allergic patients (MacDonald et al. 1995). However, a number of subsequent studies revealed that the 'scope' of TCTP/HRF as an extracellular 'signalling molecule' in immunological reactions is much wider, in that its activity is not IgE-dependent and that it is able to modulate the release of various cytokines in a number of different cell types involved in immune functions (MacDonald 2012a, b). These include B-cells (IL-1, IL-8 release), basophils (IL-4, IL-13 release), T-cells (inhibition of IL-2, IL-13 release), bronchial epithelial cells (IL-8, GM-CSF release) and GM-CSF-primed eosinophils (IL-8 release) (Macdonald 2012b). The potential binding of TCTP to IgE antibodies was a matter of debate (Wantke et al. 1999), but a more recent paper showed that the dimerised form of TCTP is able to bind to a subset of IgE and IgG molecules (Kashiwakura et al. 2012).
2. *Intracellular signalling events induced by TCTP/HRF in immune cells*. In order to gain insight into the intracellular signalling pathways/events that are involved in mediating the TCTP/HRF-dependent release of histamine and cytokines from immune cells, the MacDonald group studied signalling events in basophiles that were triggered by human recombinant TCTP/HRF for histamine release. They found that the activity of the inositol 5′ phosphatase SHIP-1, which inhibits the PI3-kinase signalling pathway, is inversely correlated with the HRF-dependent histamine release in basophils from IgE(+) donors. This finding was corroborated by the demonstration that the PI3-kinase inhibitor Ly294002 also prevented the HRF-dependent histamine release in these cells (Vonakis et al. 2001). In another study, they showed that, in TCTP/HRF-responsive basophils (but not in non-responsive ones), HRF-treatment resulted in phosphorylation of the protein kinase Akt (Vonakis et al. 2008), further confirming that the PI3-kinase pathway is involved in mediating the intracellular events in these type of cells.
3. *Secretion of TCTP, pathways and regulation*. Secretion of TCTP was reported from a number cell types, largely cells involved in the immune system, such as macrophages, dendritic cells and PBMN cells, as well as a range of cancer cells (reviewed in Maeng et al. 2012). This paper also names the different agents that trigger and regulate TCTP release, and it describes the underlying mechanisms for secretion in detail, which I will only briefly summarise here. First, the protein TCTP does not have a signal sequence (Chitpatima et al. 1988; Gross et al. 1989), and no precursor protein for TCTP was detected, meaning that TCTP is not secreted through the classical secretory pathway, via the ER and Golgi apparatus. This was confirmed by the finding that its secretion was insensitive to brefeldin A or monensin, two inhibitors of this pathway (Amzallag et al. 2004). These authors proposed a non-classical pathway for the secretion of TCTP. They showed that TCTP interacts with TSAP6, a transmembrane protein, which is induced by the tumour suppressor protein p53, and that TSAP6 and TCTP co-localise to vesicular structures at the plasma membrane, indicating that TCTP is secreted through an exosomal pathway, guided by TSAP6.

Overexpression of TSAP6 did indeed result in increased TCTP levels in exosome preparations and in enhanced secretion of TCTP (Amzallag et al. 2004), whereas cells derived from TSAP6-deficient mice are severely compromised in the DNA damage-induced p53-dependent exosomal secretory pathway (Lespagnol et al. 2008). Another, also exosome-related pathway for TCTP secretion, was proposed by the Kyunglim Lee's group. –They discovered that TCTP secretion from HEK293 and U937 cells is inhibited by proton pump inhibitors (PPIs), such as omeprazole and pantoprazole, which are inhibitors of the human gastric H(+)/K(+)ATPase and are used in the treatment of gastric ulcers (Choi et al. 2009). Consistent with this, overexpression of the ATPase increased TCTP secretion from these cells, and the authors concluded that this enzyme might facilitate the secretion of TCTP via a non-classical pathway.

4.2.4.2 TCTP Dimerisation and Other Extracellular Roles

The ability of TCTP to self-interact was discovered in Dr. Lee's group as early as in 2000, using the yeast two-hybrid system and co-immunoprecipitation (Yoon et al. 2000). Later, the same group detected the dimerised protein in sera from allergic patients, and they showed that dimerisation is essential for the cytokine-like properties of TCTP/HRF. For dimerisation to happen, the protein had to undergo an N-terminal truncation of about eleven amino acids (Kim et al. 2009a). The group also identified a 7-mer peptide that was able to bind to dimerised TCTP and to inhibit its cytokine-like activity in a cellular assay system and in a mouse model (Kim et al. 2011). Based on their findings and on additional evidence, the authors developed a hypothesis describing two potential mechanisms for the activation (through dimerisation) of extracellular TCTP/HRF to acquire cytokine-like properties (Kim et al. 2013a): (1) A 'spontaneous' mechanism, involving extracellular proteases that trigger N-terminal truncation of the protein and subsequent dimerisation, driven by reactive oxygen species (ROS), which frequently occur in inflammatory processes. (2) Dimerisation by binding to HRF-reactive IgE antibodies, which would confer autoantigen properties to HRF. In each case, only the dimerised TCTP/HRF would bind to the target cell and trigger cytokine release. These ideas are discussed in detail in their review article (Kim et al. 2013a).

Dimerisation of TCTP was reported by other groups in quite different contexts. Gnanasekar et al. characterised the TCTP protein from filarial parasites and found that it occurs in multimeric forms (Gnanasekar et al. 2002). Meshnick and co-workers described TCTP from *Plasmodium falciparum* as a target protein for the antimalarial drug artemisinin (Bhisutthibhan et al. 1999). They observed that dihydroartemisinin may form adducts with both the monomeric and dimeric form of TCTP (Bhisutthibhan and Meshnick 2001). A recent detailed study on ligand binding to human TCTP revealed that binding of haemin to TCTP resulted in a conformational change of the protein and promoted its dimerisation. In contrast, Ca^{2+}-binding resulted in destabilisation of TCTP dimers formed in the presence of haemin (Lucas et al. 2014). The authors propose a 'buffer-like' function for TCTP, since it is able to bind both haemin and Ca^{2+} at higher concentrations. For example, TCTP may be able to bind intracellular free haemin and keep it in a non-toxic state, preventing the formation of ROS. This paper critically evaluates the earlier literature on Ca^{2+}-binding and on dimerisation of TCTP.

Extracellular 'functions' of TCTP were also described in the context of apoptosis. Apoptotic cell death includes a paracrine function, aimed at promoting tissue repair in the vicinity of the dying cell. To this end, apoptotic cells release nanovesicles containing a set of proteins, different from that of apoptotic blebs. Sirois et al. studied such nanovesicles released from apoptotic endothelial cells and identified TCTP as a prominent anti-apoptotic protein in these vesicles (Sirois et al. 2011). They went on to show that these nanovesicles induce an anti-apoptotic phenotype in vascular smooth muscle cells (VSMCs), and that this activity is abolished in nanovesicles treated with TCTP siRNA. Another, more recent paper studied serum TCTP (fortilin) levels in mice and humans using a newly developed ELISA. The authors found that TCTP/ fortilin levels are significantly elevated in sera from patients with solid cancers, in response to chemo- or radiation therapy, and concluded that TCTP is suitable as a biomarker for apoptosis in vivo (Sinthujaroen et al. 2014).

4.3 Involvement of TCTP in Disease Processes

4.3.1 TCTP in Human Cancer

Other chapters in this book presumably deal with the 'TCTP and Cancer' topic in more detail; I will here just summarise this important part of the TCTP story under three subheadings.

4.3.1.1 Overexpression of TCTP in Human Tumours

As mentioned at the beginning, the name 'translationally controlled tumour protein' was derived from the fact that the cDNA for the first sequence of human TCTP, published in 1989 (Gross et al. 1989), was obtained from a mammary carcinoma. During the ensuing decade, some doubts were voiced on the term 'tumour protein', since TCTP protein and mRNA were detected in essentially all eukaryotic cells and tissues (Thiele et al. 2000; Hinojosa-Moya et al. 2008; Sanchez et al. 1997). However, since the beginning of this century, a substantial body of evidence has been accumulated, demonstrating that TCTP is indeed overexpressed in cancer cells and in human tumours. A number of review articles have summarised these efforts (Telerman and Amson 2009; Bommer 2012; Acunzo et al. 2014; Amson et al. 2013; Chan et al. 2012a; Koziol and Gurdon 2012), and the reader is referred to these for papers reporting overexpression of TCTP in cancer cell lines.

Table 4.2 provides a compilation of the types of human cancers, where overexpression of TCTP was demonstrated in tumour tissues. The table lists only

Table 4.2 TCTP overexpression in human cancers

Cancer type	Methods employed	Clinical associations	References
Brain tumour: Gliomas	Immunohistochemistry, Western blot	High-grade gliomas, poor survival rate	Miao et al. (2013)
	Immunohistochemistry, RT-PCR (mRNA levels)	High-grade gliomas, poor survival rate	Gu et al. (2014)
Breast cancer	Proteomics methods		Deng et al. (2006)
	Immunohistochemistry	Agressive tumours; poor survival rate	Amson et al. (2012)
Colorectal cancer	Northern blot (mRNA)		Chung et al. (2000)
	Proteomics methods		Friedman et al. (2004)
	Microarray analysis (mRNA)		Slaby et al. (2009)
	Immunohistochemistry	(Cellular drug resistance)	Bommer et al. (2017)
	Immunohistochemistry	High grades, metastases, poor survival rate	Xiao et al. (2016)
Kidney and Renal Cell Cancer (RCC)	RT-PCR, Western blot, Immunohistochemistry		Ambrosio et al. (2015)
Leukaemia (CLL)	Immunohistochemistry		Yagci et al. (2013)
Lymphomas (DLBCL, NHL, FL, NK/T-Cell)	Immunohistochemistry		He et al. (2015)
Liver cancer	RT-PCR (mRNA levels)		Zhu et al. (2008)
Hepatocellular carcinoma (HCC)	Immunohistochemistry	Advanced tumours; poor survival rate	Chan et al. (2012b)
Lung cancer	Western blot		Kim et al. (2008b)
NF1-associated tumours	Immunohistochemistry	(Cell tumorigenicity)	Kobayashi et al. (2014)
Neuroblastomas	Immunohistochemistry	Advanced tumours, poor survival rate	Ramani et al. (2015)
Oral cancer	Proteomics methods		Lo et al. (2012)

Osteosarcoma	RT-PCR (mRNA levels)	(Cell tumorigenicity)	Shen et al. (2016)
Ovarian cancer	Immunohistochemistry	Poor prognosis, Cisplatin resistance	Chen et al. (2015)
Pancreatic cancer	Immunohistochemistry	High TCTP and Pim-3 in advanced tumours	Zhang et al. (2013)
Prostate cancer	Immunohistochemistry	Castration resistance	Baylot et al. (2012), Kaarbo et al. (2013)

those papers, which have dealt specifically with one specific type of cancer. There are two additional papers, where the authors have screened a range of cancers for TCTP overexpression, compared to normal tissue, either by western blotting (Tuynder et al. 2002) or using proteomics methods (Kuramitsu and Nakamura 2006). From this screening, it emerged that, apparently, TCTP was not overexpressed in cancers of the stomach and the pancreas (Tuynder et al. 2002; Kuramitsu and Nakamura 2006), of the oesophagus (Kuramitsu and Nakamura 2006) and the cervix (Tuynder et al. 2002), whereas overexpression was confirmed for most other types of cancers investigated. However for pancreatic cancer, a later, more specific study found high TCTP levels, particularly in advanced tumours (Table 4.2). This example shows that for each type of cancer, we have to await more detailed investigations, to confirm (or rule out) TCTP overexpression for a particular cancer type. Currently, the picture emerges that TCTP is overexpressed in a large number of cancer types, certainly in the majority of those investigated so far. Also, where clinical associations have been established, high TCTP levels were typically associated with advanced tumours and poor patient outcomes (Table 4.2).

4.3.1.2 Mechanistic Involvement in Cancer Progression

The importance of TCTP in cancer was originally demonstrated through studies on the 'tumour reversion model' in Adam Telerman's group. This approach investigates the rare events of tumour cells reverting back from the malignant to the normal phenotype (Telerman and Amson 2009). The group showed that TCTP is one of the important proteins, whose expression is down-regulated in this process. Knockdown of TCTP resulted in suppression of the malignant phenotype (Tuynder et al. 2002) and it led to a dramatically increased number of spontaneous revertants (Tuynder et al. 2004). Another group used a proteomics approach to identify proteins that are down-regulated in a reversion model of multiple myeloma. They identified STAT3 and TCTP, among other proteins, as being down-regulated in revertant cells (Ge et al. 2011). The tumour reversion model was instrumental to demonstrate the importance of TCTP in cancer, but the important question remained: What are the mechanisms, through which this protein is able to promote cancer? The following possibilities have emerged from the discussion in recent years.

1. *Anti-apoptotic activity of TCTP*. Overexpression of anti-apoptotic proteins is a common feature and part of the survival strategy of cancer cells. It is therefore not surprising that TCTP is often found to be overexpressed in cancer, alongside other 'classical' anti-apoptotic proteins, such as Bcl-2, Bcl-XL or Mcl-1. Since the discovery of the anti-apoptotic activity of TCTP (fortilin) (Li et al. 2001), several mechanistic aspects of this property have been reported (reviewed in Bommer 2012; Acunzo et al. 2014; Chan et al. 2012a). These were described earlier in Sect. 4.2.1 of this chapter and do not need to be outlined again. However, one of these anti-apoptotic mechanisms ought to be mentioned here, i.e. the antagonism of TCTP to the tumour suppressor protein p53, which is crucial for TCTP to exert its

cancer-promoting activity (Amson et al. 2012; Chen et al. 2011). The cytoprotective role of TCTP in cancer cells is well documented, for example in conditions of oxidative stress (Lucibello et al. 2011), or of treatment with various anticancer drugs, the latter often resulting in chemoresistance of these cells (Li et al. 2001; Jung et al. 2014; Graidist et al. 2004; Bommer et al. 2017; He et al. 2015), which is a frequent problem in cancer chemotherapy.

2. *Promotion of mitosis.* A very specific role of TCTP in cancer promotion was described by Chan et al. in 2012, who studied the chromodomain helicase/ATPase DNA-binding protein 1-like gene (CHD1L), which is a specific oncogene in human hepatocellular carcinoma (HCC) (Chan et al. 2012b). They found that CHD1L acts as a transcriptional activator of TCTP and that the resulting overexpression of TCTP contributed to the mitotic defects of tumour cells. Moreover, TCTP promoted the ubiquitin–proteasome-dependent degradation of the phosphatase Cdc25C during mitotic progression, resulting in a sudden drop of Cdk1 activity in mitosis. The activity of the cyclin-dependent kinase 1 (Cdk1) is normally maintained by Cdc25C and is required for an orderly mitotic exit. The sudden drop of Cdk1 activity caused by TCTP overexpression resulted in a faster mitotic exit and chromosome miss-segregation, which in turn led to chromosomal instability (Chan et al. 2012b). The observation that TCTP is involved in regulating the degradation of Cdc25C through the ubiquitin–proteasome pathway is consistent with other studies reporting a role of TCTP in regulating the ubiquitin–proteasome-dependent degradation of specific proteins (see Sect. 4.2.3). Further support comes from a proteomics study on LoVo colon cancer cells reporting that TCTP knockdown results in the alteration predominantly of components of the ubiquitin–proteasome system (Ma et al. 2010).
3. *TCTP and growth signalling in cancer.* There are several reports describing the effects of alteration of cellular TCTP levels on growth signalling pathways. These were covered earlier (Sect. 4.2.3). Briefly, TCTP overexpression was found to activate the Ras/Raf/ERK, the PLC-γ and the PI3K/Akt growth signalling pathways in HeLa cells (Kim et al. 2009b). Also, Hsu et al. described TCTP in *Drosophila* as an activator of the small GTPase Rheb, an upstream activator of the mTORC1 pathway, and based on this, implicated TCTP in growth regulation (Hsu et al. 2007). However, other follow-up studies could not confirm some of the findings in mammalian cells (Wang et al. 2008; Rehmann et al. 2008). Thus, while some studies indicate that TCTP could act as a growth-regulatory protein, these ideas and the mechanisms involved need further consolidation.
4. *TCTP in specific stages of cancer progression.* Since TCTP is active in promoting cellular growth and proliferation, one would expect the protein also to be involved in the early stages of cancerous growth. Yet, there are not many observations published on TCTP levels specifically in early cancer development. Colorectal cancer (CRC) with its distinct morphology of precancerous lesions, such as adenomas, is amenable for such investigations. Recently, we (Bommer et al. 2017) and a Chinese group (Xiao et al. 2016) demonstrated that indeed TCTP levels (assessed by immunohistochemistry) increase early in CRC, already

at the adenoma stage. This is consistent with our finding that during growth-induction of both HeLa cells and HT29 colon cancer cells, TCTP synthesis is translationally up-regulated through the PI3K/Akt/mTORC1 signalling pathway (Bommer et al. 2015), given the earlier observation that this pathway is activated early in the development of colorectal cancer (Zhang et al. 2009).

Epithelial-to-mesenchymal transition (EMT) of cancer cells is a crucial step in the development of invasiveness and metastasis of tumours. In 2015, Bae et al. published a paper showing that TCTP is able to promote EMT, cell migration and invasiveness (Bae et al. 2015). They also used murine melanoma cells to demonstrate that depletion of TCTP suppresses the development of pulmonary metastasis in a mouse model. Several other studies also demonstrated a role for TCTP in promoting cell migration and invasiveness of cancer cells, and the formation of metastases in animal models. Examples are glioma cells (Jin et al. 2015) and a model for liver metastasis in SCID mice (Chan et al. 2012a); however the cells most frequently studied in this context are those of colorectal cancer. Knockdown of TCTP inhibited proliferation, migration and invasion activities of LoVo colon cancer cells (Ma et al. 2010; Chu et al. 2011). Xiao and co-workers reported that in colon cancer patients, TCTP expression levels were higher in liver metastases, compared to primary tumours (Xiao et al. 2016). They also showed that extracellular TCTP promoted migration and invasiveness of CRC cells in vitro and contributed to distant liver metastasis in vivo. Another example of TCTP being involved in advanced state malignant disease came from a study on prostate cancer (PC). Baylot et al. demonstrated that TCTP is particularly expressed in the castration-resistant form of PC and in metastases of the bone, liver and lymph nodes resulting from this (Baylot et al. 2012). Their results also show that knockdown of TCTP inhibits growth of prostate cancer cells, progression of castration-resistant tumours in mice and reduces their chemoresistance to docetaxel.

5. *Anticancer drug resistance*. A frequent problem in cancer chemotherapy is the development of drug resistance, and some recent reports indicate that TCTP is likely to be involved in this aspect of cancer as well. We have shown that HCT116 colon cancer cells respond to treatment with 5-fluorouracil (5-FU) or with oxaliplatin, two agents frequently used in CRC treatment, with increased TCTP expression. This is probably part of a cellular stress response, since increased TCTP levels protected these cells against the cytotoxicity exerted by these drugs (Bommer et al. 2017). Similarly, the contribution of TCTP (HRF) to the development of cell adhesion and chemoresistance was reported in non-Hodgkin lymphomas (He et al. 2015). Lucibello et al. showed that in breast cancer cells, inhibition of TCTP by dihydro-artemisinin resulted in increased sensitivity to chemotherapy and that phospho-TCTP levels, an indicator of mitotic activity (Yarm 2002), are particularly increased in breast tumours that are resistant against treatment with the trastuzumab antibody (Lucibello et al. 2015).

   Reports on the involvement of TCTP in anticancer drug resistance started to appear as early as 2000. Sinha and co-workers performed a proteomics analysis on human melanoma cell lines that were resistant to drugs like vindesine, cisplatin and etoposide. They identified TCTP as one of four proteins overexpressed in those cell lines, compared to their no-resistant parental cells (Sinha et al. 2000). Fujise's laboratory demonstrated TCTP (fortilin)-dependent drug resistance against etoposide (Li et al. 2001) and 5-FU (Graidist et al. 2004) in U2OS cells. In a more recent study, Kyunglim Lee's laboratory explored the mechanism underlying the TCTP-dependent resistance against etoposide in HeLa cells. They demonstrated that TCTP interacts with the apoptotic protease-activating factor (Apaf-1), associates with the apoptosome and inhibits activation of caspase 3 and execution of apoptosis (Jung et al. 2014).

4.3.1.3 TCTP as an Anticancer Target

Given the well-established role of TCTP in cancer as summarised in this section, it is not surprising that the idea of exploring TCTP as an anticancer target protein is not new, and this interesting topic has already been covered in six review articles (Telerman and Amson 2009; Bommer 2012; Acunzo et al. 2014; Amson et al. 2013; Efferth 2005, 2006). Here I will briefly summarise these efforts, according to the groups of drugs investigated to target TCTP:

1. *Artemisinin*: Since the early discovery that TCTP of *Plasmodium falciparum* is a target protein of the antimalarial drug artemisinin (Bhisutthibhan et al. 1998), this drug interaction has been studied in more detail (Chae et al. 2006). Artemisinin not only displays antimalarial activity but was also found to be a potent anticancer agent in cellular assay (review in Efferth 2005), and screening for potential targets in this context has confirmed that TCTP is indeed a target for this drug (Efferth 2005, 2006). The interaction of artemisinin with TCTP was then studied in cancer cells, and it was found that drug treatment results in degradation of TCTP (fortilin) via the ubiquitin–proteasome pathway (Fujita et al. 2008). More recent publications described proof-of-principle investigations for the use of artemisinin as an anticancer drug in cellular models of lung cancer (Liu et al. 2014), of Neurofibromatosis type 1 (NF1)-associated tumours (Kobayashi et al. 2014) and of breast cancer (Lucibello et al. 2015). Specifically, the latter report demonstrated that dihydroartemisinin enhances the anticancer effect of doxorubicin in triple-negative breast cancer cells and acts synergistically with the antibody Trastuzumab, which is used for the treatment of HER2/neu positive breast cancers, to induce apoptosis of tumour cells.
2. *Antihistaminics and antidepressants*. Based on the identification of TCTP as histamine-releasing factor HRF (MacDonald et al. 1995), the Telerman laboratory tested a panel of antihistaminic drugs for their effect on human leukaemia and breast cancer cells (Tuynder et al. 2004). Of these ones, hydroxyzine and promethazine displayed growth-inhibiting activity on these cells. These antihistaminics are structurally related to the antidepressants sertraline and thioridazine, which also inhibited tumour growth, both in vitro and in vivo (Tuynder et al. 2004). All these drugs bind to TCTP and disrupt its interaction with other partner proteins, eventually leading to increased release of the protein from the cell and to

lowered intracellular TCTP levels (Telerman and Amson 2009). Since the demonstration of the reciprocal repressive feedback loop between TCTP and the tumour suppressor P53 by the same laboratory (Amson et al. 2012), the mechanism of action of these drugs can be described in more detail as follows (Amson et al. 2013): Binding of sertraline and thioridazine to TCTP prevents its binding to MDM2 and consequently the destabilisation of P53. The resulting increased P53 levels lead to additional transcriptional inhibition of TCTP expression. An additional mechanism is based on the finding that both sertraline (Lin et al. 2010) and thioridazine (Kang et al. 2012) inhibit the mTOR signalling pathway. Blocking this pathway in itself will inhibit TCTP expression, since TCTP mRNA translation is regulated through the mTOR pathway (Bommer et al. 2015). In *Drosophila*, TCTP was described as a positive regulator of the mTOR pathway via the small GTPase Rheb (Hsu et al. 2007); it is therefore possible that TCTP and mTOR are in a positive feedback loop (Kobayashi et al. 2014), which would be disrupted by treatment with sertraline or thioridazine. In a very recent paper, two additional antihistaminic drugs, buclizine and levomepromazine, were reported to inhibit cancer cell growth by binding to TCTP and by inducing cell differentiation (Seo and Efferth 2016).

3. *Antisense oligonucleotides (ASOs) and anti-TCTP peptide*. Synthetic agents that target TCTP were first used by Baylot and colleagues, who patented an ASO against TCTP (Baylot et al. 2012). In their study, they used a mouse model to show that treatment with the TCTP-ASO inhibits tumour growth in castration-resistant prostate cancer, enhances docetaxel chemotherapy and delays cancer progression in vivo. This effect is associated with an increase in P53 levels (Acunzo et al. 2014; Baylot et al. 2012). In addition, recently a peptide aptamer (WGQWPYHC) targeting TCTP was tested and found to display specific cytotoxicity towards TCTP expressing tumour cells, without affecting normal cells (Kadioglu and Efferth 2016).

In summary, both existing drugs that are in use for treatment of other disease groups (antimalarials, antihistaminics, antidepressants) and synthetic agents designed to specifically target TCTP show promising results in proof-of-concept studies as potential approaches for anticancer treatment.

4.3.2 Participation in Immunological Reactions

Since the discovery of TCTP as an 'IgE-dependent histamine-releasing factor (HRF)' in biological fluids of allergic patients (MacDonald et al. 1995), many papers have been published, relating to the involvement of TCT/HRF in allergic diseases and other activities of the immune system. The molecular and cellular aspects of these extracellular activities of TCTP/HRF have been summarised in the preceding Sect. 4.2.4. Here, I will just touch on some of the clinical aspects, based on the reviews and some recent papers:

As mentioned in Sect. 4.2.4, the extracellular function of TCTP/HRF consists of a cytokine-like activity, in that it is able to trigger several types of immune cells to release cytokines or other signalling molecules, not only the IgE-dependent histamine release from basophils, as described originally (reviewed in Macdonald 2012b; Kawakami et al. 2012). Although the association of TCTP/HRF with human allergic disease has been widely accepted, its exact role in the clinical context still awaits further clarification (Kawakami et al. 2012). Several small-group clinical investigations have established a correlation between the TCTP/HRF status and a range of clinical parameters typical in allergic diseases. Such clinical parameters include the following: (1) the sensitivity of basophils from a subpopulation of allergic donor patients to TCTP/HRF, (2) the intensity of symptoms in the late phase reaction of the allergic response, (3) bronchial hyper-reactivity and sensitivity to histamine, (4) the clinical status of food allergy and atopic dermatitis (reviewed in Macdonald 2012b). However, not all clinical studies were able to detect such correlations (see, e.g. Budde et al. 2002). The observation that the TCTP proteins of two highly allergenic fungal species [*Cladosporium herbarum* (Rid et al. 2008) and *Alternaria alternata* (Rid et al. 2009)] are able to cause histamine release from human basophils and to compete with human HRF is also of interest in this context.

Apart from establishing a clinical correlation, the next question is whether there is indeed a causal relationship between HRF/TCTP levels (or activity) with the disease symptoms. The generation of an inducible transgenic mouse model with overexpressing HRF/TCTP targeted to lung epithelial cells was a first step to address this question. These mice have increased HRF protein levels both in the lung epithelium and extracellularly in the BAL fluid; moreover, HRF exacerbates the allergic, asthmatic responses in these animals after ovalbumin challenge (Yeh et al. 2010). An inherent problem for delineating the precise role of TCTP/HRF in the allergic disease processes lies in its multifunctionality that entails many intracellular roles in addition to the extracellular ones, which makes it difficult to exclude off-target effects (Kawakami et al. 2012).

There was some debate in the literature, whether or not TCTP/HRF binds directly to IgE antibodies and in this way triggers histamine release from mast cell and basophils (Kawakami et al. 2012). A recent detailed study demonstrated that a subset of both IgE and IgG antibodies are able to bind TCTP/HRF (Kashiwakura et al. 2012). Dimerisation of TCTP as a prerequisite for its cytokine-like activity was shown before (Kim et al. 2009a, 2013a). Kawakami and colleagues demonstrated in their paper that it is indeed dimerised TCTP/HRF that binds to these antibodies. They mapped the binding site on TCTP and used corresponding peptides to block its interaction with the 'cognate' IgE antibody. They also showed that these Ig-interacting TCTP/HRF peptides inhibited IgE/HRF-induced mast cell activation in vitro and other allergenic symptoms in a mouse model in vivo (Kashiwakura et al. 2012). This finding demonstrates that TCTP/HRF has a proinflammatory role in asthma and skin hypersensitivity and also that it can be considered a potential therapeutic target. This idea was not entirely new, since just before, Dr. Lee's group identified a 7-mer peptide that binds preferentially to dimerised TCTP and is able to block IL-8 release from BEAS-2B cells triggered by dimeric TCTP. This peptide reduced eosinophil infiltration and other symptoms in a mouse rhinitis model (Kim et al. 2011), and it is non-toxic (Kim et al. 2013b). These two examples indicate that TCTP/HRF may indeed be a valid target for the treatment of symptoms in allergic disease.

4.3.3 TCTP in Lower Animals and Parasitic Infections

There is a large number of reports on TCTP proteins and their biological function in lower animals and plants. The high degree of conservation was mentioned in the introduction (for review see Hinojosa-Moya et al. 2008). Here, I will focus on those papers, which deal with the involvement of such TCTP proteins in infectious or immune response processes or are otherwise potentially involved in human disease.

4.3.3.1 TCTP in Protozoans

The first TCTP protein described in a protozoan was from *Plasmodium falciparum (Pf)*, one of three parasites causing malaria. It was discovered as one of the target proteins for the antimalarial drug artemisinin (Bhisutthibhan et al. 1998). This paper demonstrated that TCTP reacts with artemisinin in situ and in vitro in the presence of haemin and that it binds haemin itself. The interactions of Pf-TCTP with artemisinin (Chae et al. 2006) and of human TCTP with haemin (Lucas et al. 2014) were subsequently studied in more detail, as was the subcellular location and calcium-binding activity in *Plasmodium* (Bhisutthibhan et al. 1999).

The role of malarial TCTP in the host–parasite interaction was studied in three papers: MacDonald et al. showed that malarial TCTP is secreted by the parasite and can be detected in the blood of infected individuals (MacDonald et al. 2001). They also showed that, like human TCTP/HRF, Pf-TCTP is able to stimulate histamine release from basophils and IL-8 secretion from eosinophils, but is much less efficient to do so. Thus, malarial TCTP could compete with human HRF for the binding sites on the immune cells and in this way dim the host immune response. Similarly, another paper demonstrated that Pf-TCTP is incorporated into mouse splenic B-cells at a much higher rate than human TCTP, but has a much lower proliferative effect on B-cells than its human counterpart (Calderon-Perez et al. 2014). A recent study even explored the potential of malarial TCTP as a vaccine to reduce parasitaemia in mice. In two trials, a significant reduction in parasitaemia in the early stages of infection was observed in BALB/c mice (Taylor et al. 2015).

4.3.3.2 TCTP in Parasitic Worms

Reports on the biology of TCTP in parasitic worms typically revolve around the topic of cyto-protection, defence and host–parasite interaction. Two papers characterise the calcium-binding properties of TCTP from the filarial parasites *Brugia malayi* and *Wuchereria bancrofti* (Gnanasekar et al. 2002) and from *Schistosoma mansoni* (Rao et al. 2002). They detected the parasite TCTP protein in the bloodstream of infected mice and described its histamine-releasing activity and the promotion of allergic inflammatory responses associated with filarial infections in these mice. The same group also characterised *Brugia malayi* TCTP as an antioxidant protein (Gnanasekar and Ramaswamy 2007) and TCTP from *Schistosoma mansoni* as a heat-shock protein (Gnanasekar et al. 2009). Similarly, in two *Trichinella* species, TCTP was reported to be heat-induced (Mak et al. 2001, 2007), indicating that it is involved in cyto-protection under heat-shock conditions.

A study on TCTP in two nematode species, *Ostertagia ostertagi*, a parasitic nematode in cattle, and in the free-living nematode *Caenorhabditis elegans* revealed that in both species, TCTP was predominantly located in the eggs of the animals (Meyvis et al. 2009). Interestingly, knock-down of TCTP in *C. elegans* reduced the number of eggs laid by the hermaphrodite in the F0 generation by 90%, indicating that TCTP plays a pivotal role in nematode reproduction.

4.3.3.3 TCTP in Crustaceans and Other Waterborne Animals

Among crustaceans, the TCTP protein most widely studied was in several species of shrimp, usually by researchers from the Southeast Asian region, predominantly Thailand. This is because the shrimp industry plays a substantial role in the economy of this region. One of the major threats to shrimps is the white spot syndrome virus (WSSV), and several papers show that shrimp TCTP is involved in immune defence of the animals against the virus. One of the earliest papers on shrimp TCTP reported already that TCTP levels are severely down-regulated in the advanced stages of viral infection, compared to early infection stages and uninfected animals (Bangrak et al. 2004). Subsequently, the same group demonstrated that injection of shrimps with recombinant TCTP/fortilin after infection with WSSV resulted in 80–100% survival, and it severely reduced the virus load (Tonganunt et al. 2008). Consistent with this, TCTP/ fortilin of the shrimp *Penaeus monodon* inhibited the expression of early and late genes of the WSSV virus in an insect cell model (Nupan et al. 2011). The group also identified a novel binding partner to TCTP/fortilin, fortilin-binding protein (FBP1), which might be involved in the immune defence as well (Panrat et al. 2012).

Recently, a Chinese group working on the TCTP protein of another shrimp species, *Litopenaeus vannamei*, reported that TCTP expression was significantly up-regulated at 16 h and 48 h following infection with the WSSV virus. Silencing of TCTP with dsRNA led to a significant increase in WSSV loads (Wu et al. 2013). Yet another Chinese group studied the TCTP protein from the sea cucumber

Stichopus monotuberculatus. They demonstrated the anti-oxidation and heat-shock protein properties of recombinant TCTP protein, and their data suggested that the sea cucumber TCTP may also play an important role in the innate immune defence against bacterial and viral infections (Ren et al. 2014). Most recently, the characterisation of the TCTP protein of the scallop *Chlamys farreri* (Cf) was published (Jia et al. 2017). Its expression levels are highly regulated during embryonic development of the mollusc, and in response to stimulation with PAMPs (pathogen-associated molecular patterns). Recombinant CfTCTP could induce the release of histamine from BT-549 cells. These results indicate that TCTP plays a pivotal role in the embryonic development and immune protection of scallops.

4.3.3.4 TCTP in Arthropods

Similar to the situation in shrimps, the TCTP protein of the silkworm *Bombyx mori* (Bm) has attracted special attention, particularly from Chinese researchers, although the initial characterisation of the mRNA and the gene structure of BmTCTP was published by a Japanese group (Lee et al. 2004). Later, the Chinese group studied the role of BmTCTP in gut immunity of the silkworm in more detail (Wang et al. 2013). They found that BmTCTP is produced in intestinal epithelial cells and released into the lumen of the gut. The production increases at early time points during oral microbial infection, but declines later. BmTCTP acts as a multi-ligand-binding protein; it also functions as an opsonin that promotes phagocytosis of microorganisms. TCTP induces the production of an antimicrobial peptide via a signalling pathway, which involves activation of ERK. The authors conclude that TCTP is a dual-function protein involved in both the cellular and the humoral immune response of the silkworm. In support of this, the group recently studied the effect of silencing of TCTP by RNAi in a transgenic silkworm (Hu et al. 2015). They reported that the antimicrobial capacity of the silkworm decreased, since the expression of the gut antimicrobial peptide was not sufficiently induced during microbial challenge. This led to the suppression of the innate intestinal immunity, as result of RNAi-mediated knockdown of TCTP.

In summary, the results described in the last two sections show that TCTP is involved not only in cellular defence mechanisms, such as protection against oxidation or heat-shock, but also in innate immunity, of both mammals and the lower taxa of the animal kingdom. A recent, very unusual addition to the various biological roles of the TCTP protein represents a more ‘aggressive’ one, i.e. the participation in the deadly cocktail of spider venoms. Two papers reported that translationally controlled tumour protein is a component of the venom from the brown spider *Loxosceles intermedia* (Gremski et al. 2014; Sade et al. 2012). Another one investigated the spit (used to ‘glue’ the prey to a solid surface) and the venom of the spider *Scytodes thoracica* using transcriptomic and proteomic analyses. In these secretions, they detected TCTP alongside 19 different groups of toxic peptides (Zobel-Thropp et al. 2014).

4.3.4 *TCTP in Other (Patho)physiological Processes*

Apart from being involved in cancer and, as an extracellular protein, in inflammatory and immune reactions, TCTP also participates in other physiological processes and, if deregulated, in pathologic derailments of these. Known examples will be summarised in this section.

4.3.4.1 Metabolic Regulation and Diabetes

About five years ago, we studied the regulation and protective role of TCTP in pancreatic β-cells and demonstrated that TCTP levels are up-regulated in response to stimulatory glucose concentrations, but down-regulated in stress conditions induced by fatty acids (palmitate). Overexpression of TCTP prevented cell death induced by palmitate (Diraison et al. 2011). These results imply that TCTP protects β-cells against stress induced by hyperglycaemia and by high concentrations of fatty acids. More recently, Tsai and colleagues studied the effect of TCTP levels on β-cell proliferation in mice (Tsai et al. 2014). They found that (1) TCTP expression levels are increased under conditions of enhanced β-cell proliferation, i.e. in the perinatal development period and in insulin-resistant states (induced by high-fat diet); (2) TCTP-knockout resulted in decreased β-cell proliferation and cell mass, and in reduced insulin production, eventually leading to hyperglycaemia. Together, these two papers highlight the importance of TCTP for maintaining the homeostasis of pancreatic β-cells and reducing the risk of developing hyperglycaemia and eventually type 2 diabetes.

Whilst in the case of β-cells, the growth-promoting activity of TCTP helps to maintain metabolic homeostasis; in other cell types it may contribute to pathological alterations caused by diabetes. An example is provided through a paper by Kim et al. on podocyte hypertrophy, one of the renal pathologies induced by diabetes (Kim et al. 2012b). They reported that TCTP levels are increased in the glomeruli of diabetic mice, compared to control animals. Knockdown of TCTP led to reduced activity of the mTORC1 pathway in diabetic glomeruli; it reduced the size of the podocytes and prevented the development of diabetic nephropathy.

Very recently, Goodman and colleagues studied the involvement of TCTP and the mTOR signalling pathway in physiological models of skeletal muscle hypertrophy and atrophy (Goodman et al. 2017). Their results show that TCTP and mTOR signalling are up-regulated in both hypertrophy and atrophy of skeletal muscle. The increase in TCTP observed under these conditions occurred in part via an mTOR-dependent mechanism. However, the overexpression of TCTP was not sufficient to activate mTOR signalling. The authors provide preliminary evidence to show that TCTP may act through inhibiting protein degradation, rather than activation of protein synthesis.

4.3.4.2 Blood Circulation

There are also pathologies of the cardiovascular system that are being promoted through TCTP. First, a paper by Kyunglim Lee's group reported that a transgenic mouse overexpressing TCTP develops systemic hypertension (Kim et al. 2008a). These authors had previously shown that TCTP inhibits the Na/K-ATPase, and they proposed that promotion of hypertension by TCTP operates through this mechanism.

A different disease, despite its similar name, is pulmonary arterial hypertension (PAH). It is a lethal disease, caused by excessive proliferation of pulmonary vascular endothelial cells. The hereditary form (HPAH) is often caused by mutations in the bone morphogenetic protein receptor type 2 gene (BMPR2). Lavoie et al., through a proteomics screen, comparing HPAH patients with BMPR2 mutations with healthy control subjects, identified TCTP as one of 22 significantly altered proteins (Lavoie et al. 2014). They reported that TCTP is markedly up-regulated in remodelled blood vessels of complex lesions in lungs from patients with PAH. Silencing of TCTP expression increased apoptosis and abrogated the hyperproliferative phenotype of epithelial cells from patients with HPAH.

Also in the case of atherosclerosis, TCTP seems to play a disease-promoting role, albeit through a different mechanism. Ken Fujise's group has generated a mouse model with heterozygous deficiency of TCTP/fortilin in a background of hypercholesterolemia, which develops atherosclerotic characteristics, similar to those in humans (Pinkaew et al. 2013). Studying this animal model, they arrived at the conclusion that TCTP/fortilin acts by reducing apoptosis in macrophages, one of the main players in the development of atherosclerosis. On the other hand, based on experiments using TCTP overexpression in ApoE-knockout mice, Kyunglim Lee's group proposed that TCTP enhances the severity of atherosclerotic lesions through the induction of hypertension (Cho et al. 2012).

Taken together, the examples of TCTP's involvement in diseases given in Sect. 4.3.4 are by and large based on the growth-promoting effect of TCTP in quite different cellular settings. In one case (pancreatic β-cells), the net effect of TCTP is beneficial; in most other cases, it is disease promoting. All examples—inclusive of cancer—show that, whilst TCTP is generally a cytoprotective protein, its excessive up-regulation is likely to cause disease. Therefore, understanding the regulation of TCTP at the cellular level (see following section) is essential for exploring the mechanisms of such diseases and eventually for our ability to modulate them.

4.4 Regulation of Cellular TCTP Levels

4.4.1 Cell Physiologic Conditions That Result in Regulation of TCTP Levels

Considering the sheer number and range of cell biological processes TCTP is involved in (see Sect. 4.2), it is not surprising that cellular levels of the protein are highly regulated, in response to many different environmental cues and also through a variety of regulatory mechanisms. Presumably, a large number of researchers, who today work on TCTP, originally discovered the protein either in an interaction screen (see Table 4.1), or in search of genes/proteins that are regulated in defined alterations of physiologic conditions, or during transition to a disease state. Typically, TCTP is among those proteins, which display the most prominent changes.

Table 4.3 provides a list of publications, which reported alterations of intracellular TCTP levels in response to altered cell physiologic conditions. The Table lists the type of signals for adaptation ('Stimulus'), the cellular system observed and the likely regulatory mechanism involved. Within the scope of this chapter, it is impossible to discuss each of these cases, and the reader is referred to the individual reference for further details. For this table, I have included only those specific cases of cancer, where a specific mechanism of TCTP up-regulation has been described. The other examples of TCTP overexpression in cancer are listed in Table 4.2.

Two general points emerge from this compilation: (1) The type of physiological settings/adaptations that result in regulation of cellular TCTP levels largely correspond to those conditions, where TCTP was shown to be involved, either as promoting or as protective protein (compare Sect. 4.2.1 to 4.2.3). (2) A whole range of regulatory mechanisms may be involved in regulating TCTP levels, i.e. transcriptional or translational regulation, or stability regulation of TCTP mRNA or protein. However, since not all of the papers listed went into detail to prove or disprove the one or other option, the overall picture is incomplete and may look biased towards transcriptional or translational control. Further work will be necessary to refine this list and confirm or disprove certain types of regulatory mechanisms.

4.4.2 Mechanisms Involved in Regulation of Cellular TCTP Levels

In this section, I will only discuss those papers which explored the *mechanisms* and signalling pathways that underlie the adaptation of TCTP levels in specific cell physiologic settings, not those which just report altered TCTP protein or mRNA levels.

Table 4.3 Cell physiologic conditions resulting in alterations of TCTP levels

Stimulus	Cell/tissue type	Change	Mechanism involved	References
Growth signals: Serum	Swiss 3T3 cells	Up	Translational control	Thomas et al. (1981), Thomas and Thomas (1986)
Serum	Ehrlich ascites cells	Up	Translational control	Bohm et al. (1989, 1991)
Serum	NIH 3T3 cells	Up	Translational control (eIF4E)	Bommer et al. (1994)
Serum	HeLa, HT29 cells	Up	Translational (mTORC1)	Bommer et al. (2015)
Liver regeneration	Rat liver	Up	Transcription activation	Zhu et al. (2008)
Serum starvation	MEF cells	Down	Translation inhibition (PKR)	Bommer et al. (2002)
Hypertrophy/ Atrophy	Mouse muscle	Up	Translational (mTORC1)	Goodman et al. (2017)
Cell cycle: Mitotic exit	Xenopus embryos	Down	Proteasomal degradation	Kubiak et al. (2008)
Cell signalling: Phorbol ester	T24 carcinoma cells	Up	Transcription activation	Andree et al. (2006)
M-CSF	Mouse macrophages	Up	Increased mRNA and protein	Teshima et al. (1998)
Cell differentiation	Mouse erythro-leukaemia cells	Up	Transcription activation	Yenofsky et al. (1983)
Nutrients: Glucose	Pancreatic β-cells	Up	TCTP protein levels	Diraison et al. (2011)
Fatty acids	Pancreatic β-cells	Down	TCTP protein levels	Diraison et al. (2011)
Ammonium starvation	Yeast	Down	Transcription repression	Bonnet et al. (2000)
Heat shock	Prostate cancer cells	(Up)	Hsp27 protects TCTP protein	Baylot et al. (2012)
	Schistosoma; human	Up	Transcription activation	Gnanasekar et al. (2009)
	Trichinella spiralis	Up	Transcription activation	Mak et al. (2001)
	Trichinella pseudospiralis	Up	Translational control	Mak et al. (2007)
Oxidative stress: strong: ATO	Cancer cells	Down	TCTP protein levels	Lucibello et al. (2011)
(mild) H_2O_2	Cancer cells	Up	TCTP protein levels	Lucibello et al. (2011)
(mild) H_2O_2	Human keratinocytes	Up	Transcription activation (VDR)	Rid et al. (2010)
H_2O_2	Plant stresses	Up	Transcription activation	Chen et al. (2014b)

(continued)

Table 4.3 (continued)

Stimulus	Cell/tissue type	Change	Mechanism involved	References
Ca(2+)-Stress: A23187; Thapsigargin	Cos-7 cells	Up	Transcription and translation	Xu et al. (1999)
A23187; Thapsigargin	MEF cells; β-cells	Down	Translation inhibition (PKR)	Bommer et al. (2010)
Thapsigargin	Pancreatic β-cells	Down	TCTP protein levels	Diraison et al. (2011)
DNA damage: γ-Rays	Human cells	Up	TCTP protein levels	Zhang et al. (2012)
Specific cancer types[a]	Hepatocellular cancer (HCC)	Up	Transcription activ. (CHD1L)	Chan et al. (2012b)
	Oral cancer cells	Down	mRNA repression (miRNA-27b)	Lo et al. (2012)
	NF1-associated Tumours	Up	Translational (mTORC1)	Kobayashi et al. (2014)
Alzheimers; Down Syndrome	Brain regions	Down	Protein levels decreased	Kim et al. (2001)
Apoptotic factors: P53	Murine cells	Up	Transcription activation	Chen et al. (2013b)
P53	Breast cancer cells	Down	P53 inhibits transcription	Amson et al. (2012)
P53	RTL6 cells; Mouse erythro-leukemia cells	Down	P53 reduces TCTP protein	Tuynder et al. (2002), Bommer et al. (2010)
Depletion of Mcl-1	U2OS cells	Down	Protein degradation	Zhang et al. (2002)
Heavy metals: Copper (Cu)	Calu-6 and Cos-7 cells	Up	Transcription and translation	Schmidt et al. (2007)
Cobalt, Nickel	Calu-6 and Cos-7 cells	Up	mRNA stabilisation	Schmidt et al. (2007)
Cd, Cu, Pb, Zn	Earth worms	Up	Transcription activation	Sturzenbaum et al. (1998)
Uranium nitrate	Mouse kidneys	Up	Transcription and translation	Taulan et al. (2006)
Aluminium stress	Soybean cultivars	Up	Transcription activation	Ermolayev et al. (2003)
Salt stress	Cassava plant	Up	Transcription activation	Santa Brigida et al. (2014)
Mercury Stress	Rice roots	Up	TCTP protein levels	Wang et al. (2012)
Toxins: Dioxin	Mouse ES cells	Up	Transcription activation	Oikawa et al. (2002)
Dioxin	Calu-6 and Cos-7 cells	Up	Transcription activation	Schmidt et al. (2007)
Naphthenic acids	Earth worms	Up	Transcription activation	Wang et al. (2015)

(continued)

Table 4.3 (continued)

Stimulus	Cell/tissue type	Change	Mechanism involved	References
Drugs: 5-Fluorouracil	Colon cancer cells	Up	Translational (mTORC1)	Bommer et al. (2017)
Oxaliplatin	Colon cancer cells	Up	Translational (mTORC1)	Bommer et al. (2017), Yao et al. (2009)
Ursolic acid	Hepatocellular carcinoma cells	Down	Reduced PI3K signalling	Chuang et al. (2016)
Dihydro-artemisinin	Several cell lines	Down	Proteasomal degradation	Fujita et al. (2008)
Dihydro-artemisinin	Breast cancer cells	Down	Proteasomal degradation	Lucibello et al. (2015)
Sertraline, Thioridazine	Breast cancer cells	Down	Increased p53; mTOR inhibition	Amson et al. (2013)

[a]Only those cancer types are listed, where the mechanism of TCTP regulation has been explored

4.4.2.1 Transcriptional Regulation of TPT1 Gene Expression

The most comprehensive analysis of the gene structure and transcriptional regulation of the mammalian TPT1 gene came from the Thiele laboratory in Berlin. The gene and mRNA structures were described in Sect. 4.1.2; here I will just summarise the results on mechanisms of transcriptional regulation of TCTP expression. Andree et al. (2006) performed predictions of potential transcription factor-binding sites in the 5′-flanking region of the TPT1 gene of five mammalian species, which revealed the conservation of a cluster of five such binding sites within the first 170 nucleotides 5′-terminal to the transcription start site of these genes. These comprised two binding sites each for the transcription factors ETS1 and CREB and one binding site for MZF1. In this paper, they confirmed experimentally that TCTP expression is indeed regulated by phorbol ester (PMA) and forskolin through the cAMP-PKA signalling pathway via transcription factor CREB.

Other confirmed examples of transcriptional regulation of TCTP synthesis include the following:

Differentiation of mouse erythroleukaemia (MEL) cells is induced by treatment of cells with DMSO. It belongs to the very early observations on TCTP (P21) mRNA that its synthesis increases early in this process, even though its translation rate decreases after exposure to DMSO (Yenofsky et al. 1983).

Other examples of specific *transcriptional regulation of the TPT1 gene* are *relevant to cancer*: (1) The tumour suppressor protein P53 acts as a transcription factor and, as an antagonist to TCTP, it binds to a p53-response element upstream of the TPT1 gene and represses its transcription in human cells (Amson et al. 2012). This inhibitory activity is alleviated in cancers, which bear mutations in the p53 gene, as it is frequently the case. (2) In hepatocellular cancer (HCC), a transcription factor called CHD1L has been identified as a specific oncogene. CHD1L binds to the promoter region of the TPT1 gene and activates its transcription (Chan et al. 2012b). (3) Transcriptional regulation of TCTP by HIF-1α was reported by Xiao et al. in colon cancer cells (Xiao et al. 2016). (4) In rat liver regeneration, a model of actively proliferating tissue, the expression of TCTP mRNA is transiently up-regulated, from 3 to 12 h after partial hepatectomy (Zhu et al. 2008).

There are also examples of *TCTP being transcriptionally regulated in stress conditions*: (1) Mild oxidative stress induced by hydrogen peroxide was reported to induce transcription activation of TCTP expression through the vitamin D3 receptor (VDR) in keratinocytes (Rid et al. 2010) and plants (maize) under flood stress conditions (Chen et al. 2014b). (2) Transcriptional induction of TCTP synthesis by heavy metals has been observed very early in earthworms (Sturzenbaum et al. 1998). In this case, copper and cadmium led to the highest rates of TCTP mRNA synthesis. A more detailed study on the regulation of TCTP by heavy metals was subsequently performed in the Thiele laboratory (Schmidt et al. 2007). They found that copper induced TCTP synthesis at both the transcriptional and the translational level, whereas cobalt and nickel seem to result in TCTP mRNA stabilisation. These authors also showed that the potent toxin dioxin (TCDD) transcriptionally activates TCTP synthesis in human and Calu-6 and Cos-7 cells, which mirrors an earlier report on mouse embryonic stem cells (Oikawa et al. 2002).

Further examples of transcriptional activation of TCTP synthesis are given in Table 4.3.

4.4.2.2 Translational Regulation of TCTP Synthesis

The early demonstration that TCTP is a translationally controlled protein was based on the following evidence: (1) The rate of synthesis of the proteins 'Q23' (Thomas et al. 1981) and 'P23' (Bohm et al. 1989) were found to be increased very early after serum induction of Swiss 3T3 fibroblasts and mouse Ehrlich ascites tumour cells, respectively. This rate increase was visible within 10 min, and it was resistant to inhibition by the transcription inhibitor actinomycin D. (2) These two proteins were later shown to be identical to yet another one called 'P21', the mRNA of which was found abundantly in cytoplasmic untranslated mRNP particles in mouse sarcoma ascites cells (Yenofsky et al. 1982). These mRNP particles were considered reserve pools of untranslated mRNAs.

At the beginning of the 1990s, mechanisms of translational control were just being unveiled, and the importance of translation initiation factors in cellular regulation was recognised (reviewed in Clemens and Bommer 1999). In particular, the ability of the cap-binding protein eIF4E to trigger the malignant transformation of cells was discovered at this time (Lazaris-Karatzas et al. 1990). The prevailing hypothesis to explain this crucial role of eIF4E was that certain mRNAs that code for growth-related proteins are highly structured and are therefore poorly translated. Overexpression of eIF4E results in more efficient translation of these mRNAs leading to promotion of cell proliferation. In collaboration with the Sonenberg lab, we were able to show that the mRNA of TCTP (P23) is one of the mRNAs, the translational efficiency of which is highly dependent on eIF4E (Bommer et al. 1994). Consistent with this, we later demonstrated that TCTP(P23) mRNA is indeed a highly structured molecule (Bommer et al. 2002).

It was only late in the 1990s that the link was established between the PI3K/Akt/mTORC1 growth signalling pathway and the activation of eIF4E via phosphorylation of the inhibitory eIF4E-binding proteins, 4E-BPs (Clemens and Bommer 1999). Also, it became increasingly clear that a subset of mRNAs, i.e. the ones bearing a 5′-terminal oligopyrimidine tract (5′-TOP), is specifically regulated through this pathway. The main representatives of this group of mRNAs are those which encode components of the translational apparatus (Meyuhas and Kahan 2015). Since the TCTP mRNA also features a 5′-TOP (Yamashita et al. 2008), we recently revisited this topic and showed that the growth factor-dependent induction of TCTP synthesis is indeed regulated through the PI3K/Akt/mTORC1 pathway and eIF4E (Bommer et al. 2015). We also found that treatment of colon cancer cells with DNA-damaging anticancer drugs leads to a four-fold up-regulation of TCTP levels through the mTORC1 pathway (Bommer et al. 2017). Similarly, Goodman et al. reported an increase in TCTP levels in muscle hypertrophy and atrophy and its regulation through mTOR signalling (Goodman et al. 2017).

An important negative translational control mechanism operates via phosphorylation of translation initiation factor eIF2α. This phosphorylation event prevents the recycling of the guanosine nucleotide exchange factor for eIF2, eIF2B, and results in the shutdown of protein synthesis. There are four specific protein kinases that are able to phosphorylate eIF2α, each one being activated under very specific cell stress conditions (Clemens and Bommer 1999). One of these, the dsRNA-dependent kinase PKR, is an antiviral enzyme that is activated by double-stranded RNA, which is often formed during viral replication. However, there are also few cellular mRNAs, which are highly structured and able to activate PKR. We have shown that the mRNA for TCTP is one of those mRNA molecules. Due to its structure, it can activate PKR locally and prevent its own translation (Bommer et al. 2002). We have also demonstrated that TCTP mRNA translation is indeed inhibited by PKR in cell stress conditions, as, e.g. in serum starvation (Bommer et al. 2002) or under calcium stress (Bommer et al. 2010).

These findings are consistent with reports on the regulation of other anti-apoptotic proteins, such as Mcl-1, Bcl-XL and survivin. They are all translationally regulated through the mTORC1 pathway [see references in Bommer et al. (2015)], and Mcl-1 is also regulated through PKR (Fritsch et al. 2007).

4.4.2.3 Other Post-transcriptional Regulation

1. *Regulation by Micro-RNAs*. The expression of genes involved in fundamental biological processes, such as development, apoptosis or cancer, are often subject to an additional layer of regulation, by micro-RNAs. This also applies to anti-apoptotic proteins, as they are mentioned above. It is, therefore, appropriate to assume that TCTP would be regulated by micro-RNAs as well. However, reports

on this topic are scarce in the literature; in fact, there are only two papers from 2012 reporting the regulation of TCTP levels by micro-RNAs.

Lo et al. performed a proteomics study on genes differentially expressed in oral cancer patients compared to normal individuals; they found TCTP to be up-regulated in the cancer patient group (Lo et al. 2012). At the same time, the miRNA miR-27b was down-regulated in oral cancer, and expression of miR-27b in two oral cancer cell lines resulted in TCTP levels being reduced to about 30%. In a study to validate an assay for miRNA target identification, Gäken and colleagues used miR-130a as an example. They reported TCTP as the one of the five target mRNAs newly identified by this method, which is regulated to the highest extent (Gaken et al. 2012).

We ourselves attempted to identify micro-RNAs that may target TCTP mRNA. *In-silico* searches revealed several potential target sites in the 3′-UTR of this mRNA; however, our attempts to validate some of these in luciferase reporter gene and in cellular assays did not yield consistent results (UA Bommer, J Clancy and T. Preiss, unpublished observations), and we did not pursue this project further. Thus, the only two reported cases of miRNAs targeting TCTP mRNA are miR-27b and miR-130a.

2. *Regulation of mRNA stability.* In their study on the regulation of TCTP expression by heavy metals, Schmidt et al. found that cobalt and nickel moderately increase TCTP expression through stabilisation of its mRNA (Schmidt et al. 2007). They speculated about the potential role of AUUUA motifs in the 3′-UTR of TCTP mRNA in this process. However, these AUUUA elements do not completely match the 'classical' AU-rich elements (AREs), which in cytokine mRNAs serve to target these RNAs for regulated degradation. The role of these motifs in TCTP mRNA, if any, is yet to be elucidated. Overall, TCTP mRNA was found to be generally fairly stable and abundant in mammalian cells (Yenofsky et al. 1983), although the abundance varies depending on the tissue type (Thiele et al. 2000). Our finding that TCTP mRNA has a high degree of structure (Bommer et al. 2002) is also consistent with the notion that it is a fairly stable RNA molecule.
3. *Protein stability regulation.* There are a few reports, which indicate that in specific instances TCTP protein may be subject to regulated degradation. Two studies described the ability of another anti-apoptotic protein, Mcl-1 (Zhang et al. 2002), and heat-shock protein Hsp27 (Baylot et al. 2012) to stabilise the TCTP protein, which implies that it may be destabilised if those proteins are absent. Kubiak et al. reported that TCTP, a protein involved in maintaining the integrity of the mitotic spindle, is partially degraded during mitotic exit (Kubiak et al. 2008). A novel mechanism for regulated degradation of TCTP protein was published in a very recent paper (see 'Note Added in Proof').

In summary, TCTP levels can be regulated at all levels of gene expression, inclusive of protein stability, although transcriptional and translational regulation of protein expression seems to be the most commonly reported regulatory mechanisms.

4.5 Synopsis

During the 35 years since its original discovery, the translationally controlled tumour protein TCTP has attracted an ever-increasing level of attention. Through a body of well over 300 publications as of today, a considerable number of functional associations of this protein have been established.

Although TCTP is largely a cellular protein, it has also extracellular functions, and its importance in fundamental biological processes often has implications for homeostasis at the whole-organism level. In trying to summarise the functional importance of this protein, the following key words come to mind: cell growth and proliferation, early development, cyto-protection and defence. Many examples studied by today confirm that TCTP is involved in the biological defence against a wide range of cell stresses, in nearly all eukaryotic kingdoms. The specifics of the defence reactions, in which TCTP is engaged, may be quite varied between lower and higher organisms or between plants and animals. What we need now is a deeper understanding of the mechanisms, by which TCTP exerts all these effects.

Naturally, the involvement in these important processes requires a high degree of regulation of both levels and activity of the protein. We do understand some of the underlying mechanisms, but by far not yet all details. Such understanding is of particular importance, since the frequent participation of TCTP in disease processes, such as in cancer, is often due to deregulation of TCTP or due to the exertion of its effects at the wrong time in the wrong place. Promising initial discoveries have been made, identifying TCTP as a potential target in anticancer (or other disease) strategies. We have to learn more about some of the functional interactions of this protein, as well as its regulation, in order eventually to be able to translate this knowledge into meaningful clinical application.

Note Added in Proof

Since the original completion of this manuscript (December 2016) more than half a year has passed, and quite a few new papers on the TCTP protein were published in the meantime. Of these, I just wish to mention the following five publications, which added some interesting new aspects to our understanding of the biology of this protein. (My apology to those colleagues, whose work was not considered here, or even in the main manuscript!)

1. As pointed out in Sect. 4.2.1, there are a large number of publications reporting TCTP as a cyto-protective protein, able to protect cells against a wide range of cytotoxic stresses, inclusive of Ca^{2+}-stress. However, so far, an involvement of TCTP in modulating **ER-stress** and the unfolded protein response (UPR) was not yet described. A recent paper by Pinkaew et al. fills this gap (Pinkaew et al. 2017), by reporting that TCTP (fortilin) binds to the cytoplasmic domain of the ER-stress sensor IRE1α, inhibiting its endonuclease (RNase) and protein kinase activities, and in this way protecting cells against apoptotic cell death.
2. The involvement of TCTP in the cellular process of **autophagy** was studied in a recent paper by the group of Kyunglim Lee (Bae et al. 2017). They found that TCTP interferes with this process in both the mTORC1/AMPK-dependent and

the mTORC1-independent pathway. Thus, TCTP inhibits macroautophagy both at early stages and later at the step of autophagosome maturation. This conclusion contrasts that of another paper (Chen et al. 2014a), cited in Sect. 4.2.3. A possible explanation for this discrepancy is given in the discussion of the more recent paper (Bae et al. 2017).

3. In the very last section of the main text, I mentioned that relatively little is known about the mechanisms involved in **regulation of TCTP degradation**. This gap was closed by a very recent paper by Bonhoure et al. (2017), who showed that TCTP protein can be degraded through the chaperone-mediated autophagy (CMA) pathway. This pathway is different from the process of macroautophagy mentioned in the preceding paragraph. CMA involves the targeting of individual cytoplasmic proteins for lysosomal degradation. In the case of TCTP, it involves the acetylation of the protein at Lysine 19 and its binding to Hsc70, and it requires the activity of lysosome-associated membrane protein type 2A (LAMP-2A). The authors show that this is an underlying mechanism for the down-regulation of TCTP levels under serum-starvation conditions (Bonhoure et al. 2017), which is likely to be corroborated by a block in TCTP mRNA translation (Bommer et al. 2002, 2015).
4. Previous studies on the **role of TCTP in plants**, other than in stress reactions, showed that plant TCTP is important for cell division, growth and development (Sect. 4.2.3). A new study extended these investigations. De Carvalho and colleagues expressed tomato TCTP in tobacco plants and, by transcriptomics analysis, studied the pathways that are differentially regulated as a result of TCTP overexpression (de Carvalho et al. 2017). They observed that genes involved in photosynthesis, fatty acid metabolism and water transport are up-regulated, while genes involved in the synthesis of the phytohormone ethylene were down-regulated. TCTP overexpression also promoted biomass production and it protected plants against salt and osmotic stress. These observations are also consistent with an earlier study that demonstrated the ability of *Arabidopsis thaliana* TCTP to protect tobacco leafs against induction of programmed cell death (Hoepflinger et al. 2013).

Acknowledgements Work in my laboratory was supported by project grants from The Wellcome Trust (UK), by a Short-Term Fellowship from the Human Frontier Science Program (Strasbourg) and by small grants from the Cancer Prevention Research Trust (London, UK). I received small grants from the Illawarra Health and Medical Research Institute and from the Graduate School of Medicine, University of Wollongong, NSW, Australia. I wish to thank all friends and colleagues, who contributed to our work.

References

Acunzo J, Baylot V, So A, Rocchi P (2014) TCTP as therapeutic target in cancers. Cancer Treat Rev 40(6):760–769

Ambrosio MR, Rocca BJ, Barone A, Onorati M, Mundo L, Crivelli F, Di Nuovo F, De Falco G, del Vecchio MT, Tripodi SA et al (2015) Expression of translationally controlled tumor protein in human kidney and in renal cell carcinoma. Biomed Res Int 2015:730390

Amson R, Kubiak JZ, Van Montagu M, Telerman A (2011) Could TCTP contribute to Armin Braun's paradigm of tumor reversion in plants? Cell Cycle 10(1):1

Amson R, Pece S, Lespagnol A, Vyas R, Mazzarol G, Tosoni D, Colaluca I, Viale G, Rodrigues-Ferreira S, Wynendaele J et al (2012) Reciprocal repression between P53 and TCTP. Nat Med 18(1):91–99

Amson R, Karp JE, Telerman A (2013) Lessons from tumor reversion for cancer treatment. Curr Opin Oncol 25(1):59–65

Amzallag N, Passer BJ, Allanic D, Segura E, Thery C, Goud B, Amson R, Telerman A (2004) TSAP6 facilitates the secretion of translationally controlled tumor protein/histamine-releasing factor via a nonclassical pathway. J Biol Chem 279(44):46104–46112

Andree H, Thiele H, Fahling M, Schmidt I, Thiele BJ (2006) Expression of the human TPT1 gene coding for translationally controlled tumor protein (TCTP) is regulated by CREB transcription factors. Gene 380(2):95–103

Arcuri F, Papa S, Carducci A, Romagnoli R, Liberatori S, Riparbelli MG, Sanchez J-C, Tosi P, del Vecchio MT (2004) Translationally controlled tumor protein (TCTP) in the human prostate and prostate cancer cells: expression, distribution, and calcium binding activity. Prostate 60 (2):130–140

Arcuri F, Papa S, Meini A, Carducci A, Romagnoli R, Bianchi L, Riparbelli MG, Sanchez J-C, Palmi M, Tosi P et al (2005) The translationally controlled tumor protein is a novel calcium binding protein of the human placenta and regulates calcium handling in trophoblast cells. Biol Reprod 73(4):745–751

Bae SY, Kim HJ, Lee KJ, Lee K (2015) Translationally controlled tumor protein induces epithelial to mesenchymal transition and promotes cell migration, invasion and metastasis. Sci Rep 5:8061

Bae SY, Byun S, Bae SH, Min DS, Woo HA, Lee K (2017) TPT1 (tumor protein, translationally-controlled 1) negatively regulates autophagy through the BECN1 interactome and an MTORC1-mediated pathway. Autophagy 13(5):820–833

Bangrak P, Graidist P, Chotigeat W, Phongdara A (2004) Molecular cloning and expression of a mammalian homologue of a translationally controlled tumor protein (TCTP) gene from Penaeus monodon shrimp. J Biotechnol 108(3):219–226

Baylot V, Katsogiannou M, Andrieu C, Taieb D, Acunzo J, Giusiano S, Fazli L, Gleave M, Garrido C, Rocchi P (2012) Targeting TCTP as a new therapeutic strategy in castration-resistant prostate cancer. Mol Ther 20(12):2244–2256

Bazile F, Pascal A, Arnal I, Le Clainche C, Chesnel F, Kubiak JZ (2009) Complex relationship between TCTP, microtubules and actin microfilaments regulates cell shape in normal and cancer cells. Carcinogenesis 30(4):555–565

Berkowitz O, Jost R, Pollmann S, Masle J (2008) Characterization of TCTP, the translationally controlled tumor protein, from Arabidopsis thaliana. Plant Cell 20(12):3430–3447

Bhisutthibhan J, Meshnick SR (2001) Immunoprecipitation of [(3)H]dihydroartemisinin translationally controlled tumor protein (TCTP) adducts from Plasmodium falciparum-infected erythrocytes by using anti-TCTP antibodies. Antimicrob Agents Chemother 45(8):2397–2399

Bhisutthibhan J, Pan XQ, Hossler PA, Walker DJ, Yowell CA, Carlton J, Dame JB, Meshnick SR (1998) The Plasmodium falciparum translationally controlled tumor protein homolog and its reaction with the antimalarial drug artemisinin. J Biol Chem 273(26):16192–16198

Bhisutthibhan J, Philbert MA, Fujioka H, Aikawa M, Meshnick SR (1999) The Plasmodium falciparum translationally controlled tumor protein: subcellular localization and calcium binding. Eur J Cell Biol 78(9):665–670

Bohm H, Benndorf R, Gaestel M, Gross B, Nurnberg P, Kraft R, Otto A, Bielka H (1989) The growth-related protein P23 of the Ehrlich ascites tumor: translational control, cloning and primary structure. Biochem Int 19(2):277–286

Bohm H, Gross B, Gaestel M, Bommer UA, Ryffel G, Bielka H (1991) The 5'-untranslated region of p23 mRNA from the Ehrlich ascites tumor is involved in translation control of the growth related protein p23. Biomed Biochim Acta 50(12):1193–1203

Bommer UA (2012) Cellular function and regulation of the translationally controlled tumour protein TCTP. Open Allergy J 5:19–32

Bommer UA, Thiele BJ (2004) The translationally controlled tumour protein (TCTP). Int J Biochem Cell Biol 36(3):379–385

Bommer UA, Lazaris-Karatzas A, De Benedetti A, Nurnberg P, Benndorf R, Bielka H, Sonenberg N (1994) Translational regulation of the mammalian growth-related protein P23: involvement of eIF-4E. Cell Mol Biol Res 40(7-8):633–641

Bommer UA, Borovjagin AV, Greagg MA, Jeffrey IW, Russell P, Laing KG, Lee M, Clemens MJ (2002) The mRNA of the translationally controlled tumor protein P23/TCTP is a highly structured RNA, which activates the dsRNA-dependent protein kinase PKR. RNA 8 (4):478–496

Bommer UA, Heng C, Perrin A, Dash P, Lobov S, Elia A, Clemens MJ (2010) Roles of the translationally controlled tumour protein (TCTP) and the double-stranded RNA-dependent protein kinase, PKR, in cellular stress responses. Oncogene 29(5):763–773

Bommer UA, Iadevaia V, Chen J, Knoch B, Engel M, Proud CG (2015) Growth-factor dependent expression of the translationally controlled tumour protein TCTP is regulated through the PI3-K/Akt/mTORC1 signalling pathway. Cell Signal 27(8):1557–1568

Bommer UA, Vine KL, Puri P, Engel M, Belfiore L, Fildes K, Batterham M, Lochhead A, Aghmesheh M (2017) Translationally controlled tumour protein TCTP is induced early in human colorectal tumours and contributes to the resistance of HCT116 colon cancer cells to 5-FU and oxaliplatin. Cell Commun Signal 15(1):9

Bonhoure A, Vallentin A, Martin M, Senff-Ribeiro A, Amson R, Telerman A, Vidal M (2017) Acetylation of translationally controlled tumor protein promotes its degradation through chaperone-mediated autophagy. Eur J Cell Biol 96(2):83–98

Bonnet C, Perret E, Dumont X, Picard A, Caput D, Lenaers G (2000) Identification and transcription control of fission yeast genes repressed by an ammonium starvation growth arrest. Yeast 16(1):23–33

Brioudes F, Thierry AM, Chambrier P, Mollereau B, Bendahmane M (2010) Translationally controlled tumor protein is a conserved mitotic growth integrator in animals and plants. Proc Natl Acad Sci U S A 107(37):16384–16389

Budde IK, Lopuhaa CE, de Heer PG, Langdon JM, MacDonald SM, van der Zee JS, Aalberse RC (2002) Lack of correlation between bronchial late allergic reaction to Dermatophagoides pteronyssinus and in vitro immunoglobulin E reactivity to histamine-releasing factor derived from mononuclear cells. Ann Allergy Asthma Immunol 89(6):606–612

Burgess A, Labbe JC, Vigneron S, Bonneaud N, Strub JM, Van Dorsselaer A, Lorca T, Castro A (2008) Chfr interacts and colocalizes with TCTP to the mitotic spindle. Oncogene 27 (42):5554–5566

Calderon-Perez B, Xoconostle-Cazares B, Lira-Carmona R, Hernandez-Rivas R, Ortega-Lopez J, Ruiz-Medrano R (2014) The Plasmodium falciparum translationally controlled tumor protein (TCTP) is incorporated more efficiently into B cells than its human homologue. PLoS One 9 (1):e85514

Cans C, Passer BJ, Shalak V, Nancy-Portebois V, Crible V, Amzallag N, Allanic D, Tufino R, Argentini M, Moras D et al (2003) Translationally controlled tumor protein acts as a guanine

nucleotide dissociation inhibitor on the translation elongation factor eEF1A. Proc Natl Acad Sci U S A 100(24):13892–13897
Cao B, Lu Y, Chen G, Lei J (2010) Functional characterization of the translationally controlled tumor protein (TCTP) gene associated with growth and defense response in cabbage. Plant Cell Tissue Org Cult 103:217–226
Chae J, Choi I, Kim C (2006) Homology modeling and molecular docking study of translationally controlled tumor protein and artemisinin. Arch Pharm Res 29(1):50–58
Chan TH, Chen L, Guan XY (2012a) Role of translationally controlled tumor protein in cancer progression. Biochem Res Int 2012:369384
Chan TH, Chen L, Liu M, Hu L, Zheng BJ, Poon VK, Huang P, Yuan YF, Huang JD, Yang J et al (2012b) Translationally controlled tumor protein induces mitotic defects and chromosome missegregation in hepatocellular carcinoma development. Hepatology 55(2):491–505
Chattopadhyay A, Pinkaew D, Doan HQ, Jacob RB, Verma SK, Friedman H, Peterson AC, Kuyumcu-Martinez MN, McDougal OM, Fujise K (2016) Fortilin potentiates the peroxidase activity of Peroxiredoxin-1 and protects against alcohol-induced liver damage in mice. Sci Rep 6:18701
Chen SH, Wu P-S, Chou C-H, Yan Y-T, Liu H, Weng S-Y, Yang-Yen H-F (2007a) A knockout mouse approach reveals that TCTP functions as an essential factor for cell proliferation and survival in a tissue- or cell type-specific manner. Mol Biol Cell 18(7):2525–2532
Chen Z, Zhang H, Yang H, Huang X, Zhang X, Zhang P (2007b) The expression of AmphiTCTP, a TCTP orthologous gene in amphioxus related to the development of notochord and somites. Comp Biochem Physiol B Biochem Mol Biol 147(3):460–465
Chen Y, Fujita T, Zhang D, Doan H, Pinkaew D, Liu Z, Wu J, Koide Y, Chiu A, Lin CC et al (2011) Physical and functional antagonism between tumor suppressor protein p53 and fortilin, an anti-apoptotic protein. J Biol Chem 286:32575–32585
Chen K, Chen S, Huang C, Cheng H, Zhou R (2013a) TCTP increases stability of hypoxia-inducible factor 1alpha by interaction with and degradation of the tumour suppressor VHL. Biol Cell 105(5):208–218
Chen W, Wang H, Tao S, Zheng Y, Wu W, Lian F, Jaramillo M, Fang D, Zhang DD (2013b) Tumor protein translationally controlled 1 is a p53 target gene that promotes cell survival. Cell Cycle 12(14):2321–2328
Chen K, Huang C, Yuan J, Cheng H, Zhou R (2014a) Long-term artificial selection reveals a role of TCTP in autophagy in mammalian cells. Mol Biol Evol 31(8):2194–2211
Chen Y, Chen X, Wang H, Bao Y, Zhang W (2014b) Examination of the leaf proteome during flooding stress and the induction of programmed cell death in maize. Proteome Sci 12:33
Chen C, Deng Y, Hua M, Xi Q, Liu R, Yang S, Liu J, Zhong J, Tang M, Lu S et al (2015) Expression and clinical role of TCTP in epithelial ovarian cancer. J Mol Histol 46(2):145–156
Cheng X, Li J, Deng J, Li Z, Meng S, Wang H (2012) Translationally controlled tumor protein (TCTP) downregulates Oct4 expression in mouse pluripotent cells. BMB Rep 45(1):20–25
Chitpatima ST, Makrides S, Bandyopadhyay R, Brawerman G (1988) Nucleotide sequence of a major messenger RNA for a 21 kilodalton polypeptide that is under translational control in mouse tumor cells. Nucleic Acids Res 16(5):2350
Cho Y, Maeng J, Ryu J, Shin H, Kim M, Oh GT, Lee MY, Lee K (2012) Hypertension resulting from overexpression of translationally controlled tumor protein increases the severity of atherosclerosis in apolipoprotein E knock-out mice. Transgenic Res 21(6):1245–1254
Choi S, Min HJ, Kim M, Hwang ES, Lee K (2009) Proton pump inhibitors exert anti-allergic effects by reducing TCTP secretion. PLoS One 4(6):e5732
Chou M, Xia C, Feng Z, Sun Y, Zhang D, Zhang M, Wang L, Wei G (2016) A translationally controlled tumor protein gene Rpf41 is required for the nodulation of Robinia pseudoacacia. Plant Mol Biol 90(4-5):389–402
Chu ZH, Liu L, Zheng CX, Lai W, Li SF, Wu H, Zeng YJ, Zhao HY, Guan YF (2011) Proteomic analysis identifies translationally controlled tumor protein as a mediator of phosphatase of

regenerating liver-3-promoted proliferation, migration and invasion in human colon cancer cells. Chin Med J 124(22):3778–3785

Chuang WL, Lin PY, Lin HC, Chen YL (2016) The apoptotic effect of ursolic acid on SK-Hep-1 cells is regulated by the PI3K/Akt, p38 and JNK MAPK signaling pathways. Molecules 21 (4):460

Chung S, Kim M, Choi W, Chung J, Lee K (2000) Expression of translationally controlled tumor protein mRNA in human colon cancer. Cancer Lett 156(2):185–190

Clemens MJ, Bommer UA (1999) Translational control: the cancer connection. Int J Biochem Cell Biol 31(1):1–23

Cucchi U, Gianellini LM, De Ponti A, Sola F, Alzani R, Patton V, Pezzoni A, Troiani S, Saccardo MB, Rizzi S et al (2010) Phosphorylation of TCTP as a marker for polo-like kinase-1 activity in vivo. Anticancer Res 30(12):4973–4985

de Carvalho M, Acencio ML, AVN L, de Araujo LM, de Lara Campos Arcuri M, do Nascimento LC, Maia IG (2017) Impacts of the overexpression of a tomato translationally controlled tumor protein (TCTP) in tobacco revealed by phenotypic and transcriptomic analysis. Plant Cell Rep 36(6):887–900

Deng S-S, Xing T-Y, Zhou H-Y, Xiong R-H, Lu Y-G, Wen B, Liu S-Q, Yang H-J (2006) Comparative proteome analysis of breast cancer and adjacent normal breast tissues in human. Genomics Proteomics Bioinformatics 4(3):165–172

Deng Z, Chen J, Leclercq J, Zhou Z, Liu C, Liu H, Yang H, Montoro P, Xia Z, Li D (2016) Expression profiles, characterization and function of HbTCTP in rubber tree (Hevea brasiliensis). Front Plant Sci 7:789

Diraison F, Hayward K, Sanders KL, Brozzi F, Lajus S, Hancock J, Francis JE, Ainscow E, Bommer UA, Molnar E et al (2011) Translationally controlled tumour protein (TCTP) is a novel glucose-regulated protein that is important for survival of pancreatic beta cells. Diabetologia 54(2):368–379

Dong X, Yang B, Li Y, Zhong C, Ding J (2009) Molecular basis of the acceleration of the GDP-GTP exchange of human ras homolog enriched in brain by human translationally controlled tumor protein. J Biol Chem 284(35):23754–23764

Efferth T (2005) Mechanistic perspectives for 1,2,4-trioxanes in anti-cancer therapy. Drug Resist Updat 8(1–2):85–97

Efferth T (2006) Molecular pharmacology and pharmacogenomics of artemisinin and its derivatives in cancer cells. Curr Drug Targets 7(4):407–421

Ermolayev V, Weschke W, Manteuffel R (2003) Comparison of Al-induced gene expression in sensitive and tolerant soybean cultivars. J Exp Bot 54(393):2745–2756

Fadeel B, Ottosson A, Pervaiz S (2008) Big wheel keeps on turning: apoptosome regulation and its role in chemoresistance. Cell Death Differ 15(3):443–452

Feng Y, Liu D, Yao H, Wang J (2007a) Solution structure and mapping of a very weak calcium-binding site of human translationally controlled tumor protein by NMR. Arch Biochem Biophys 467(1):48–57

Feng Z, Hu W, de Stanchina E, Teresky AK, Jin S, Lowe S, Levine AJ (2007b) The regulation of AMPK beta1, TSC2, and PTEN expression by p53: stress, cell and tissue specificity, and the role of these gene products in modulating the IGF-1-AKT-mTOR pathways. Cancer Res 67 (7):3043–3053

Fiucci G, Lespagnol A, Stumptner-Cuvelette P, Beaucourt S, Duflaut D, Susini L, Amson R, Telerman A (2003) Genomic organization and expression of mouse Tpt1 gene. Genomics 81 (6):570–578

Fleischer TC, Weaver CM, McAfee KJ, Jennings JL, Link AJ (2006) Systematic identification and functional screens of uncharacterized proteins associated with eukaryotic ribosomal complexes. Genes Dev 20(10):1294–1307

Friedman DB, Hill S, Keller JW, Merchant NB, Levy SE, Coffey RJ, Caprioli RM (2004) Proteome analysis of human colon cancer by two-dimensional difference gel electrophoresis and mass spectrometry. Proteomics 4(3):793–811

Fritsch RM, Schneider G, Saur D, Scheibel M, Schmid RM (2007) Translational repression of MCL-1 couples stress-induced eIF2 alpha phosphorylation to mitochondrial apoptosis initiation. J Biol Chem 282(31):22551–22562

Fujita T, Felix K, Pinkaew D, Hutadilok-Towatana N, Liu Z, Fujise K (2008) Human fortilin is a molecular target of dihydroartemisinin. FEBS Lett 582(7):1055–1060

Funston G, Goh W, Wei SJ, Tng QS, Brown C, Jiah Tong L, Verma C, Lane D, Ghadessy F (2012) Binding of translationally controlled tumour protein to the N-terminal domain of HDM2 is inhibited by nutlin-3. PLoS One 7(8):e42642

Gachet Y, Tournier S, Lee M, Lazaris-Karatzas A, Poulton T, Bommer UA (1999) The growth-related, translationally controlled protein P23 has properties of a tubulin binding protein and associates transiently with microtubules during the cell cycle. J Cell Sci 112(Pt 8):1257–1271

Gaken J, Mohamedali AM, Jiang J, Malik F, Stangl D, Smith AE, Chronis C, Kulasekararaj AG, Thomas NS, Farzaneh F et al (2012) A functional assay for microRNA target identification and validation. Nucleic Acids Res 40(10):e75

Ge F, Zhang L, Tao SC, Kitazato K, Zhang ZP, Zhang XE, Bi LJ (2011) Quantitative proteomic analysis of tumor reversion in multiple myeloma cells. J Proteome Res 10(2):845–855

Gnanasekar M, Ramaswamy K (2007) Translationally controlled tumor protein of Brugia malayi functions as an antioxidant protein. Parasitol Res 101(6):1533–1540

Gnanasekar M, Rao KVN, Chen L, Narayanan RB, Geetha M, Scott AL, Ramaswamy K, Kaliraj P (2002) Molecular characterization of a calcium binding translationally controlled tumor protein homologue from the filarial parasites Brugia malayi and Wuchereria bancrofti. Mol Biochem Parasitol 121(1):107–118

Gnanasekar M, Dakshinamoorthy G, Ramaswamy K (2009) Translationally controlled tumor protein is a novel heat shock protein with chaperone-like activity. Biochem Biophys Res Commun 386(2):333–337

Goodman CA, Coenen AM, Frey JW, You JS, Barker RG, Frankish BP, Murphy RM, Hornberger TA (2017) Insights into the role and regulation of TCTP in skeletal muscle. Oncotarget 8 (12):18754–18772

Graidist P, Phongdara A, Fujise K (2004) Antiapoptotic protein partners fortilin and MCL1 independently protect cells from 5-fluorouracil-induced cytotoxicity. J Biol Chem 279 (39):40868–40875

Graidist P, Yazawa M, Tonganunt M, Nakatomi A, Lin CC, Chang JY, Phongdara A, Fujise K (2007) Fortilin binds Ca2+ and blocks Ca2+-dependent apoptosis in vivo. Biochem J 408 (2):181–191

Gremski LH, Trevisan-Silva D, Ferrer VP, Matsubara FH, Meissner GO, Wille AC, Vuitika L, Dias-Lopes C, Ullah A, de Moraes FR et al (2014) Recent advances in the understanding of brown spider venoms: from the biology of spiders to the molecular mechanisms of toxins. Toxicon 83:91–120

Gross B, Gaestel M, Bohm H, Bielka H (1989) cDNA sequence coding for a translationally controlled human tumor protein. Nucleic Acids Res 17(20):8367

Gu X, Yao L, Ma G, Cui L, Li Y, Liang W, Zhao B, Li K (2014) TCTP promotes glioma cell proliferation in vitro and in vivo via enhanced beta-catenin/TCF-4 transcription. Neuro-Oncology 16(2):217–227

Guillaume E, Pineau C, Evrard B, Dupaix A, Moertz E, Sanchez JC, Hochstrasser DF, Jegou B (2001) Cellular distribution of translationally controlled tumor protein in rat and human testes. Proteomics 1(7):880–889

Gutierrez-Galeano DF, Toscano-Morales R, Calderon-Perez B, Xoconostle-Cazares B, Ruiz-Medrano R (2014) Structural divergence of plant TCTPs. Front Plant Sci 5:361

Haghighat NG, Ruben L (1992) Purification of novel calcium binding proteins from Trypanosoma brucei: properties of 22-, 24- and 38-kilodalton proteins. Mol Biochem Parasitol 51(1):99–110

Hao S, Qin Y, Yin S, He J, He D, Wang C (2016) Serum translationally controlled tumor protein is involved in rat liver regeneration after hepatectomy. Hepatol Res 46(13):1392–1401

He S, Huang Y, Wang Y, Tang J, Song Y, Yu X, Ma J, Wang S, Yin H, Li Q et al (2015) Histamine-releasing factor/translationally controlled tumor protein plays a role in induced cell adhesion, apoptosis resistance and chemoresistance in non-Hodgkin lymphomas. Leuk Lymphoma 56 (7):2153–2161

Hinojosa-Moya J, Xoconostle-Cazares B, Piedra-Ibarra E, Mendez-Tenorio A, Lucas WJ, Ruiz-Medrano R (2008) Phylogenetic and structural analysis of translationally controlled tumor proteins. J Mol Evol 66(5):472–483

Hoepflinger MC, Reitsamer J, Geretschlaeger AM, Mehlmer N, Tenhaken R (2013) The effect of translationally controlled tumour protein (TCTP) on programmed cell death in plants. BMC Plant Biol 13:135

Hong ST, Choi KW (2013) TCTP directly regulates ATM activity to control genome stability and organ development in Drosophila melanogaster. Nat Commun 4:2986

Hong ST, Choi KW (2016) Antagonistic roles of Drosophila Tctp and Brahma in chromatin remodelling and stabilizing repeated sequences. Nat Commun 7:12988

Hsu Y-C, Chern JJ, Cai Y, Liu M, Choi K-W (2007) Drosophila TCTP is essential for growth and proliferation through regulation of dRheb GTPase. Nature 445(7129):785–788

Hu C, Wang F, Ma S, Li X, Song L, Hua X, Xia Q (2015) Suppression of intestinal immunity through silencing of TCTP by RNAi in transgenic silkworm, Bombyx mori. Gene 574 (1):82–87

Jaglarz MK, Bazile F, Laskowska K, Polanski Z, Chesnel F, Borsuk E, Kloc M, Kubiak JZ (2012) Association of TCTP with centrosome and microtubules. Biochem Res Int 2012:541906

Jeon HJ, You SY, Park YS, Chang JW, Kim JS, Oh JS (2016) TCTP regulates spindle microtubule dynamics by stabilizing polar microtubules during mouse oocyte meiosis. Biochim Biophys Acta 1863(4):630–637

Jia Z, Wang M, Yue F, Wang X, Wang L, Song L (2017) The immunomodulation of a maternal translationally controlled tumor protein (TCTP) in Zhikong scallop Chlamys farreri. Fish Shellfish Immunol 60:141–149

Jin H, Zhang X, Su J, Teng Y, Ren H, Yang L (2015) RNA interference mediated knockdown of translationally controlled tumor protein induces apoptosis, and inhibits growth and invasion in glioma cells. Mol Med Rep 12(5):6617–6625

Johansson H, Simonsson S (2010) Core transcription factors, Oct4, Sox2 and Nanog, individually form complexes with nucleophosmin (Npm1) to control embryonic stem (ES) cell fate determination. Aging (Albany, NY) 2(11):815–822

Johansson H, Vizlin-Hodzic D, Simonsson T, Simonsson S (2010a) Translationally controlled tumor protein interacts with nucleophosmin during mitosis in ES cells. Cell Cycle 9(11):2160–2169

Johansson H, Svensson F, Runnberg R, Simonsson T, Simonsson S (2010b) Phosphorylated nucleolin interacts with translationally controlled tumor protein during mitosis and with Oct4 during interphase in ES cells. PLoS One 5(10):e13678

Jung J, Kim M, Kim M-J, Kim J, Moon J, Lim J-S, Kim M, Lee K (2004) Translationally controlled tumor protein interacts with the third cytoplasmic domain of Na,K-ATPase alpha subunit and inhibits the pump activity in HeLa cells. J Biol Chem 279(48):49868–49875

Jung J, Kim HY, Kim M, Sohn K, Lee K (2011) Translationally controlled tumor protein induces human breast epithelial cell transformation through the activation of Src. Oncogene 30 (19):2264–2274

Jung J, Kim HY, Maeng J, Kim M, Shin DH, Lee K (2014) Interaction of translationally controlled tumor protein with Apaf-1 is involved in the development of chemoresistance in HeLa cells. BMC Cancer 14:165

Kaarbo M, Storm ML, Qu S, Waehre H, Risberg B, Danielsen HE, Saatcioglu F (2013) TCTP is an androgen-regulated gene implicated in prostate cancer. PLoS One 8(7):e69398

Kadioglu O, Efferth T (2016) Peptide aptamer identified by molecular docking targeting translationally controlled tumor protein in leukemia cells. Investig New Drugs 34(4):515–521

Kang S, Dong SM, Kim BR, Park MS, Trink B, Byun HJ, Rho SB (2012) Thioridazine induces apoptosis by targeting the PI3K/Akt/mTOR pathway in cervical and endometrial cancer cells. Apoptosis 17(9):989–997

Kashiwakura JC, Ando T, Matsumoto K, Kimura M, Kitaura J, Matho MH, Zajonc DM, Ozeki T, Ra C, MacDonald SM et al (2012) Histamine-releasing factor has a proinflammatory role in mouse models of asthma and allergy. J Clin Invest 122(1):218–228

Kawakami T, Ando T, Kawakami Y (2012) HRF-interacting molecules. Open Allergy J 5:41–46

Kim M, Jung Y, Lee K, Kim C (2000) Identification of the calcium binding sites in translationally controlled tumor protein. Arch Pharm Res 23(6):633–636

Kim SH, Cairns N, Fountoulakisc M, Lubec G (2001) Decreased brain histamine-releasing factor protein in patients with Down syndrome and Alzheimer's disease. Neurosci Lett 300(1):41–44

Kim M-J, Kwon J-S, Suh SH, Suh J-K, Jung J, Lee S-N, Kim Y-H, Cho M-C, Oh GT, Lee K (2008a) Transgenic overexpression of translationally controlled tumor protein induces systemic-hypertension via repression of Na+, K+-ATPase. J Mol Cell Cardiol 44(1):151–159

Kim JE, Koo KH, Kim YH, Sohn J, Park YG (2008b) Identification of potential lung cancer biomarkers using an in vitro carcinogenesis model. Exp Mol Med 40(6):709–720

Kim M, Min HJ, Won HY, Park H, Lee JC, Park HW, Chung J, Hwang ES, Lee K (2009a) Dimerization of translationally controlled tumor protein is essential for its cytokine-like activity. PLoS One 4(7):e6464

Kim M, Jung J, Lee K (2009b) Roles of ERK, PI3 kinase, and PLC-gamma pathways induced by overexpression of translationally controlled tumor protein in HeLa cells. Arch Biochem Biophys 485(1):82–87

Kim M, Chung J, Lee C, Jung J, Kwon Y, Lee K (2011) A peptide binding to dimerized translationally controlled tumor protein modulates allergic reactions. J Mol Med (Berl) 89 (6):603–610

Kim YM, Han YJ, Hwang OJ, Lee SS, Shin AY, Kim SY, Kim JI (2012a) Overexpression of Arabidopsis translationally controlled tumor protein gene AtTCTP enhances drought tolerance with rapid ABA-induced stomatal closure. Mol Cells 33(6):617–626

Kim DK, Nam BY, Li JJ, Park JT, Lee SH, Kim DH, Kim JY, Kang HY, Han SH, Yoo TH et al (2012b) Translationally controlled tumour protein is associated with podocyte hypertrophy in a mouse model of type 1 diabetes. Diabetologia 55(4):1205–1217

Kim M, Maeng J, Lee K (2013a) Dimerization of TCTP and its clinical implications for allergy. Biochimie 95(4):659–666

Kim M, Jin YB, Lee K, Lee YS (2013b) A new antiallergic agent that binds to dimerized translationally controlled tumor protein and inhibits allergic symptoms is nontoxic. Hum Exp Toxicol 32(11):1119–1125

Kobayashi D, Hirayama M, Komohara Y, Mizuguchi S, Wilson Morifuji M, Ihn H, Takeya M, Kuramochi A, Araki N (2014) Translationally controlled tumor protein is a novel biological target for neurofibromatosis type 1 (NF1)-associated tumors. J Biol Chem 289(38):26314–26326

Koide Y, Kiyota T, Tonganunt M, Pinkaew D, Liu Z, Kato Y, Hutadilok-Towatana N, Phongdara A, Fujise K (2009) Embryonic lethality of fortilin-null mutant mice by BMP-pathway overactivation. Biochim Biophys Acta 1790(5):326–338

Koziol MJ, Gurdon JB (2012) TCTP in development and cancer. Biochem Res Int 2012:105203

Koziol MJ, Garrett N, Gurdon JB (2007) Tpt1 activates transcription of oct4 and nanog in transplanted somatic nuclei. Curr Biol 17(9):801–807

Kubiak JZ, Bazile F, Pascal A, Richard-Parpaillon L, Polanski Z, Ciemerych MA, Chesnel F (2008) Temporal regulation of embryonic M-phases. Folia Histochem Cytobiol 46(1):5–9

Kuramitsu Y, Nakamura K (2006) Proteomic analysis of cancer tissues: shedding light on carcinogenesis and possible biomarkers. Proteomics 6(20):5650–5661

Langdon JM, Vonakis BM, MacDonald SM (2004) Identification of the interaction between the human recombinant histamine releasing factor/translationally controlled tumor protein and elongation factor-1 delta (also known as eElongation factor-1B beta). Biochim Biophys Acta 1688(3):232–236

Laplante M, Sabatini DM (2012) mTOR signaling in growth control and disease. Cell 149(2):274–293
Lavoie JR, Ormiston ML, Perez-Iratxeta C, Courtman DW, Jiang B, Ferrer E, Caruso P, Southwood M, Foster WS, Morrell NW et al (2014) Proteomic analysis implicates translationally controlled tumor protein as a novel mediator of occlusive vascular remodeling in pulmonary arterial hypertension. Circulation 129(21):2125–2135
Lazaris-Karatzas A, Montine KS, Sonenberg N (1990) Malignant transformation by a eukaryotic initiation factor subunit that binds to mRNA 5′ cap. Nature 345(6275):544–547
Le TP, Vuong LT, Kim AR, Hsu YC, Choi KW (2016) 14-3-3 proteins regulate Tctp-Rheb interaction for organ growth in Drosophila. Nat Commun 7:11501
Lee JM, Kusakabe T, Kawaguchi Y, Miyagawa Y, Takahashi M, Mon H, Nho S-K, Koga K (2004) Molecular cloning and characterization of the translationally controlled tumor protein gene in Bombyx mori. Comp Biochem Physiol B Biochem Mol Biol 139(1):35–43
Lespagnol A, Duflaut D, Beekman C, Blanc L, Fiucci G, Marine JC, Vidal M, Amson R, Telerman A (2008) Exosome secretion, including the DNA damage-induced p53-dependent secretory pathway, is severely compromised in TSAP6/Steap3-null mice. Cell Death Differ 15(11):1723–1733
Li F, Zhang D, Fujise K (2001) Characterization of fortilin, a novel antiapoptotic protein. J Biol Chem 276(50):47542–47549
Li S, Chen X, Ding Y, Liu X, Wang Y, He J (2011) Expression of translationally controlled tumor protein (TCTP) in the uterus of mice of early pregnancy and its possible significance during embryo implantation. Hum Reprod 26(11):2972–2980
Li D, Deng Z, Liu X, Qin B (2013) Molecular cloning, expression profiles and characterization of a novel translationally controlled tumor protein in rubber tree (Hevea brasiliensis). J Plant Physiol 170(5):497–504
Li S, Chen M, Xiong Q, Zhang J, Cui Z, Ge F (2016) Characterization of the translationally controlled tumor protein (TCTP) interactome reveals novel binding partners in human cancer cells. J Proteome Res 15(10):3741–3751
Lin CJ, Robert F, Sukarieh R, Michnick S, Pelletier J (2010) The antidepressant sertraline inhibits translation initiation by curtailing mammalian target of rapamycin signaling. Cancer Res 70 (8):3199–3208
Liu H, Peng H-W, Cheng Y-S, Yuan HS, Yang-Yen H-F (2005) Stabilization and enhancement of the antiapoptotic activity of mcl-1 by TCTP. Mol Cell Biol 25(8):3117–3126
Liu LK, Wu HF, Guo ZR, Chen XJ, Yang D, Shu YQ, Zhang JN (2014) Targeted efficacy of dihydroartemisinin for translationally controlled protein expression in a lung cancer model. Asian Pac J Cancer Prev 15(6):2511–2515
Lo WY, Wang HJ, Chiu CW, Chen SF (2012) miR-27b-regulated TCTP as a novel plasma biomarker for oral cancer: from quantitative proteomics to post-transcriptional study. J Proteome 77:154–166
Lucas AT, Fu X, Liu J, Brannon MK, Yang J, Capelluto DG, Finkielstein CV (2014) Ligand binding reveals a role for heme in translationally-controlled tumor protein dimerization. PLoS One 9(11):e112823
Lucibello M, Gambacurta A, Zonfrillo M, Pierimarchi P, Serafino A, Rasi G, Rubartelli A, Garaci E (2011) TCTP is a critical survival factor that protects cancer cells from oxidative stress-induced cell-death. Exp Cell Res 317(17):2479–2489
Lucibello M, Adanti S, Antelmi E, Dezi D, Ciafre S, Carcangiu ML, Zonfrillo M, Nicotera G, Sica L, De Braud F et al (2015) Phospho-TCTP as a therapeutic target of Dihydroartemisinin for aggressive breast cancer cells. Oncotarget 6(7):5275–5291
Ma Q, Geng Y, Xu W, Wu Y, He F, Shu W, Huang M, Du H, Li M (2010) The role of translationally controlled tumor protein in tumor growth and metastasis of colon adenocarcinoma cells. J Proteome Res 9(1):40–49
MacDonald SM (2012a) Histamine releasing factor/translationally controlled tumor protein: history, functions and clinical implications. Open Allergy J 5:12–18
Macdonald SM (2012b) Potential role of histamine releasing factor (HRF) as a therapeutic target for treating asthma and allergy. J Asthma Allergy 5:51–59
MacDonald SM, Rafnar T, Langdon J, Lichtenstein LM (1995) Molecular identification of an IgE-dependent histamine-releasing factor. Science 269(5224):688–690

MacDonald SM, Paznekas WA, Jabs EW (1999) Chromosomal localization of tumor protein, translationally-controlled 1 (TPT1) encoding the human histamine releasing factor (HRF) to 13q12-->q14. Cytogenet Cell Genet 84(1-2):128–129

MacDonald SM, Bhisutthibhan J, Shapiro TA, Rogerson SJ, Taylor TE, Tembo M, Langdon JM, Meshnick SR (2001) Immune mimicry in malaria: Plasmodium falciparum secretes a functional histamine-releasing factor homolog in vitro and in vivo. Proc Natl Acad Sci U S A 98 (19):10829–10832

Maeng J, Kim M, Lee K (2012) On the mechanisms underlying the secretion and export of translationally controlled tumor protein/histamine releasing factor (TCTP/HRF). Open Allergy J 5:33–40

Mak CH, Su KW, Ko RC (2001) Identification of some heat-induced genes of Trichinella spiralis. Parasitology 123(Pt 3):293–300

Mak CH, Poon MW, Lun HM, Kwok PY, Ko RC (2007) Heat-inducible translationally controlled tumor protein of Trichinella pseudospiralis: cloning and regulation of gene expression. Parasitol Res 100(5):1105–1111

Meyuhas O (2000) Synthesis of the translational apparatus is regulated at the translational level. Eur J Biochem 267(21):6321–6330

Meyuhas O, Kahan T (2015) The race to decipher the top secrets of TOP mRNAs. Biochim Biophys Acta 1849(7):801–811

Meyvis Y, Houthoofd W, Visser A, Borgonie G, Gevaert K, Vercruysse J, Claerebout E, Geldhof P (2009) Analysis of the translationally controlled tumour protein in the nematodes Ostertagia ostertagi and Caenorhabditis elegans suggests a pivotal role in egg production. Int J Parasitol 39(11):1205–1213

Miao X, Chen YB, Xu SL, Zhao T, Liu JY, Li YR, Wang J, Zhang J, Guo GZ (2013) TCTP overexpression is associated with the development and progression of glioma. Tumour Biol 34 (6):3357–3361

Nagano-Ito M, Banba A, Ichikawa S (2009) Functional cloning of genes that suppress oxidative stress-induced cell death: TCTP prevents hydrogen peroxide-induced cell death. FEBS Lett 583(8):1363–1367

Nupan B, Phongdara A, Saengsakda M, Leu JH, Lo CF (2011) Shrimp Pm-fortilin inhibits the expression of early and late genes of white spot syndrome virus (WSSV) in an insect cell model. Dev Comp Immunol 35(4):469–475

Oikawa K, Ohbayashi T, Mimura J, Fujii-Kuriyama Y, Teshima S, Rokutan K, Mukai K, Kuroda M (2002) Dioxin stimulates synthesis and secretion of IgE-dependent histamine-releasing factor. Biochem Biophys Res Commun 290(3):984–987

Panrat T, Sinthujaroen P, Nupan B, Wanna W, Tammi MT, Phongdara A (2012) Characterization of a novel binding protein for Fortilin/TCTP--component of a defense mechanism against viral infection in Penaeus monodon. PLoS One 7(3):e33291

Pinkaew D, Le RJ, Chen Y, Eltorky M, Teng BB, Fujise K (2013) Fortilin reduces apoptosis in macrophages and promotes atherosclerosis. Am J Physiol Heart Circ Physiol 305(10):H1519–H1529

Pinkaew D, Chattopadhyay A, King MD, Chunhacha P, Liu Z, Stevenson HL, Chen Y, Sinthujaroen P, McDougal OM, Fujise K (2017) Fortilin binds IRE1alpha and prevents ER stress from signaling apoptotic cell death. Nat Commun 8(1):18

Ramani P, Nash R, Sowa-Avugrah E, Rogers C (2015) High levels of polo-like kinase 1 and phosphorylated translationally controlled tumor protein indicate poor prognosis in neuroblastomas. J Neuro-Oncol 125(1):103–111

Rao KVN, Chen L, Gnanasekar M, Ramaswamy K (2002) Cloning and characterization of a calcium-binding, histamine-releasing protein from Schistosoma mansoni. J Biol Chem 277 (34):31207–31213

Rehmann H, Bruning M, Berghaus C, Schwarten M, Kohler K, Stocker H, Stoll R, Zwartkruis FJ, Wittinghofer A (2008) Biochemical characterisation of TCTP questions its function as a guanine nucleotide exchange factor for Rheb. FEBS Lett 582(20):3005–3010

Ren C, Chen T, Jiang X, Wang Y, Hu C (2014) The first characterization of gene structure and biological function for echinoderm translationally controlled tumor protein (TCTP). Fish Shellfish Immunol 41(2):137–146

Rho SB, Lee JH, Park MS, Byun HJ, Kang S, Seo SS, Kim JY, Park SY (2011) Anti-apoptotic protein TCTP controls the stability of the tumor suppressor p53. FEBS Lett 585(1):29–35

Rid R, Simon-Nobbe B, Langdon J, Holler C, Wally V, Poll V, Ebner C, Hemmer W, Hawranek T, Lang R et al (2008) Cladosporium herbarum translationally controlled tumor protein (TCTP) is an IgE-binding antigen and is associated with disease severity. Mol Immunol 45(2):406–418

Rid R, Onder K, MacDonald S, Lang R, Hawranek T, Ebner C, Hemmer W, Richter K, Simon-Nobbe B, Breitenbach M (2009) Alternaria alternata TCTP, a novel cross-reactive ascomycete allergen. Mol Immunol 46(16):3476–3487

Rid R, Onder K, Trost A, Bauer J, Hintner H, Ritter M, Jakab M, Costa I, Reischl W, Richter K et al (2010) H_2O_2-dependent translocation of TCTP into the nucleus enables its interaction with VDR in human keratinocytes: TCTP as a further module in calcitriol signalling. J Steroid Biochem Mol Biol 118(1–2):29–40

Rinnerthaler M, Jarolim S, Heeren G, Palle E, Perju S, Klinger H, Bogengruber E, Madeo F, Braun RJ, Breitenbach-Koller L et al (2006) MMI1 (YKL056c, TMA19), the yeast orthologue of the translationally controlled tumor protein (TCTP) has apoptotic functions and interacts with both microtubules and mitochondria. Biochim Biophys Acta 1757(5–6):631–638

Rinnerthaler M, Lejskova R, Grousl T, Stradalova V, Heeren G, Richter K, Breitenbach-Koller L, Malinsky J, Hasek J, Breitenbach M (2013) Mmi1, the yeast homologue of mammalian TCTP, associates with stress granules in heat-shocked cells and modulates proteasome activity. PLoS One 8(10):e77791

Roque CG, Wong HH, Lin JQ, Holt CE (2016) Tumor protein Tctp regulates axon development in the embryonic visual system. Development 143(7):1134–1148

Sade YB, Boia-Ferreira M, Gremski LH, da Silveira RB, Gremski W, Senff-Ribeiro A, Chaim OM, Veiga SS (2012) Molecular cloning, heterologous expression and functional characterization of a novel translationally-controlled tumor protein (TCTP) family member from Loxosceles intermedia (brown spider) venom. Int J Biochem Cell Biol 44(1):170–177

Sanchez JC, Schaller D, Ravier F, Golaz O, Jaccoud S, Belet M, Wilkins MR, James R, Deshusses J, Hochstrasser D (1997) Translationally controlled tumor protein: a protein identified in several nontumoral cells including erythrocytes. Electrophoresis 18(1):150–155

Santa Brigida AB, dos Reis SP, Costa Cde N, Cardoso CM, Lima AM, de Souza CR (2014) Molecular cloning and characterization of a cassava translationally controlled tumor protein gene potentially related to salt stress response. Mol Biol Rep 41(3):1787–1797

Schmidt I, Fahling M, Nafz B, Skalweit A, Thiele BJ (2007) Induction of translationally controlled tumor protein (TCTP) by transcriptional and post-transcriptional mechanisms. FEBS J 274 (20):5416–5424

Seo EJ, Efferth T (2016) Interaction of antihistaminic drugs with human translationally controlled tumor protein (TCTP) as novel approach for differentiation therapy. Oncotarget 7 (13):16818–16839

Seo J, Maeng J, Kim HJ (2016) Translationally controlled tumor protein stimulates dopamine release from PC12 cells via Ca2+-independent phospholipase A(2) pathways. Int J Mol Sci 17(10)

Shen JH, Qu CB, Chu HK, Cui MY, Wang YL, Sun YX, Song YD, Li G (2016) Shi FJ: siRNA targeting TCTP suppresses osteosarcoma cell growth and induces apoptosis in vitro and in vivo. Biotechnol Appl Biochem 63(1):5–14

Sinha P, Kohl S, Fischer J, Hutter G, Kern M, Kottgen E, Dietel M, Lage H, Schnolzer M, Schadendorf D (2000) Identification of novel proteins associated with the development of chemoresistance in malignant melanoma using two-dimensional electrophoresis. Electrophoresis 21(14):3048–3057

Sinthujaroen P, Wanachottrakul N, Pinkaew D, Petersen JR, Phongdara A, Sheffield-Moore M, Fujise K (2014) Elevation of serum fortilin levels is specific for apoptosis and signifies cell death in vivo. BBA Clin 2:103–111

Sirois I, Raymond MA, Brassard N, Cailhier JF, Fedjaev M, Hamelin K, Londono I, Bendayan M, Pshezhetsky AV, Hebert MJ (2011) Caspase-3-dependent export of TCTP: a novel pathway for antiapoptotic intercellular communication. Cell Death Differ 18(3):549–562

Slaby O, Sobkova K, Svoboda M, Garajova I, Fabian P, Hrstka R, Nenutil R, Sachlova M, Kocakova I, Michalek J et al (2009) Significant overexpression of Hsp110 gene during colorectal cancer progression. Oncol Rep 21(5):1235–1241

Sturzenbaum SR, Kille P, Morgan AJ (1998) Identification of heavy metal induced changes in the expression patterns of the translationally controlled tumour protein (TCTP) in the earthworm Lumbricus rubellus1. Biochim Biophys Acta 1398(3):294–304

Susini L, Besse S, Duflaut D, Lespagnol A, Beekman C, Fiucci G, Atkinson AR, Busso D, Poussin P, Marine JC et al (2008) TCTP protects from apoptotic cell death by antagonizing bax function. Cell Death Differ 15(8):1211–1220

Takahashi T, Yano T, Zhu J, Hwang GW, Naganuma A (2010) Overexpression of FAP7, MIG3, TMA19, or YLR392c confers resistance to arsenite on Saccharomyces cerevisiae. J Toxicol Sci 35(6):945–946

Tani T, Shimada H, Kato Y, Tsunoda Y (2007) Bovine oocytes with the potential to reprogram somatic cell nuclei have a unique 23-kDa protein, phosphorylated transcriptionally controlled tumor protein (TCTP). Cloning Stem Cells 9(2):267–280

Tao JJ, Cao YR, Chen HW, Wei W, Li QT, Ma B, Zhang WK, Chen SY, Zhang JS (2015) Tobacco translationally controlled tumor protein interacts with ethylene receptor tobacco histidine kinase1 and enhances plant growth through promotion of cell proliferation. Plant Physiol 169(1):96–114

Taulan M, Paquet F, Argiles A, Demaille J, Romey MC (2006) Comprehensive analysis of the renal transcriptional response to acute uranyl nitrate exposure. BMC Genomics 7:2

Taylor KJ, Van TT, MacDonald SM, Meshnick SR, Fernley RT, Macreadie IG, Smooker PM (2015) Immunization of mice with Plasmodium TCTP delays establishment of Plasmodium infection. Parasite Immunol 37(1):23–31

Telerman A, Amson R (2009) The molecular programme of tumour reversion: the steps beyond malignant transformation. Nat Rev Cancer 9(3):206–216

Teshima S, Rokutan K, Nikawa T, Kishi K (1998) Macrophage colony-stimulating factor stimulates synthesis and secretion of a mouse homolog of a human IgE-dependent histamine-releasing factor by macrophages in vitro and in vivo. J Immunol 161(11):6356–6366

Thaw P, Baxter NJ, Hounslow AM, Price C, Waltho JP, Craven CJ (2001) Structure of TCTP reveals unexpected relationship with guanine nucleotide-free chaperones. Nat Struct Biol 8 (8):701–704

Thayanithy V (2005) Evolution and expression of translationally controlled tumour protein (TCTP) of fish. Comp Biochem Physiol B Biochem Mol Biol 142(1):8–17

Thebault S, Agez M, Chi X, Stojko J, Cura V, Telerman SB, Maillet L, Gautier F, Billas-Massobrio I, Birck C et al (2016) TCTP contains a BH3-like domain, which instead of inhibiting, activates Bcl-xL. Sci Rep 6:19725

Thiele H, Berger M, Lenzner C, Kuhn H, Thiele BJ (1998) Structure of the promoter and complete sequence of the gene coding for the rabbit translationally controlled tumor protein (TCTP) P23. Eur J Biochem 257(1):62–68

Thiele H, Berger M, Skalweit A, Thiele BJ (2000) Expression of the gene and processed pseudogenes encoding the human and rabbit translationally controlled tumour protein (TCTP). Eur J Biochem 267(17):5473–5481

Thomas G, Thomas G (1986) Translational control of mRNA expression during the early mitogenic response in Swiss mouse 3T3 cells: identification of specific proteins. J Cell Biol 103(6 Pt 1):2137–2144

Thomas G, Thomas G, Luther H (1981) Transcriptional and translational control of cytoplasmic proteins after serum stimulation of quiescent Swiss 3T3 cells. Proc Natl Acad Sci U S A 78 (9):5712–5716

Tonganunt M, Nupan B, Saengsakda M, Suklour S, Wanna W, Senapin S, Chotigeat W, Phongdara A (2008) The role of Pm-fortilin in protecting shrimp from white spot syndrome virus (WSSV) infection. Fish Shellfish Immunol 25(5):633–637

Tsai MJ, Yang-Yen HF, Chiang MK, Wang MJ, Wu SS, Chen SH (2014) TCTP is essential for beta-cell proliferation and mass expansion during development and beta-cell adaptation in response to insulin resistance. Endocrinology 155(2):392–404

Tsarova K, Yarmola EG, Bubb MR (2011) Identification of a cofilin-like actin-binding site on translationally controlled tumor protein (TCTP). FEBS Lett 584(23):4756–4760

Tuynder M, Susini L, Prieur S, Besse S, Fiucci G, Amson R, Telerman A (2002) Biological models and genes of tumor reversion: cellular reprogramming through tpt1/TCTP and SIAH-1. Proc Natl Acad Sci U S A 99(23):14976–14981

Tuynder M, Fiucci G, Prieur S, Lespagnol A, Geant A, Beaucourt S, Duflaut D, Besse S, Susini L, Cavarelli J et al (2004) Translationally controlled tumor protein is a target of tumor reversion. Proc Natl Acad Sci U S A 101(43):15364–15369

Vedadi M, Lew J, Artz J, Amani M, Zhao Y, Dong A, Wasney GA, Gao M, Hills T, Brokx S et al (2007) Genome-scale protein expression and structural biology of Plasmodium falciparum and related Apicomplexan organisms. Mol Biochem Parasitol 151(1):100–110

Vonakis BM, Gibbons S Jr, Sora R, Langdon JM, MacDonald SM (2001) Src homology 2 domain-containing inositol 5' phosphatase is negatively associated with histamine release to human recombinant histamine-releasing factor in human basophils. J Allergy Clin Immunol 108 (5):822–831

Vonakis BM, Macglashan DW Jr, Vilarino N, Langdon JM, Scott RS, MacDonald SM (2008) Distinct characteristics of signal transduction events by histamine-releasing factor/translationally controlled tumor protein (HRF/TCTP)-induced priming and activation of human basophils. Blood 111(4):1789–1796

Wang X, Fonseca BD, Tang H, Liu R, Elia A, Clemens MJ, Bommer UA, Proud CG (2008) Re-evaluating the roles of proposed modulators of mammalian target of rapamycin complex 1 (mTORC1) signaling. J Biol Chem 283(45):30482–30492

Wang F, Shang Y, Yang L, Zhu C (2012) Comparative proteomic study and functional analysis of translationally controlled tumor protein in rice roots under Hg2+ stress. J Environ Sci (China) 24(12):2149–2158

Wang F, Hu C, Hua X, Song L, Xia Q (2013) Translationally controlled tumor protein, a dual functional protein involved in the immune response of the silkworm, Bombyx mori. PLoS One 8(7):e69284

Wang J, Cao X, Sun J, Chai L, Huang Y, Tang X (2015) Transcriptional responses of earthworm (Eisenia fetida) exposed to naphthenic acids in soil. Environ Pollut 204:264–270

Wantke F, MacGlashan DW, Langdon JM, MacDonald SM (1999) The human recombinant histamine releasing factor: functional evidence that it does not bind to the IgE molecule. J Allergy Clin Immunol 103(4):642–648

Wu W, Wu B, Ye T, Huang H, Dai C, Yuan J, Wang W (2013) TCTP is a critical factor in shrimp immune response to virus infection. PLoS One 8(9):e74460

Wu H, Gong W, Yao X, Wang J, Perrett S, Feng Y (2015) Evolutionarily conserved binding of translationally controlled tumor protein to eukaryotic elongation factor 1B. J Biol Chem 290 (14):8694–8710

Xiao B, Chen D, Luo S, Hao W, Jing F, Liu T, Wang S, Geng Y, Li L, Xu W et al (2016) Extracellular translationally controlled tumor protein promotes colorectal cancer invasion and metastasis through Cdc42/JNK/ MMP9 signaling. Oncotarget 7(31):50057–50073

Xu A, Bellamy AR, Taylor JA (1999) Expression of translationally controlled tumour protein is regulated by calcium at both the transcriptional and post-transcriptional level. Biochem J 342 (Pt 3):683–689

Yagci M, Yegin ZA, Akyurek N, Kayhan H, Ozkurt ZN, Sucak GT, Haznedar R (2013) TCTP/ HRF pathway and angiogenesis: a feasible intercourse in chronic lymphocytic leukemia. Leuk Res 37(6):665–670

Yamashita R, Suzuki Y, Takeuchi N, Wakaguri H, Ueda T, Sugano S, Nakai K (2008) Comprehensive detection of human terminal oligo-pyrimidine (TOP) genes and analysis of their characteristics. Nucleic Acids Res 36(11):3707–3715

Yang Y, Yang F, Xiong Z, Yan Y, Wang X, Nishino M, Mirkovic D, Nguyen J, Wang H, Yang X-F (2005) An N-terminal region of translationally controlled tumor protein is required for its antiapoptotic activity. Oncogene 24(30):4778–4788

Yao Y, Jia XY, Tian HY, Jiang YX, Xu GJ, Qian QJ, Zhao FK (2009) Comparative proteomic analysis of colon cancer cells in response to oxaliplatin treatment. Biochim Biophys Acta 1794 (10):1433–1440

Yarm FR (2002) Plk phosphorylation regulates the microtubule-stabilizing protein TCTP. Mol Cell Biol 22(17):6209–6221

Yeh YC, Xie L, Langdon JM, Myers AC, Oh SY, Zhu Z, Macdonald SM (2010) The effects of overexpression of histamine releasing factor (HRF) in a transgenic mouse model. PLoS One 5 (6):e11077

Yenofsky R, Bergmann I, Brawerman G (1982) Messenger RNA species partially in a repressed state in mouse sarcoma ascites cells. Proc Natl Acad Sci U S A 79(19):5876–5880

Yenofsky R, Cereghini S, Krowczynska A, Brawerman G (1983) Regulation of mRNA utilization in mouse erythroleukemia cells induced to differentiate by exposure to dimethyl sulfoxide. Mol Cell Biol 3(7):1197–1203

Yoon T, Jung J, Kim M, Lee KM, Choi EC, Lee K (2000) Identification of the self-interaction of rat TCTP/IgE-dependent histamine-releasing factor using yeast two-hybrid system. Arch Biochem Biophys 384(2):379–382

Yoon T, Kim M, Lee K (2006) Inhibition of Na,K-ATPase-suppressive activity of translationally controlled tumor protein by sorting nexin 6. FEBS Lett 580(14):3558–3564

Yubero N, Esteso G, Cardona H, Morera L, Garrido JJ, Barbancho M (2009) Molecular cloning, expression analysis and chromosome localization of the Tpt1 gene coding for the pig translationally controlled tumor protein (TCTP). Mol Biol Rep 36(7):1957–1965

Zhang D, Li F, Weidner D, Mnjoyan ZH, Fujise K (2002) Physical and functional interaction between myeloid cell leukemia 1 protein (MCL1) and fortilin. The potential role of MCL1 as a fortilin chaperone. J Biol Chem 277(40):37430–37438

Zhang YJ, Dai Q, Sun DF, Xiong H, Tian XQ, Gao FH, Xu MH, Chen GQ, Han ZG, Fang JY (2009) mTOR signaling pathway is a target for the treatment of colorectal cancer. Ann Surg Oncol 16(9):2617–2628

Zhang J, de Toledo SM, Pandey BN, Guo G, Pain D, Li H, Azzam EI (2012) Role of the translationally controlled tumor protein in DNA damage sensing and repair. Proc Natl Acad Sci U S A 109(16):E926–E933

Zhang F, Liu B, Wang Z, Yu XJ, Ni QX, Yang WT, Mukaida N, Li YY (2013) A novel regulatory mechanism of Pim-3 kinase stability and its involvement in pancreatic cancer progression. Mol Cancer Res 11(12):1508–1520

Zhu WL, Cheng HX, Han N, Liu DL, Zhu WX, Fan BL, Duan FL (2008) Messenger RNA expression of translationally controlled tumor protein (TCTP) in liver regeneration and cancer. Anticancer Res 28(3A):1575–1580

Zobel-Thropp PA, Correa SM, Garb JE, Binford GJ (2014) Spit and venom from scytodes spiders: a diverse and distinct cocktail. J Proteome Res 13(2):817–835

Zoncu R, Efeyan A, Sabatini DM (2011) mTOR: from growth signal integration to cancer, diabetes and ageing. Nat Rev Mol Cell Biol 12(1):21–35

Chapter 5
Current Understanding of the TCTP Interactome

Siting Li and Feng Ge

Abstract Evolutionarily conserved and pleiotropic, the translationally controlled tumor protein (TCTP) is a housekeeping protein present in eukaryotic organisms. It plays an important role in regulating many fundamental processes, such as cell proliferation, cell death, immune responses, and apoptosis. As a result of the pioneer work by Adam Telerman and Robert Amson, the critical role of TCTP in tumor reversion was revealed. Moreover, TCTP has emerged as a regulator of cell fate determination and a promising therapeutic target for cancers. The multifaceted action of TCTP depends on its ability to interact with different proteins. Through this interaction network, TCTP regulates diverse physiological and pathological processes in a context-dependent manner. Complete mapping of the entire sets of TCTP protein interactions (interactome) is essential to understand its various cellular functions and to lay the foundation for the rational design of TCTP-based therapeutic approaches. So far, the global profiling of the interacting partners of TCTP has rarely been performed, but many interactions have been identified in small-scale studies in a specific biological system. This chapter, based on information from protein interaction databases and the literature, illustrates current knowledge of the TCTP interactome.

5.1 Introduction

Translationally controlled tumor protein (TCTP), also termed as tumor protein translationally controlled 1 (TPT1), histamine releasing factor (HRF), p23, or fortilin, was initially discovered in the tumor cells in mice by researchers working on translationally regulated genes (Yenofsky et al. 1983; Macdonald et al. 1995). It

S. Li
Key Laboratory of Algal Biology, Institute of Hydrobiology, Chinese Academy of Sciences, Wuhan 430072, China

University of Chinese Academy of Sciences, Beijing 100049, China

F. Ge (✉)
Key Laboratory of Algal Biology, Institute of Hydrobiology, Chinese Academy of Sciences, Wuhan 430072, China
e-mail: gefeng@ihb.ac.cn

A. Telerman, R. Amson (eds.), *TCTP/tpt1 - Remodeling Signaling from Stem Cell to Disease*, Results and Problems in Cell Differentiation 64,
DOI 10.1007/978-3-319-67591-6_5

was subsequently found that TCTP is an evolutionarily conserved protein that shares a high degree of homology with the protein from plants to mammals (Acunzo et al. 2014; Amson et al. 2013; Bommer and Thiele 2004). Numerous studies have shown that TCTP plays indispensable roles in various physiological processes, including cell proliferation (Gu et al. 2014; Hsu et al. 2007), cell death (Chen et al. 2014; Lucibello et al. 2011; Susini et al. 2008), cell cycle (Burgess et al. 2008; Johnson et al. 2008), the cytoskeleton (Jeon et al. 2016; Jaglarz et al. 2012; Bazile et al. 2009), protein synthesis (Chen et al. 2013a; Rho et al. 2011), immune responses (Tsai et al. 2014; Kaarbo et al. 2013), malignant transformation (Huang et al. 2015), and nuclear reprogramming (Amson et al. 2013; Roque et al. 2016; Wu et al. 2012; Cheng et al. 2012; Sirois et al. 2011). Telerman and colleagues demonstrated that TCTP plays an important role in tumor reversion, which is defined as the process by which cancer cells lose their malignant phenotype (Tuynder et al. 2002, 2004). Our previous study also revealed that downregulation of TCTP in multiple myeloma cells can lead to tumor reversion (Ge et al. 2011). Importantly, increasing evidences suggest that TCTP is a promising therapeutic target for cancer prevention and intervention (Acunzo et al. 2014; Lucibello et al. 2015; Baylot et al. 2012).

The role of TCTP in many cellular functions is the result of its dynamic interactions with numerous cellular proteins. It is well known that tumorigenesis is the consequence of multiple genetic and epigenetic events that induce cell proliferation and the progression of tumor growth. TCTP was implicated in diverse cellular functions due to interactions with other proteins related to tumorigenesis. Therefore, identification and the characterization of the TCTP interacting proteins on a large scale is important for the understanding of its regulatory mechanisms and revealing its functions in tumorigenesis.

5.2 Global Interactome Profiling Methods

Individual proteins perform their functions through interactions with other proteins and these interactions are crucial for all cellular processes. The knowledge about the entire set of protein interactions (interactome) is essential for our understanding of both the function of individual proteins and the functional organization of the whole cell (Lage 2014). Many experimental high-throughput (HTP) approaches have been developed to determine the protein interactomes in various organisms on a large scale. Through their integration with other "omics" data, interactome datasets have provided valuable information to uncover the functional cellular protein networks and the origin of many diseases. In this chapter, we discuss the emerging and established techniques currently employed to identify the interactome with a particular focus on yeast two-hybrid screens (Y2H) and mass spectrometry (MS)-based approaches. Excellent in-depth reviews on HTP approaches are already available (Mehta and Trinkle-Mulcahy 2016; Lievens et al. 2010).

The yeast two-hybrid (Y2H) assay has been used for over 25 years and remains the most popular choice for researchers investigating interactomes (Rajagopala 2015). This assay is a genetic complementation technique where the proteins to be tested for interaction (referred to as "bait" and "prey") are fused to the DNA-binding domain and the activation domain of the transcription factor (Bruckner et al. 2009). The proteins are co-expressed in a yeast strain reconstituting transcription factor activity, which drives the expression of a reporter gene (Lentze and Auerbach 2008). Large-scale Y2H strategies have been applied to map the human interactome and to generate protein interactome in a number of model organisms (Rajagopala 2015; Zhang et al. 2010; Yu et al. 2008; Li et al. 2004). There are many different variations of the Y2H assay developed such as the recruitment of the bait and prey to the cytosol, plasma membrane, and endoplasmic reticulum or using multiple baits (Koegl and Uetz 2007; Stellberger et al. 2010).

MS-based proteomics has become a widely used technology to identify protein–protein interactions (PPIs) during the past decade (Smits and Vermeulen 2016). The workflows of the commonly used MS-based interaction proteomics are based on affinity-purification MS (AP-MS) of the protein of interest using specific antibodies. The application of quantitative proteomics such as quantitative immunoprecipitation combined with knockdown (QUICK) for protein enrichments from crude lysates to discriminate bona fide interactors from background proteins has proved to be particularly useful (Ge et al. 2010; Selbach and Mann 2006; Zheng et al. 2012; Chen et al. 2013b). Recently, many different MS-based global interactome profiling approaches have been developed, such as proximity-ligation technology based on engineered ascorbate peroxidase (APEX) labeling and global interactome profiling based on the co-behavior of proteins in biochemical fractionations (Havugimana et al. 2012; Kristensen et al. 2012) or perturbation experiments (Christoforou et al. 2016).

An alternative method that has proved invaluable in protein interactome research is protein arrays, which are miniaturized parallel assay systems that contain small amounts of purified proteins in a high-density format (Phizicky et al. 2003). They allow the simultaneous determination of a variety of analytes from small amounts of samples in a single experiment (Tao et al. 2007; Tao and Zhu 2006). This technique has undergone considerable developments since it was first introduced (Chen et al. 2013b; Yang et al. 2016). This has become one of the most powerful multiplexed detection platforms, which can be used for identification of protein interactome, antibody classification, and protein functional analysis.

5.3 The Current Knowledge of the TCTP Interactome

The results of TCTP interactome analysis imply that TCTP interacting proteins belong to a range of functional groups (Li et al. 2016), including nucleic acid-binding proteins, cytoskeletal proteins, chaperones, enzyme modulators, and transferases, which is consistent with the multifunctional nature of TCTP.

5.3.1 Chaperone Proteins

Chaperone proteins from the highly conserved HSP70 family were identified to be interacting partners of TCTP, and HSPA9 was subsequently confirmed as a TCTP-binding partner (Li et al. 2016). HSP70 family members have key regulatory roles in a variety of cellular stress responses (Murphy 2013; Daugaard et al. 2007). The study of over-expression of HSPA9 in tumor cells demonstrated its function in protecting cells from oxidative damage (Liu et al. 2005; Orsini et al. 2004). Interestingly, TCTP is upregulated during oxidative stress and has been implicated as an antioxidant protein (Lucibello et al. 2011; Oikawa et al. 2002; Rupec et al. 1998). Thus, it is likely that the HSPA9–TCTP complex can function together in resistance to intracellular oxidative stress. The complex of TCTP and HSP70 family may also be involved in anti-apoptotic processes (Fig. 5.1a), which is similar to the HSP27–TCTP complex (Baylot et al. 2012).

5.3.2 Nucleic Acid-Binding Proteins

A large number of nucleic acid-binding proteins are the regulatory proteins of transcription and translation. Among them, XRCC6 (Ku70) has been demonstrated to play a critical role in genomic stability maintenance by binding to TCTP and XRCC5 (Ku80) (Zhang et al. 2012; Wang et al. 2015; Gullo et al. 2006). When DNA double-strand breaks (DSB) occur, TCTP accumulates at the damage sites, co-localizing with XRCC6 and XRCC5 (Ku80) and forms complexes for DSB repairs (Fig. 5.1b). However, the levels of XRCC5 and XRCC6 in nuclei are reduced in the absence of TCTP (Zhang et al. 2012), suggesting that TCTP can act as a chaperone. Moreover, Gurdon et al. demonstrated that TCTP directly binds to the promoter region of *oct4* and acts as a transcription factor for this gene (Koziol et al. 2007). They further showed that TCTP also indirectly activates *nanog* transcription by binding to a distant site from its promoter (Koziol et al. 2007). Together, the interactome analysis further confirmed that TCTP is a critical transcription and translation regulator and may fulfill its functions by binding to other proteins.

5.3.3 Cytoskeletal Proteins

The cytoskeleton is a highly dynamic system comprising of different groups of structural proteins including tubulin, actin, and intermediate filaments to form polymers and associated proteins with diverse regulatory functions (Petrasek and Schwarzerova 2009). TCTP has been reported to be associated with cytoskeleton proteins and many related cellular processes (Bazile et al. 2009; Gachet et al. 1999;

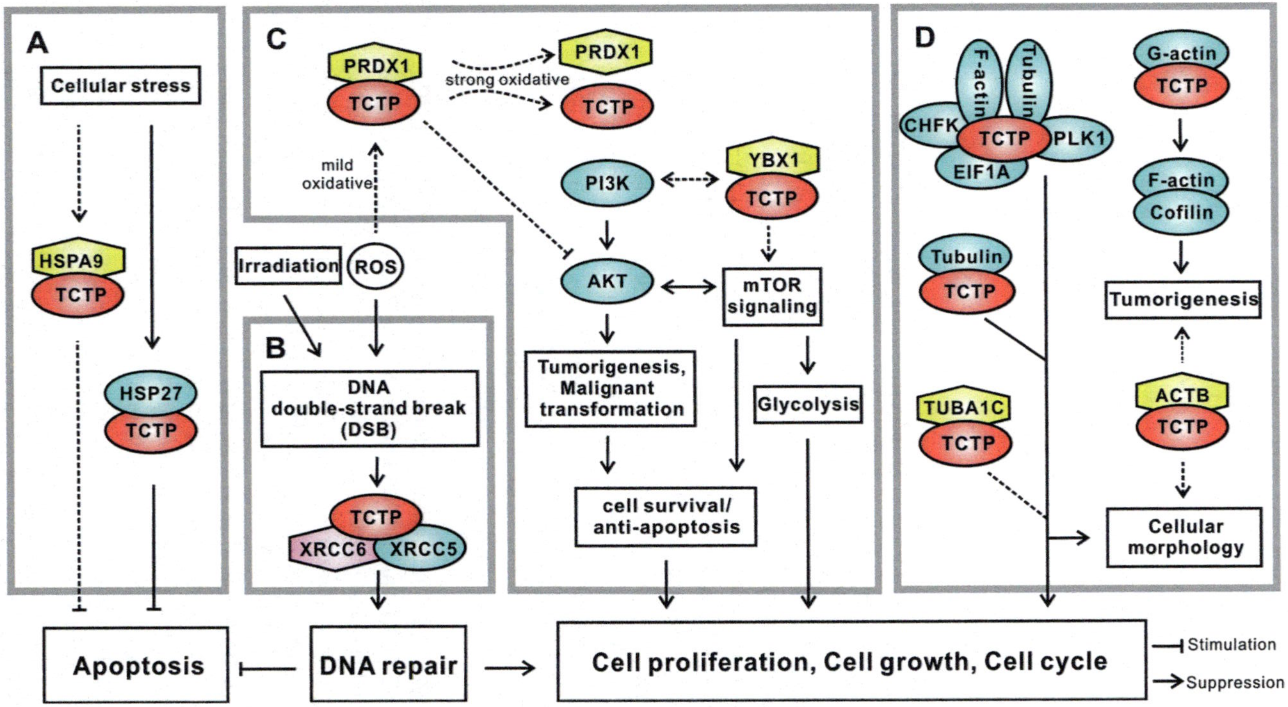

Fig. 5.1 Proposed model depicting the molecular mechanism of the validated TCTP-binding proteins. (**a**) When exposed to cellular stress, HSPA9–TCTP may form complexes to resist apoptosis, which is similar to HSP27–TCTP complex. (**b**) When exposed to irradiation or ROS, which may result in DSB, the TCTP–XRCC6–XRCC5 complexes will form and exert their function in DSB sensing and repairing. (**c**) The TCTP–PRDX1 complex may prevent Akt-driven transformation by similar mechanism as TCTP–PTEN in a mild oxidative stress. However, under high doses of oxidative stress, TCTP may also dissociate from oxidized PRDX1 and inactivated. The TCTP–YBX1 complexes are associated with PI3K–Akt signaling pathway which regulates tumorigenesis, malignant transformation, and mTOR signaling. The complexes may also activate mTOR signaling directly and further regulate cell proliferation, cell growth, and cell cycle. (**d**) The ACTB–TCTP and TUBA1C–TCTP complexes may probably have similar functions as the interaction of TCTP between F-actin, G-actin, and tubulin and exert their function in cell morphology, tumorigenesis, cell proliferation, cell growth, and cell cycle. *Red*: TCTP protein. *Yellow*: the novel TCTP-binding proteins; *Pink*: TCTP-binding proteins that identified previously. *Blue*: molecules contribute to those functional pathways. The *solid line* indicates the reported regulation relationships, while the *dash line* indicates the conjectural regulation relationships

Tsarova et al. 2010). For example, TCTP is involved in regulating cell shape, probably via complex interactions with both F-actin and the microtubule cytoskeleton (Bazile et al. 2009). TCTP also associates with microtubules during specific phases of the cell cycle by binding to tubulin (Gachet et al. 1999). TCTP can release the binding of cofilin to G-actin and transfer the active cofilin to F-actin, increasing the cofilin-activity cycle in invasive tumor cells (Tsarova et al. 2010). The proteomics analysis reveals 15 TCTP interacting cytoskeleton proteins and sheds new light on the role of TCTP in cytoskeleton-related functions like cell morphology, tumorigenesis, cell proliferation, cell growth, and cell cycle (Fig. 5.1d).

5.3.4 Other Functions

PRDX1 and YBX1 were also validated to be TCTP-binding partners. PRDX1 was the first antioxidant protein reported to protect other proteins from inactivation through interaction (Neumann et al. 2009). When exposed to mild oxidative stress, PRDX1 is upregulated and binds to phosphatase and tensin homolog (PTEN) to protect it from oxidation-induced inactivation (Neumann et al. 2009). However, under high doses of oxidative stress, PTEN irreversibly dissociates from oxidized PRDX1 and becomes inactivated, resulting in hyperactivation of Akt signaling (Neumann et al. 2009; Stambolic et al. 1998, 2000; Backman et al. 2004). Notably, TCTP upregulation has been detected in surviving cells after oxidative stress (Lucibello et al. 2011). Conversely, when exposed to a strong oxidative stress, cancer cells caused a downregulation of TCTP, followed by cell death (Lucibello et al. 2011). Therefore, the binding of TCTP and PRDX1 may also be involved in antioxidant pathways, and the TCTP–PRDX1 complex may prevent Akt-driven transformation by a similar mechanism as PTEN under mild oxidative stress (Fig. 5.1c).

YBX1 has been implicated in numerous cellular processes similar to TCTP. The pleiotropic functions of YBX1 and TCTP indicate that the TCTP–YBX1 complex may be involved in vital signaling pathways. YBX1 is closely related to the PI3K/Akt/mTOR signaling pathway (Dazert and Hall 2011). It transcriptionally activates the expression of *PIK3CA* in basal-like breast cancer cells (Astanehe et al. 2009). Serine phosphorylation of the YBX1 102 residue relies on Akt kinase activity (Sutherland et al. 2005; Basaki et al. 2007; Sinnberg et al. 2012). The inhibition of the PI3K pathway can also reduce the expression of YBX1 (Sinnberg et al. 2012). Through experiments of YBX1 silencing, Lee et al. confirmed that the reduction of YBX1 resulted in decreasing of mTOR protein levels (Lee et al. 2008). Interestingly, the translation of *TCTP* mRNA is regulated by PI3-K/Akt/mTOR signaling, and a positive feedback loop between TCTP and mTOR contributes to tumor formation (Bommer et al. 2015; Kobayashi et al. 2014). By interacting with each other, TCTP and YBX1 may work cooperatively in the PI3K/Akt/mTOR pathway, which regulates tumorigenesis, malignant transformation, and mTOR signaling. The complexes may also be directly related to mTOR signaling and may further influence the glycolysis pathway to regulate cell proliferation, cell growth, and cell cycle (Fig. 5.1c).

5.4 Concluding Remarks

As described above, the TCTP interactome is complex, multifunctional, and has many missing pieces. There are still many unexplained functions that have not yet been attributed to a specific TCTP interactor. Much work is still required to understand the TCTP interactome and its physiological significance. Large-scale approaches, such as next generation Y2H, tandem-affinity purification coupled to MS, and protein arrays might lead to the identification of the new interactors of TCTP and the entire TCTP interactome. The identification of bona fide TCTP interactors can reveal novel functional properties of TCTP. A complete identification of the TCTP interactome will give us a better understanding of the role of TCTP, its mechanism of action, and its associations with the interacting proteins to affect diverse biological and pathological processes. It is critical to identify the key hubs and nodes in the TCTP interaction networks as well as to obtain a detailed molecular characterization of the TCTP interactome. This knowledge will facilitate the development of agents to perturb (or mimic) these interactions. As an example of the potential of this approach, the identification of the interaction between TCTP and YBX1 gives us an idea to find the therapeutic small molecule compound (inhibitor) which can specifically block the interaction (Li et al. 2016). Therefore, elucidating the mechanisms of action of TCTP interactome should provide substantial benefits for the discovery of novel drug targets and biomarkers of disease.

Acknowledgements This work was supported by the National Key Research and Development Program (2016YFA0501304), the Strategic Priority Research Program of the Chinese Academy of Sciences (Grant No. XDB14030202).

References

Acunzo J et al (2014) TCTP as therapeutic target in cancers. Cancer Treat Rev 40:760–769
Amson R et al (2013) TPT1/TCTP-regulated pathways in phenotypic reprogramming. Trends Cell Biol 23:37–46
Astanehe A et al (2009) The transcriptional induction of PIK3CA in tumor cells is dependent on the oncoprotein Y-box binding protein-1. Oncogene 28:2406–2418
Backman SA et al (2004) Early onset of neoplasia in the prostate and skin of mice with tissue-specific deletion of pten. Proc Natl Acad Sci USA 101:1725–1730
Basaki Y et al (2007) Akt-dependent nuclear localization of Y-box-binding protein 1 in acquisition of malignant characteristics by human ovarian cancer cells. Oncogene 26:2736–2746
Baylot V et al (2012) Targeting TCTP as a new therapeutic strategy in castration-resistant prostate cancer. Mol Ther 20:2244–2256

Bazile F et al (2009) Complex relationship between TCTP, microtubules and actin microfilaments regulates cell shape in normal and cancer cells. Carcinogenesis 30:555–565
Bommer UA, Thiele BJ (2004) The translationally controlled tumour protein (TCTP). Int J Biochem Cell Biol 36:379–385
Bommer UA et al (2015) Growth-factor dependent expression of the translationally controlled tumour protein TCTP is regulated through the PI3-K/akt/mTORC1 signalling pathway. Cell Signal 27:1557–1568
Bruckner A et al (2009) Yeast two-hybrid, a powerful tool for systems biology. Int J Mol Sci 10:2763–2788
Burgess A et al (2008) Chfr interacts and colocalizes with TCTP to the mitotic spindle. Oncogene 27:5554–5566
Chen K et al (2013a) TCTP increases stability of hypoxia-inducible factor 1alpha by interaction with and degradation of the tumour suppressor VHL. Biol Cell 105:208–218
Chen Y et al (2013b) Bcl2-associated athanogene 3 interactome analysis reveals a new role in modulating proteasome activity. Mol Cell Proteomics 12:2804–2819
Chen K et al (2014) Long-term artificial selection reveals a role of TCTP in autophagy in mammalian cells. Mol Biol Evol 31:2194–2211
Cheng X et al (2012) Translationally controlled tumor protein (TCTP) downregulates Oct4 expression in mouse pluripotent cells. BMB Rep 45:20–25
Christoforou A et al (2016) A draft map of the mouse pluripotent stem cell spatial proteome. Nat Commun 7:8992
Daugaard M et al (2007) The heat shock protein 70 family: highly homologous proteins with overlapping and distinct functions. FEBS Lett 581:3702–3710
Dazert E, Hall MN (2011) mTOR signaling in disease. Curr Opin Cell Biol 23:744–755
Gachet Y et al (1999) The growth-related, translationally controlled protein P23 has properties of a tubulin binding protein and associates transiently with microtubules during the cell cycle. J Cell Sci 112(Pt 8):1257–1271
Ge F et al (2010) Identification of novel 14-3-3zeta interacting proteins by quantitative immunoprecipitation combined with knockdown (QUICK). J Proteome Res 9:5848–5858
Ge F et al (2011) Quantitative proteomic analysis of tumor reversion in multiple myeloma cells. J Proteome Res 10:845–855
Gu X et al (2014) TCTP promotes glioma cell proliferation in vitro and in vivo via enhanced beta-catenin/TCF-4 transcription. Neuro-Oncology 16:217–227
Gullo C et al (2006) The biology of ku and its potential oncogenic role in cancer. Biochim Biophys Acta Rev Cancer 1765:223–234
Havugimana PC et al (2012) A census of human soluble protein complexes. Cell 150:1068–1081
Hsu YC et al (2007) Drosophila TCTP is essential for growth and proliferation through regulation of dRheb GTPase. Nature 445:785–788
Huang H et al (2015) Poly(ADP-ribose) glycohydrolase silencing down-regulates TCTP and cofilin-1 associated with metastasis in benzo(a)pyrene carcinogenesis. Am J Cancer Res 5:155–167
Jaglarz MK et al (2012) Association of TCTP with centrosome and microtubules. Biochem Res Int 2012:541906
Jeon HJ et al (2016) TCTP regulates spindle microtubule dynamics by stabilizing polar microtubules during mouse oocyte meiosis. Biochim Biophys Acta 1863:630–637
Johnson TM et al (2008) Plk1 activation by Ste20-like kinase (slk) phosphorylation and polo-box phosphopeptide binding assayed with the substrate translationally controlled tumor protein (TCTP). Biochemistry 47:3688–3696
Kaarbo M et al (2013) TCTP is an androgen-regulated gene implicated in prostate cancer. PLoS One 8:e69398
Kobayashi D et al (2014) Translationally controlled tumor protein is a novel biological target for neurofibromatosis type 1-associated tumors. J Biol Chem 289:26314–26326

Koegl M, Uetz P (2007) Improving yeast two-hybrid screening systems. Brief Funct Genomics Proteomics 6:302–312
Koziol MJ et al (2007) Tpt1 activates transcription of oct4 and nanog in transplanted somatic nuclei. Curr Biol 17:801–807
Kristensen AR et al (2012) A high-throughput approach for measuring temporal changes in the interactome. Nat Methods 9:907–909
Lage K (2014) Protein-protein interactions and genetic diseases: the interactome. Biochim Biophys Acta 1842:1971–1980
Lee C et al (2008) Targeting YB-1 in HER-2 overexpressing breast cancer cells induces apoptosis via the mTOR/STAT3 pathway and suppresses tumor growth in mice. Cancer Res 68:8661–8666
Lentze N, Auerbach D (2008) The yeast two-hybrid system and its role in drug discovery. Expert Opin Ther Targets 12:505–515
Li S et al (2004) A map of the interactome network of the metazoan C. elegans. Science 303:540–543
Li S et al (2016) Characterization of the translationally controlled tumor protein (TCTP) interactome reveals novel binding partners in human cancer cells. J Proteome Res 15:3741–3751
Lievens S et al (2010) Large-scale protein interactome mapping: strategies and opportunities. Expert Rev Proteomics 7:679–690
Liu Y et al (2005) Effect of GRP75/mthsp70/PBP74/mortalin overexpression on intracellular ATP level, mitochondrial membrane potential and ROS accumulation following glucose deprivation in PC12 cells. Mol Cell Biochem 268:45–51
Lucibello M et al (2011) TCTP is a critical survival factor that protects cancer cells from oxidative stress-induced cell-death. Exp Cell Res 317:2479–2489
Lucibello M et al (2015) Phospho-TCTP as a therapeutic target of dihydroartemisinin for aggressive breast cancer cells. Oncotarget 6:5275–5291
Macdonald SM et al (1995) Molecular-identification of an ige-dependent histamine-releasing factor. Science 269:688–690
Mehta V, Trinkle-Mulcahy L (2016) Recent advances in large-scale protein interactome mapping. F1000Res 5
Murphy ME (2013) The HSP70 family and cancer. Carcinogenesis 34:1181–1188
Neumann CA et al (2009) Peroxiredoxin 1 and its role in cell signaling. Cell Cycle 8:4072–4078
Oikawa K et al (2002) Dioxin stimulates synthesis and secretion of IgE-dependent histamine-releasing factor. Biochem Biophys Res Commun 290:984–987
Orsini F et al (2004) The life span determinant p66Shc localizes to mitochondria where it associates with mitochondrial heat shock protein 70 and regulates trans-membrane potential. J Biol Chem 279:25689–25695
Petrasek J, Schwarzerova K (2009) Actin and microtubule cytoskeleton interactions. Curr Opin Plant Biol 12:728–734
Phizicky E et al (2003) Protein analysis on a proteomic scale. Nature 422:208–215
Rajagopala SV (2015) Mapping the protein-protein interactome networks using yeast two-hybrid screens. Adv Exp Med Biol 883:187–214
Rho SB et al (2011) Anti-apoptotic protein TCTP controls the stability of the tumor suppressor p53. FEBS Lett 585:29–35
Roque CG et al (2016) Tumor protein Tctp regulates axon development in the embryonic visual system. Development 143(7):1134–1148
Rupec RA et al (1998) Isolation of a hypoxia-induced cDNA with homology to the mammalian growth-related protein p23. Oncol Res 10:69–74
Selbach M, Mann M (2006) Protein interaction screening by quantitative immunoprecipitation combined with knockdown (QUICK). Nat Methods 3:981–983
Sinnberg T et al (2012) MAPK and PI3K/AKT mediated YB-1 activation promotes melanoma cell proliferation which is counteracted by an autoregulatory loop. Exp Dermatol 21:265–270

Sirois I et al (2011) Caspase-3-dependent export of TCTP: a novel pathway for antiapoptotic intercellular communication. Cell Death Differ 18:549–562
Smits AH, Vermeulen M (2016) Characterizing protein-protein interactions using mass spectrometry: challenges and opportunities. Trends Biotechnol 34:825–834
Stambolic V et al (1998) Negative regulation of PKB/akt-dependent cell survival by the tumor suppressor PTEN. Cell 95:29–39
Stambolic V et al (2000) High incidence of breast and endometrial neoplasia resembling human cowden syndrome in pten(+/−) mice. Cancer Res 60:3605–3611
Stellberger T et al (2010) Improving the yeast two-hybrid system with permutated fusions proteins: the varicella zoster virus interactome. Proteome Sci 8:8
Susini L et al (2008) TCTP protects from apoptotic cell death by antagonizing bax function. Cell Death Differ 15:1211–1220
Sutherland BW et al (2005) Akt phosphorylates the Y-box binding protein 1 at Ser102 located in the cold shock domain and affects the anchorage-independent growth of breast cancer cells. Oncogene 24:4281–4292
Tao SC, Zhu H (2006) Protein chip fabrication by capture of nascent polypeptides. Nat Biotechnol 24:1253–1254
Tao SC et al (2007) Applications of protein microarray technology. Comb Chem High Throughput Screen 10:706–718
Tsai MJ et al (2014) TCTP is essential for beta-cell proliferation and mass expansion during development and beta-cell adaptation in response to insulin resistance. Endocrinology 155:392–404
Tsarova K et al (2010) Identification of a cofilin-like actin-binding site on translationally controlled tumor protein (TCTP). FEBS Lett 584:4756–4760
Tuynder M et al (2002) Biological models and genes of tumor reversion: cellular reprogramming through tpt1/TCTP and SIAH-1. Proc Natl Acad Sci USA 99:14976–14981
Tuynder M et al (2004) Translationally controlled tumor protein is a target of tumor reversion. Proc Natl Acad Sci USA 101:15364–15369
Wang W et al (2015) The proteomic investigation reveals interaction of mdig protein with the machinery of DNA double-strand break repair. Oncotarget 6:28269–28281
Wu D et al (2012) Upregulation of TCTP expression in human skin squamous cell carcinoma increases tumor cell viability through anti-apoptotic action of the protein. Exp Ther Med 3:437–442
Yang L et al (2016) Identification of serum biomarkers for gastric cancer diagnosis using a human proteome microarray. Mol Cell Proteomics 15:614–623
Yenofsky R et al (1983) Regulation of messenger-RNA utilization in mouse erythroleukemia-cells induced to differentiate by exposure to dimethylsulfoxide. Mol Cell Biol 3:1197–1203
Yu H et al (2008) High-quality binary protein interaction map of the yeast interactome network. Science 322:104–110
Zhang Y et al (2010) Plant protein-protein interaction network and interactome. Curr Genomics 11:40–46
Zhang J et al (2012) Role of the translationally controlled tumor protein in DNA damage sensing and repair. Proc Natl Acad Sci USA 109:E926–E933
Zheng P et al (2012) QUICK identification and SPR validation of signal transducers and activators of transcription 3 (Stat3) interacting proteins. J Proteome 75:1055–1066

Chapter 6
Role and Fate of TCTP in Protein Degradative Pathways

Michel Vidal

Abstract This chapter focuses on published studies specifically concerning TCTP and its involvement in degradation or stabilization of various proteins, and also in its own degradation in different ways. The first part relates to the inhibition of proteasomal degradation of proteins. This can be achieved by masking ubiquitination sites of specific partners, by favoring ubiquitin E3 ligase degradation, or by regulating proteasome activity. The second part addresses the ability of TCTP to favor degradation of specific proteins through proteasome or macroautophagic pathways. The third part discusses about the different ways by which TCTP has been shown to be degraded.

Abbreviations

Bre5	Brefeldin A sensitivity 5
CMA	Chaperone-mediated autophagy
DHA	Dihydroartemisinin
HIF1α	Hypoxia-inducible factor 1α
HRF	Histamine releasing factor
Hsp27	Heat shock protein 27
Mcl-1	Myeloid cell leukemia 1
Mdm2	Murine double minute 2
Mmi1	Microtubule and mitochondria interacting protein
Mss4	Mammalian suppressor of yeast Sec4
Mst-1	Mammalian sterile twenty-1
NTHK1	Tobacco histidine kinase-1
Pim-3	Serine/threonine-protein kinase Pim-3
PRX1	Peroxiredoxin-1
Rpn	26S proteasome regulatory subunit
Rpt	Proteasome regulatory particle base subunit

M. Vidal (✉)
Université Montpellier, UMR 5235, cc107, 34095 Montpellier, France
e-mail: michel.vidal@univ-montp2.fr

A. Telerman, R. Amson (eds.), *TCTP/tpt1 - Remodeling Signaling from Stem Cell to Disease*, Results and Problems in Cell Differentiation 64,
DOI 10.1007/978-3-319-67591-6_6

UPS	Ubiquitin–proteasome system
TCTP	Translationally controlled tumor protein
Ubp3	ubiquitin specific protease 3
VHL	von Hippel–Lindau protein

6.1 Introduction

Translationally controlled tumor protein (TCTP), also known as fortilin or HRF (histamine releasing factor), is a multifunctional protein implicated in diverse processes such as apoptosis (Yang et al. 2005; Liu et al. 2005; Susini et al. 2008), survival (Lucibello et al. 2011; Diraison et al. 2011), the cell cycle (Gachet et al. 1999; Cucchi et al. 2010; Burgess et al. 2008), proliferation and growth (Chen et al. 2007b; Hsu et al. 2007), development (Le et al. 2016; Roque et al. 2016), and DNA repair (Zhang et al. 2012; Hong and Choi 2013). TCTP has been shown to be subjected to various posttranslational modifications that can change its intracellular behavior. It is a cytosolic protein which was also found to be functionally associated with subcellular compartments such as mitochondria, nucleus, and microtubules (Susini et al. 2008; Hong and Choi 2013; Bazile et al. 2009). TCTP interactome was recently characterized and about a hundred interacting proteins were identified (Li et al. 2016). From a functional standpoint, TCTP often regulates protein behavior by favoring stabilization of protein partners or, on the contrary, promoting degradation of others. Regarding its own expression, TCTP has been shown to be transcriptionally and translationally regulated (Bommer et al. 2002, 2010, 2015; Amson et al. 2012). Moreover, TCTP can be unconventionally secreted in the extracellular space while exhibiting cytokine-like activity (MacDonald et al. 1995) or released within exosomes and resulting in a decrease in its intracellular levels (Amzallag et al. 2004).

6.2 TCTP as Protein Stabilizer

TCTP has often been described as a chaperone protein due to its stabilizing effect on protein partners. TCTP does not have foldase or holdase characteristics like those of classical chaperone proteins, although it has been reported that human and parasite (*Schistosoma mansoni*) TCTP can bind to a variety of denatured proteins and protect them from thermal shock (Gnanasekar et al. 2009). Note also that TCTP was shown to be structurally close to the mammalian suppressor of yeast Sec4 (Mss4) protein that binds to the GDP/GTP free form of Rab proteins, which is called the guanine nucleotide-free chaperone (Thaw et al. 2001). Moreover, TCTP has been reported to inhibit, in different ways, degradation of specific proteins by the ubiquitin–proteasome system (UPS).

6.2.1 *TCTP Masks the Ubiquitination Sites of Its Partners*

A widely described function of TCTP relates to its anti-apoptotic activity (Susini et al. 2008; Graidist et al. 2004; Baudet et al. 1998). This anti-apoptotic function is partly due to TCTP-induced p53 downregulation (Amson et al. 2012), preventing transcriptional activation of the pro-apoptotic gene Bax. However, TCTP has been shown to inhibit apoptosis in other ways, such as via stabilization of the extremely labile anti-apoptotic protein Mcl-1, thus counteracting Bax dimerization (Susini et al. 2008). TCTP interaction with Mcl-1 was found by two-hybrid screening and confirmed in GST pull-down and co-immunoprecipitation experiments (Liu et al. 2005). It was shown that a mutant form of Mcl-1 unable to bind to TCTP was much more susceptible to ubiquitination and, conversely, that TCTP overexpression blocked Mcl-1 from undergoing ubiquitination. TCTP has also been shown to be involved in protecting cells against ROS-mediated apoptosis independently of p53. TCTP protects the antioxidant enzyme peroxiredoxin-1 (PRX1) from inactivation by the kinase Mst-1 (mammalian sterile twenty-1) by masking phosphorylation sites when physically interacting with PRX1 (Chattopadhyay et al. 2016). TCTP-shielded PRX1 is enzymatically active, and it was shown that TCTP overexpression in mouse liver enhanced the peroxidase activity protecting mice against alcohol-induced ROS-mediated liver damage. Concomitantly, TCTP binding to PRX1 was also demonstrated to protect PRX1 from ubiquitin–proteasome degradation. TCTP depletion by shRNA induced PRX1 downregulation which was reversed by adding the proteasome inhibitor MG-132. Accordingly, TCTP overexpression in the U2OS cell line or in mouse liver induced a decrease in PRX1 polyubiquitination (Chattopadhyay et al. 2016). In the same line, TCTP has been identified through a yeast two-hybrid screen and shown to interact with Pim-3, a proto-oncogene with serine/threonine kinase activity (Zhang et al. 2013). TCTP is not phosphorylated by Pim-3 but modulates the protein kinase stability. Actually, depletion of TCTP by siRNA in the human pancreatic carcinoma cell line PCI55 increases Pim-3 degradation, which is abrogated by the proteasome inhibitor MG132. Accordingly, Pim-3 ubiquitination was promoted by TCTP siRNA treatment.

TCTP is a highly conserved protein identified in a wide range of eukaryotic organisms, across animal and plant kingdoms and in yeast. Cultivated tobacco (*Nicotiana tabacum*) NtTCTP has been shown to regulate seedling growth through control of cell proliferation. Ethylene is a phytohormone that inhibits vegetative growth. This inhibition is regulated by a feedback mechanism, in which ethylene-induced NtTCTP binds to and stabilizes ethylene receptor tobacco histidine kinase-1 (NTHK1) and reduces the plant response to ethylene, promoting plant growth by increasing cell proliferation (Tao et al. 2015). Interaction of NtTCTP with NTHK1 was identified by two-hybrid screening and confirmed by GST pull-down and co-immunoprecipitation experiments (Tao et al. 2015). In the presence of cycloheximide, NtTCTP overexpression in plants stabilized NTHK1 compared to wild-type plants, while this decrease in NTHK1 levels in wild-type plants was abolished by proteasome inhibitor treatment. It is thus tempting to speculate that in all of these cases (Mcl-1, PRX1, PIM-3, and NTHK1), ubiquitination sites on TCTP partners are masked by TCTP interaction, thus avoiding ubiquitination and further degradation by the proteasome.

6.2.2 TCTP Binding Leads to E3 Ligase Degradation

TCTP can also stabilize protein by inducing degradation of the specific E3-ubiquitin ligase involved, thus inhibiting ubiquitination of the protein. This has been described for TCTP stabilization of hypoxia-inducible factor 1α (HIF1α). The tumor suppressor von Hippel–Lindau protein (VHL) functions as an E3 ligase that can interact with and ubiquitinate HIF1α (Chen et al. 2013). It has been shown that TCTP specifically binds to the β-domain responsible for substrate recognition by VHL, while competing with HIF1α. However, TCTP is not ubiquitinated by VHL, as suggested by the constant TCTP protein levels observed when VHL is either overexpressed or depleted. Conversely, TCTP promotes K48-linked ubiquitination of VHL and its further degradation by proteasomes. TCTP thus competes with HIF1α for binding to VHL, reduces E3 ligase stability, inducing upregulation of the HIF1α protein level.

6.2.3 Mmi1/ScTCTP Modulates Proteasome Activity

The microtubule and mitochondria interacting protein (Mmi1), the yeast homologue of mammalian TCTP, is described as a stress sensor and stress-response regulator (Rinnerthaler et al. 2006). In high-throughput studies, Mmi1 was found to be a putative interactor of various proteasomal subunits such as Rpn1, Rpt5, and Rpn11 (Guerrero et al. 2008). Using fluorescence imaging, Mmi1-GFP was found to be uniformly distributed in the cytoplasm of exponentially growing yeast cells. However, when yeasts were submitted to robust heat stress, Mmi1-RFP partially colocalized with the proteasome (labeled using Rpn1-GFP) in the nucleus and gradually returned to its diffusely cytoplasmic location as cells recovered from the heat shock (Rinnerthaler et al. 2013). The partial relocalization of Mmi1 to the nucleus in heat-stressed cells was confirmed by immunogold electron microscopy. By comparing the proteasomal activity of WT yeast cells or Mmi1-deleted mutants, at low temperature or after heat shock, it was shown that Mmi1 slightly but consistently inhibited proteasome activity. Moreover, Mmi1 was found to interact with other components that potentially modulate proteasome degradation. Indeed, Bre5 and Ubp3 can form a complex to specifically de-ubiquitinate proteins by cleaving off the first conjugated ubiquitin, thus modifying their turnover, as shown for Sec23 (Cohen et al. 2003). This Bre5-Ubp3 de-ubiquitination complex was shown to colocalize with Mmi1 in association with stress granules in the cytoplasm of heat-stressed yeast cells (Rinnerthaler et al. 2013).

6.3 TCTP as Degradation Inducer

The tumor suppressor p53 is tightly controlled by the E3-ubiquitin ligase Mdm2 protein. Interaction with Mdm2 maintains p53 at low levels under basal conditions through different mechanisms, including proteasomal degradation after ubiquitination. In response to stress, the cellular level of p53 is elevated through a posttranslational mechanism, leading to cell cycle arrest or apoptosis. It has been demonstrated that TCTP binds to the p53–Mdm2 complex and increases the Mdm2-mediated ubiquitination and proteasome degradation of p53, concomitantly with Mdm2 autoubiquitination inhibition (Amson et al. 2012). TCTP overexpression or knock-down in HCT116 cells, respectively, resulted in a decrease or increase in p53 protein levels. Accordingly, analysis of tissues from *Tctp* heterozygous ($Tctp^{+/-}$) mice revealed readily detectable p53 levels, contrary to *Tctp* WT mice in which basal p53 levels were undetectable. TCTP-induced p53 degradation is inhibited by the proteasome inhibitor MG132 and antagonized by NUMB, a regulator of p53 that was previously shown to bind to the p53–Mdm2 complex, therefore preventing p53 ubiquitination and degradation (Colaluca et al. 2008). Co-immunoprecipitation experiments on endogenous proteins in HCT116 cells demonstrated that TCTP forms complex with p53–Mdm2 and NUMB (Amson et al. 2012). Importantly, high TCTP levels were found to be correlated with breast cancer aggressiveness, for which it is an independent prognostic factor (Amson et al. 2012).

Hepatocellular carcinoma is also a cancer in which TCTP overexpression was detected and associated with the advanced tumor stage (Chan et al. 2012). It was shown that the chromodomain helicase/ATPase DNA binding protein1-like gene binds to the promoter region of *TCTP* and activates its transcription. TCTP overexpression was shown to contribute to the mitotic defects of tumor cells by inducing a decrease in Cdc25, leading to failure in the dephosphorylation of Cdk1 and its dysfunction during mitosis. TCTP-induced Cdc25 downregulation was shown to occur at the protein rather than mRNA level. TCTP overexpression in cell lines induced Cdc25 ubiquitination and degradation, which was abolished by MG132 treatment. Conversely, TCTP depletion by shRNA increased Cdc25 levels, which in turn increased Cdk1 activity (Chan et al. 2012).

Besides regulating proteasome degradation of various proteins, as seen above, TCTP has recently been shown to be involved in macroautophagy regulation. Macroautophagy is a self-degradative process through which macromolecules and even organelles are transported to lysosomes for degradation by the prominent formation of autophagic vesicles in the cytoplasm. This process is important for balancing sources of energy at critical times in development and in response to nutrient stress. Macroautophagy is orchestrated by a series of core autophagy proteins (ATG proteins) that are evolutionarily conserved (Tsukada and Ohsumi 1993; Ohsumi 2014). By studying long-term artificial selection of pigs (wild vs domestic species), *TCTP/TPT1* genes have been found to be upregulated during artificial selection, concomitantly with an increase in female fecundity (Chen et al. 2014). It is suggested that this could be related to macroautophagy that takes place during oogenesis in ovarian granulosa and cumulus cells. TCTP was shown to regulate AMPK and mTORC1 activities, both of which were involved in macroautophagy activation under hypoxic conditions. In line with this, it was shown that the phosphatidylethanolamine-conjugated form of ATG8 (ATG8-PE, also named LC3-II), a key molecular component initiating and contributing to autophagic vacuole elongation, was increased in TCTP knockdown COS-7 cells cultured in normoxic conditions, but decreased in the same cells cultured in hypoxic conditions. Similar results were obtained under serum starvation conditions. Moreover, by co-immunoprecipitation, TCTP was found to interact with ATG5, ATG12, and ATG16, three ATG proteins forming a complex in charge of ATG8 phospholipid conjugation and autophagic vacuole formation (Walczak and Martens 2013). These data suggest that TCTP could positively regulate macroautophagy to generate energy during oogenesis.

6.4 TCTP Degradation

As seen above, numerous studies have shown that TCTP regulates protein degradation by the ubiquitin–proteasome system or through the macroautophagy pathway (Fig. 6.1). Surprisingly, very few studies have examined the specific degradation of TCTP. One explanation is that TCTP is very efficiently and finely regulated at transcriptional and translational levels (Bommer et al. 2010, 2015). Intuitively, it is difficult to imagine TCTP as a component that could regulate proteasome degradation, while itself being a specific target for UPS degradation. As a matter of fact, TCTP can physically interact with E3-ligases (e.g., Mdm2, VHL) but is not a substrate for these ubiquitin ligases, and accordingly TCTP has been shown to have a quite long half-life (Tuynder et al. 2002; Baylot et al. 2012).

However, it was reported that in some cases TCTP degradation by UPS can be upregulated. Dihydroartemisinin (DHA) is a metabolite of artemisinin, a molecule originally used as an antimalarial drug. The anticancer activity of dihydroartemisinin was examined in human ovarian cancer cells (Jiao et al. 2007). It was found that DHA induced cell growth inhibition via the induction of apoptosis with a decrease in Bcl-xL/Bcl-2 and an increase in Bax/Bad, while also blocking cell cycle progression. Subsequently, DHA was found to bind to TCTP and shorten its half-life, a process that is blocked by MG132. TCTP was shown to be ubiquitinated in a DHA-dose dependent manner, likely after a structural change induced by DHA binding (Fujita et al. 2008).

The TCTP half-life has also been shown to be shortened after depletion of the heat shock protein 27 (hsp27) in the prostate carcinoma cell line (Baylot et al. 2012). It was first shown that hsp27 knockdown using antisense oligonucleotides and siRNA induced apoptosis and enhanced chemotherapy in prostate cancer (Rocchi et al. 2006). Hsp27 is highly overexpressed in castration-resistant prostate cancer, and TCTP was found to be a client protein of the chaperone in co-immunoprecipitation experiments. Overexpression of hsp27 increased TCTP

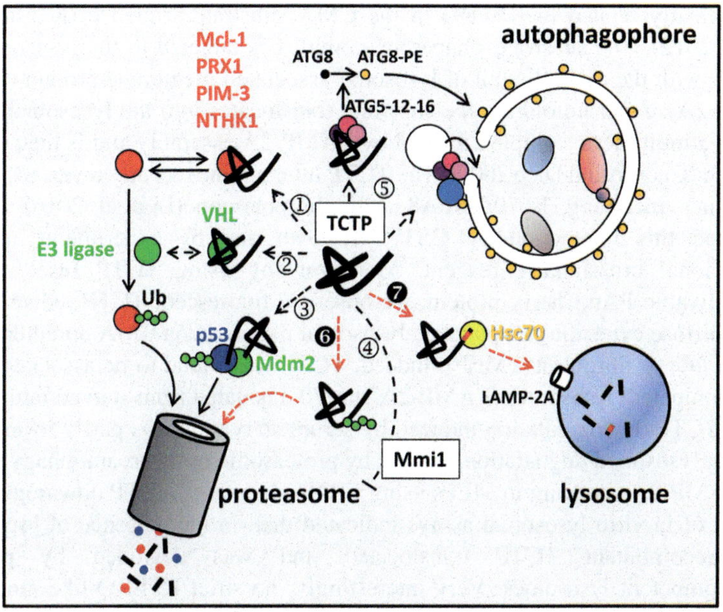

Fig. 6.1 Schematic representation of the different types of involvement of TCTP in protein degradative pathways discussed in the review. References for steps: ①: Liu et al. (2005), Susini et al. (2008), Chattopadhyay et al. (2016), Zhang et al. (2013) and Tao et al. (2015); ②: Chen et al. (2013); ③: Amson et al. (2012), Chan et al. (2012); ④: Guerrero et al. (2008) and Rinnerthaler et al. (2013); ⑤: Chen et al. (2014); ❻: Jiao et al. (2007) and Fujita et al. (2008) ; ❼: Bonhoure et al. (2017)

levels, while hsp27 knockdown led to a decrease in TCTP protein levels without affecting expression of its mRNA. Moreover, the proteasome inhibitor MG132 was found to reverse the effect of hsp27 knockdown and prolonged the TCTP half-life (Baylot et al. 2012).

These studies showed that TCTP can be degraded by the ubiquitin–proteasome system. However, these are uncommon situations which probably do not account for physiologic regulation of TCTP downregulation. As noted earlier, TCTP has been shown to be posttranslationally modified through ubiquitination but also through acetylation, phosphorylation, and sumoylation. These modifications can change the behavior of TCTP, as demonstrated for TCTP phosphorylation by polo-like kinase-1 (PLK1), thus decreasing its microtubule-stabilizing activity (Yarm 2002) and promoting its nuclear localization (Cucchi et al. 2010). Relocalization is a way to rapidly switch off or promote protein function, in addition to transcription or translation regulation. Thus, we explored the possibility of TCTP degradation promoted by posttranslational modification and found that TCTP was degraded by chaperone-mediated autophagy (CMA) after acetylation (Bonhoure et al. 2017). In contrast to regular macroautophagy, CMA allows lysosomal degradation of specific cytosolic proteins on a molecule-by-molecule basis. The selectivity of this pathway is conferred through recognition by the cytosolic chaperone hsc70 of a pentapeptide biochemically related to KFERQ in the CMA substrate sequence (Kaushik and Cuervo 2012). The substrate–chaperone complex is targeted to the lysosome and interacts with the cytosolic tail of lysosome-associated membrane protein type 2A (LAMP-2A). After unfolding, the substrate translocates into the lysosomal lumen through a multimeric complex formed by LAMP-2A assembly and is then rapidly degraded (Cuervo and Dice 2000). The TCTP interactome recently revealed various chaperones, including Hsc70/HSPA8, as binding partners (Li et al. 2016), and we confirmed this interaction by GST pull-down and by experiments using a tri-functional crosslinking reagent. Moreover, by using TCTP fused with a photoactivable PAmCherry protein, we observed fluorescent TCTP redistribution from a diffuse cytosolic to a punctate lysosomal pattern upon CMA upregulation in mouse embryo fibroblasts (MEF). Indeed, TCTP was found to be associated with CMA-competent lysosomes (LAMP-2A$^+$/hsc70$^+$) isolated from starved rat liver. In MEFs, TCTP downregulation induced by serum starvation was partly reversed by blocking lysosomal degradation, but not by proteasome or macroautophagy inhibition. LAMP-2A silencing in MEFs using siRNA decreased TCTP downregulation. The use of in vitro lysosomal assays indicated that, in the presence of hsc70 and ATP, recombinant TCTP translocated and was degraded by purified CMA-competent lysosomes. Very interestingly, no strict KFERQ-like motif was found in TCTP. However, acetylation endows Lys19 (Liu et al. 2014) with the status of critical (pseudo)glutamine in the ^{19}KIREI^{23} motif. This assumption was confirmed by showing that acetylation mimetic mutants (K19Q and K19 N) are efficiently targeted to the lysosomal (LAMP-2A$^+$) compartment, contrary to the acetylation ablative mutant (K19A). Accordingly, treatment of cells with deacetylase inhibitors decreased TCTP intracellular levels. Overall, these data show that CMA is involved in TCTP regulation, while its acetylation is critical for its degradation (Fig. 6.2).

6.5 Conclusion

TCTP relationships with protein degradative pathways thus appear to be very finely regulated, as a regulator of its partners' degradation but also of its own degradation.

It is thus tempting to speculate that acetylation/CMA sequential processing could be involved in the regulation of TCTP functioning by switching off interactions with partners and inducing its own degradation. Interestingly, the conditional KFERQ-like motif (^{19}KIREI^{23}) encompasses Arg21, which was shown to be critical for the interaction between TCTP and the Bcl-xL homolog Mcl-1 (Zhang et al. 2002). This is in keeping with interaction of TCTP with Mcl-1, which was reported to increase the TCTP half-life (Zhang et al. 2002). More broadly, the N-terminal sequence in TCTP, including the conditional KFERQ-like motif, was shown to mediate binding to the BH3 domain of Bcl-xL (Yang et al. 2005; Thebault et al. 2016).

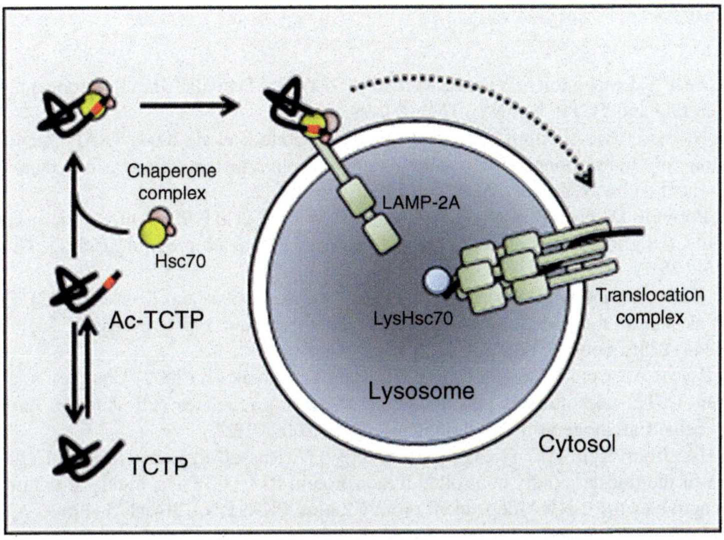

Fig. 6.2 Schematic representation of the different steps involved in TCTP degradation by chaperone-mediated autophagy

Note that Ku70 is another effector that is activated through deacetylation to enhance cell survival. Interestingly Ku70, like TCTP with which it interacts (Zhang et al. 2012), is involved in DNA double-strand break repair and apoptosis regulation in an acetylation sensitive-manner (Chen et al. 2007a; Subramanian et al. 2005). Importantly, it has been shown that CMA is upregulated in response to DNA damage and participates in the timely degradation of nuclear Chk1 subsequent to its phosphorylation by ATR kinase after DNA repair (Park et al. 2015). We could speculate a similar fate for TCTP with acetylation subsequent to DNA repair.

Acknowledgements This work was supported by grants from the CNRS, the University of Montpellier, and from the Labex LERMIT.

Compliance with Ethics Guidelines

The author declares that he has no conflict of interest with the contents of this chapter.

References

Amson R, Pece S, Lespagnol A, Vyas R, Mazzarol G, Tosoni D et al (2012) Reciprocal repression between P53 and TCTP. Nat Med 18(1):91–99

Amzallag N, Passer BJ, Allanic D, Segura E, Thery C, Goud B et al (2004) TSAP6 facilitates the secretion of translationally controlled tumor protein/histamine-releasing factor via a nonclassical pathway. J Biol Chem 279(44):46104–46112

Baudet C, Perret E, Delpech B, Kaghad M, Brachet P, Wion D et al (1998) Differentially expressed genes in C6.9 glioma cells during vitamin D-induced cell death program. Cell Death Differ 5 (1):116–125

Baylot V, Katsogiannou M, Andrieu C, Taieb D, Acunzo J, Giusiano S et al (2012) Targeting TCTP as a new therapeutic strategy in castration-resistant prostate cancer. Mol Ther 20 (12):2244–2256. Epub 2012/08/16

Bazile F, Pascal A, Arnal I, Le Clainche C, Chesnel F, Kubiak JZ (2009) Complex relationship between TCTP, microtubules and actin microfilaments regulates cell shape in normal and cancer cells. Carcinogenesis 30(4):555–565. Epub 2009/01/27

Bommer UA, Borovjagin AV, Greagg MA, Jeffrey IW, Russell P, Laing KG et al (2002) The mRNA of the translationally controlled tumor protein P23/TCTP is a highly structured RNA, which activates the dsRNA-dependent protein kinase PKR. RNA 8(4):478–496

Bommer UA, Heng C, Perrin A, Dash P, Lobov S, Elia A et al (2010) Roles of the translationally controlled tumour protein (TCTP) and the double-stranded RNA-dependent protein kinase, PKR, in cellular stress responses. Oncogene 29(5):763–773

Bommer UA, Iadevaia V, Chen J, Knoch B, Engel M, Proud CG (2015) Growth-factor dependent expression of the translationally controlled tumour protein TCTP is regulated through the PI3-K/Akt/mTORC1 signalling pathway. Cell Signal 27(8):1557–1568. Epub 2015/05/06

Bonhoure A, Vallentin A, Martin M, Senff-Ribeiro A, Amson A, Telerman A et al (2017) Acetylation of translationally controlled tumor protein promotes its degradation through chaperone-mediated autophagy. Eur. J. Cell Biol. 96: 83–98

Burgess A, Labbe JC, Vigneron S, Bonneaud N, Strub JM, Van Dorsselaer A et al (2008) Chfr interacts and colocalizes with TCTP to the mitotic spindle. Oncogene 27(42):5554–5566

Chan TH, Chen L, Liu M, Hu L, Zheng BJ, Poon VK et al (2012) Translationally controlled tumor protein induces mitotic defects and chromosome missegregation in hepatocellular carcinoma development. Hepatology 55(2):491–505. Epub 2011/09/29

Chattopadhyay A, Pinkaew D, Doan HQ, Jacob RB, Verma SK, Friedman H et al (2016) Fortilin potentiates the peroxidase activity of Peroxiredoxin-1 and protects against alcohol-induced liver damage in mice. Sci Rep 6:18701. Epub 2016/01/05

Chen CS, Wang YC, Yang HC, Huang PH, Kulp SK, Yang CC et al (2007a) Histone deacetylase inhibitors sensitize prostate cancer cells to agents that produce DNA double-strand breaks by targeting Ku70 acetylation. Cancer Res 67(11):5318–5327. Epub 2007/06/05

Chen SH, Wu PS, Chou CH, Yan YT, Liu H, Weng SY et al (2007b) A knockout mouse approach reveals that TCTP functions as an essential factor for cell proliferation and survival in a tissue- or cell type-specific manner. Mol Biol Cell 18(7):2525–2532

Chen K, Chen S, Huang C, Cheng H, Zhou R (2013) TCTP increases stability of hypoxia-inducible factor 1alpha by interaction with and degradation of the tumour suppressor VHL. Biol Cell 105 (5):208–218. Epub 2013/02/08

Chen K, Huang C, Yuan J, Cheng H, Zhou R (2014) Long-term artificial selection reveals a role of TCTP in autophagy in mammalian cells. Mol Biol Evol 31(8):2194–2211. Epub 2014/06/04

Cohen M, Stutz F, Belgareh N, Haguenauer-Tsapis R, Dargemont C (2003) Ubp3 requires a cofactor, Bre5, to specifically de-ubiquitinate the COPII protein, Sec23. Nat Cell Biol 5 (7):661–667

Colaluca IN, Tosoni D, Nuciforo P, Senic-Matuglia F, Galimberti V, Viale G et al (2008) NUMB controls p53 tumour suppressor activity. Nature 451(7174):76–80

Cucchi U, Gianellini LM, De Ponti A, Sola F, Alzani R, Patton V et al (2010) Phosphorylation of TCTP as a marker for polo-like kinase-1 activity in vivo. Anticancer Res 30(12):4973–4985

Cuervo AM, Dice JF (2000) Unique properties of lamp2a compared to other lamp2 isoforms. J Cell Sci 113(Pt 24):4441–4450

Diraison F, Hayward K, Sanders KL, Brozzi F, Lajus S, Hancock J et al (2011) Translationally controlled tumour protein (TCTP) is a novel glucose-regulated protein that is important for survival of pancreatic beta cells. Diabetologia 54(2):368–379. Epub 2010/11/11

Fujita T, Felix K, Pinkaew D, Hutadilok-Towatana N, Liu Z, Fujise K (2008) Human fortilin is a molecular target of dihydroartemisinin. FEBS Lett 582(7):1055–1060. Epub 2008/03/08

Gachet Y, Tournier S, Lee M, Lazaris-Karatzas A, Poulton T, Bommer UA (1999) The growth-related, translationally controlled protein P23 has properties of a tubulin binding protein and associates transiently with microtubules during the cell cycle. J Cell Sci 112(Pt 8):1257–1271

Gnanasekar M, Dakshinamoorthy G, Ramaswamy K (2009) Translationally controlled tumor protein is a novel heat shock protein with chaperone-like activity. Biochem Biophys Res Commun 386(2):333–337

Graidist P, Phongdara A, Fujise K (2004) Antiapoptotic protein partners fortilin and MCL1 independently protect cells from 5-fluorouracil-induced cytotoxicity. J Biol Chem 279 (39):40868–40875

Guerrero C, Milenkovic T, Przulj N, Kaiser P, Huang L (2008) Characterization of the proteasome interaction network using a QTAX-based tag-team strategy and protein interaction network analysis. Proc Natl Acad Sci USA 105(36):13333–13338. Epub 2008/09/02

Hong ST, Choi KW (2013) TCTP directly regulates ATM activity to control genome stability and organ development in Drosophila melanogaster. Nat Commun 4:2986. Epub 2013/12/20

Hsu YC, Chern JJ, Cai Y, Liu M, Choi KW (2007) Drosophila TCTP is essential for growth and proliferation through regulation of dRheb GTPase. Nature 445(7129):785–788

Jiao Y, Ge CM, Meng QH, Cao JP, Tong J, Fan SJ (2007) Dihydroartemisinin is an inhibitor of ovarian cancer cell growth. Acta Pharmacol Sin 28(7):1045–1056. Epub 2007/06/26

Kaushik S, Cuervo AM (2012) Chaperone-mediated autophagy: a unique way to enter the lysosome world. Trends Cell Biol 22(8):407–417

Le TP, Vuong LT, Kim AR, Hsu YC, Choi KW (2016) 14-3-3 Proteins regulate Tctp-Rheb interaction for organ growth in Drosophila. Nat Commun 7:11501. Epub 2016/05/07

Li S, Chen M, Xiong Q, Zhang J, Cui Z, Ge F (2016) Characterization of the translationally controlled tumor protein (TCTP) interactome reveals novel binding partners in human cancer cells. J Proteome Res. Epub 2016/09/09

Liu H, Peng HW, Cheng YS, Yuan HS, Yang-Yen HF (2005) Stabilization and enhancement of the antiapoptotic activity of mcl-1 by TCTP. Mol Cell Biol 25(8):3117–3126

Liu Z, Wang Y, Gao T, Pan Z, Cheng H, Yang Q et al (2014) CPLM: a database of protein lysine modifications. Nucleic Acids Res 42(Database issue):D531–D536. Epub 2013/11/12

Lucibello M, Gambacurta A, Zonfrillo M, Pierimarchi P, Serafino A, Rasi G et al (2011) TCTP is a critical survival factor that protects cancer cells from oxidative stress-induced cell-death. Exp Cell Res 317(17):2479–2489. Epub 2011/08/02

MacDonald SM, Rafnar T, Langdon J, Lichtenstein LM (1995) Molecular identification of an IgE-dependent histamine-releasing factor. Science 269(5224):688–690

Ohsumi Y (2014) Historical landmarks of autophagy research. Cell Res 24(1):9–23. Epub 2013/12/25

Park C, Suh Y, Cuervo AM (2015) Regulated degradation of Chk1 by chaperone-mediated autophagy in response to DNA damage. Nat Commun 6:6823. Epub 2015/04/17

Rinnerthaler M, Jarolim S, Heeren G, Palle E, Perju S, Klinger H et al (2006) MMI1 (YKL056c, TMA19), the yeast orthologue of the translationally controlled tumor protein (TCTP) has apoptotic functions and interacts with both microtubules and mitochondria. Biochim Biophys Acta 1757(5–6):631–638

Rinnerthaler M, Lejskova R, Grousl T, Stradalova V, Heeren G, Richter K et al (2013) Mmi1, the yeast homologue of mammalian TCTP, associates with stress granules in heat-shocked cells and modulates proteasome activity. PLoS One 8(10):e77791. Epub 2013/11/10

Rocchi P, Jugpal P, So A, Sinneman S, Ettinger S, Fazli L et al (2006) Small interference RNA targeting heat-shock protein 27 inhibits the growth of prostatic cell lines and induces apoptosis via caspase-3 activation in vitro. BJU Int 98(5):1082–1089. Epub 2006/08/02

Roque CG, Wong HH, Lin JQ, Holt CE (2016) Tumor protein Tctp regulates axon development in the embryonic visual system. Development 143(7):1134–1148. Epub 2016/02/24

Subramanian C, Opipari AW Jr, Bian X, Castle VP, Kwok RP (2005) Ku70 acetylation mediates neuroblastoma cell death induced by histone deacetylase inhibitors. Proc Natl Acad Sci USA 102(13):4842–4847. Epub 2005/03/22

Susini L, Besse S, Duflaut D, Lespagnol A, Beekman C, Fiucci G et al (2008) TCTP protects from apoptotic cell death by antagonizing bax function. Cell Death Differ 15(8):1211–1220

Tao JJ, Cao YR, Chen HW, Wei W, Li QT, Ma B et al (2015) Tobacco translationally controlled tumor protein interacts with ethylene receptor tobacco histidine kinase1 and enhances plant growth through promotion of cell proliferation. Plant Physiol 169(1):96–114. Epub 2015/05/06

Thaw P, Baxter NJ, Hounslow AM, Price C, Waltho JP, Craven CJ (2001) Structure of TCTP reveals unexpected relationship with guanine nucleotide-free chaperones. Nat Struct Biol 8 (8):701–704

Thebault S, Agez M, Chi X, Stojko J, Cura V, Telerman SB et al (2016) TCTP contains a BH3-like domain, which instead of inhibiting, activates Bcl-xL. Sci Rep 6:19725. Epub 2016/01/28

Tsukada M, Ohsumi Y (1993) Isolation and characterization of autophagy-defective mutants of Saccharomyces cerevisiae. FEBS Lett 333(1–2):169–174. Epub 1993/10/25

Tuynder M, Susini L, Prieur S, Besse S, Fiucci G, Amson R et al (2002) Biological models and genes of tumor reversion: cellular reprogramming through tpt1/TCTP and SIAH-1. Proc Natl Acad Sci USA 99(23):14976–14981

Walczak M, Martens S (2013) Dissecting the role of the Atg12-Atg5-Atg16 complex during autophagosome formation. Autophagy 9(3):424–425. Epub 2013/01/17

Yang Y, Yang F, Xiong Z, Yan Y, Wang X, Nishino M et al (2005) An N-terminal region of translationally controlled tumor protein is required for its antiapoptotic activity. Oncogene 24 (30):4778–4788

Yarm FR (2002) Plk phosphorylation regulates the microtubule-stabilizing protein TCTP. Mol Cell Biol 22(17):6209–6221

Zhang D, Li F, Weidner D, Mnjoyan ZH, Fujise K (2002) Physical and functional interaction between myeloid cell leukemia 1 protein (MCL1) and Fortilin. The potential role of MCL1 as a fortilin chaperone. J Biol Chem 277(40):37430–37438. Epub 2002/08/01

Zhang J, de Toledo SM, Pandey BN, Guo G, Pain D, Li H et al (2012) Role of the translationally controlled tumor protein in DNA damage sensing and repair. Proc Natl Acad Sci USA 109(16): E926–E933. Epub 2012/03/28

Zhang F, Liu B, Wang Z, Yu XJ, Ni QX, Yang WT et al (2013) A novel regulatory mechanism of Pim-3 kinase stability and its involvement in pancreatic cancer progression. Mol Cancer Res MCR 11(12):1508–1520. Epub 2013/10/30

Chapter 7
Roles of the Translationally Controlled Tumor Protein (TCTP) in Plant Development

Léo Betsch, Julie Savarin, Mohammed Bendahmane, and Judit Szecsi

Abstract The Translationally Controlled Tumor Protein (TCTP) is a conserved protein which expression was associated with several biochemical and cellular functions. Loss-of-function mutants are lethal both in animals and in plants, making the identification of its exact role difficult. Recent data using the model plant *Arabidopsis thaliana* provided the first viable adult knockout for TCTP and helped addressing the biological role of TCTP during organ development and the functional conservation between plants and animals. This chapter summarizes our up to date knowledge about the role of TCTP in plants and discuss about conserved functions and mechanisms between plants and animals.

7.1 Introduction

In plants as in animals, the growth of an organism and the determination of its final size require the tight regulation of multiple internal developmental processes that affect organ growth and allow the ultimate differentiation into functional organs and tissues. The rising of an individual requires the fine-tuning and coordination of cell proliferation, cell growth, cell differentiation, and cell death. The mechanisms and regulatory networks that control these developmental processes in plants remain largely unknown (Anastasiou and Lenhard 2007; Bögre et al. 2008; Pan 2007; Busov et al. 2008; Krizek 2009; Johnson and Lenhard 2011; Van Hautegem et al. 2015). Current findings begin to explain how the enormous variety in organ sizes and shapes appeared during evolution. They also highlighted that plants and animals share various common growth regulatory pathways during development and organ morphogenesis (Van Hautegem et al. 2015; Cook and Tyers 2007; Arya and White 2015; Niklas 2015; Rexin et al. 2015). However, there exist also major differences between plant and animal development. For example, while organs are

Authors "Léo Betsch" and "Julie Savarin" have contributed equally

L. Betsch • J. Savarin • M. Bendahmane (✉) • J. Szecsi (✉)
Laboratoire Reproduction et Développement des Plantes, Univ Lyon, ENS de Lyon, Université Claude Bernard Lyon 1, CNRS, INRA, 69342 Lyon, France
e-mail: mohammed.bendahmane@ens-lyon.fr; judit.szecsi@ens-lyon.fr

A. Telerman, R. Amson (eds.), *TCTP/tpt1 - Remodeling Signaling from Stem Cell to Disease*, Results and Problems in Cell Differentiation 64,
DOI 10.1007/978-3-319-67591-6_7

produced during embryogenesis in animals, plants have the capacity to generate organs during their whole life, thus constantly influencing their body size and shape. Identifying how growth regulatory pathways are shared among animal and plant species remains a daunting task.

Developmental processes largely depend on both genetic factors and environmental inputs in animals and plants. Molecular integrators, such as hormones and other factors, are stimulated by environmental signals to control development at the cellular and tissue levels (Lau and Deng 2010; Leivar and Monte 2014; Nibau et al. 2006; Peleg and Blumwald 2011). Integration of the environmental signals is achieved by a number of growth-promoting and growth-restricting factors. Many genes with growth-promoting and growth-restricting functions that act on cell division or on cell expansion have been identified in plants and in animals (Bögre et al. 2008; Pan 2007; Busov et al. 2008; Krizek 2009; Niklas 2015; Rexin et al. 2015; Penzo-Méndez and Stanger 2015; Lloyd 2013; Crickmore and Mann 2008; Chen et al. 2013). However, these genes act in independent genetic pathways, making it difficult to develop an integrated and comprehensive model of organ size control.

A common pathway controlling animal as well as plant organ growth has been proposed to involve the Translationally Controlled Tumor Protein (TCTP). In mammals, TCTP is associated with many cancers and was shown to have an antagonist role to the tumor suppressor p53 (Chen et al. 2013; Amson et al. 2012; Rho et al. 2011). TCTP is conserved among all eukaryotes and has been proposed to have diverse roles in developmental and defense processes, including several cellular functions such as cell proliferation, expansion, and death (Bommer and Thiele 2004; Brioudes et al. 2010; Bommer 2012).

In this chapter, we will present and discuss the up-to-date knowledge on the multiple roles of TCTP in growth and in response to various signals and stresses in plants, and how gained knowledge help understanding the TCTP's central role in plant and animal development.

7.2 Features of Plant TCTP Genes

Since the early 1990s, homologs of *TCTP* have been identified in many plant species, such as the model plant *Arabidopsis thaliana,* alfalfa, pea, *Pharbitis nil*, grape vine, oil palm, *Jatropha curcas*, cassava, cabbage, rice, strawberry, tobacco, rubber tree (Pay et al. 1992; Woo and Hawes 1997; Cao et al. 2010; Sage-Ono et al. 1998; Kang et al. 2003; Szécsi et al. 2006; Lopez and Franco 2006; Vincent et al. 2007; Nakkaew et al. 2010; Masura et al. 2011; Qin et al. 2011; Li et al. 2013; Santa Brígida et al. 2014; Tao et al. 2015; Wang et al. 2015; Zhang et al. 2013). In Eukaryotes, *TCTP* is generally present as a single gene copy in the genome, but many species carry more than one gene (Gutierrez-Galeano et al. 2014; Hinojosa-Moya et al. 2013). Mammals seem to have many *TCTP* gene copies in their genome, but only one is likely functional (Thiele et al. 2000; Chen et al. 2007). Plant species harbor one and up to five *TCTP* gene copies (Gutierrez-Galeano et al. 2014; Hinojosa-Moya et al. 2013; Pavy et al. 2005), but many of these gene copies are likely nonfunctional pseudogenes (Brioudes et al. 2010; Berkowitz et al. 2008). The role of these various *TCTP* pseudogenes and transcripts variants remain unknown, but might point toward a specialization of *TCTP* function.

Globally, the genomic structure of plant *TCTP* genes is very similar. *TCTP* gene is generally composed of five exons with conserved length and four introns with variable length (Zhang et al. 2013). In *Arabidopsis thaliana AtTCTP*, intron 3 is absent leading to fusion of the third and fourth exons (Zhang et al. 2013).

As in animals, 5′ and 3′ untranslated regions (UTRs) of variable sequence length are also present in plant *TCTPs*. These UTRs have been shown to be associated with *TCTP* mRNA stability and also to play a role in the regulation of its translation (Bommer and Thiele 2004; Brioudes et al. 2010). Like for animal *TCTPs, AtTCTP* 5′UTR contains a 5′TOP element (Terminal Oligo Pyrimidine) (Brioudes et al. 2010). 5′TOP elements are common in translationally controlled proteins (Meyuhas and Kahan 2015). However, conversely to animal *TCTPs, AtTCTP* 5′TOP is not GC-rich, suggesting a less complex secondary structure of the *AtTCTP* mRNA (Brioudes et al. 2010). *AtTCTP* 3′UTR contains classical AU-rich mRNA destabilizing elements found in short-lived mRNAs in animals and in plants (Ohme-Takagi et al. 1993; Barreau et al. 2005; Narsai et al. 2007).

AtTCTP mRNA expression was shown to be strong and ubiquitous in the model plant *A. thaliana* (Brioudes et al. 2010; Szécsi et al. 2006; Berkowitz et al. 2008). The promoter region of *AtTCTP* is located in a short 0.3 kb intergenic region between *TCTP* and a neighboring gene on the complementary strand. Within this 0.3 kb intergenic region, typical core promoter elements were found at -16 bp (Y-patch) and -34 bp (TATA box) upstream of the transcription start. This 0.3 kb promoter was shown as sufficient to insure a strong and constitutive mRNA expression using a reporter gene (Han et al. 2015).

At the protein level, the relative high degree of conservation between TCTP proteins across kingdoms is reminiscent of its important role in development and survival of eukaryotes (Thayanithy 2005). For example, *Arabidopsis* AtTCTP protein shares 53.6%, 56%, and 62% amino acid similarity with human, *Drosophila* and yeast counterparts, respectively, and about 30% amino acid identity with human hTCTP (Fig. 7.1) (Hinojosa-Moya et al. 2013; Thayanithy 2005). Within the plant phylum, the majority of TCTP proteins are composed of 167 or 168 highly conserved amino acids that share 70–95% identity (Gutierrez-Galeano et al. 2014) (Fig. 7.1).

Sequence comparison showed that numerous domains in TCTP proteins are conserved in all eukaryotes (Thayanithy 2005; Hinojosa-Moya et al. 2008). Almost all identified TCTPs contain two TCTP signatures that are highly conserved and a basic domain for tubulin and calcium-binding (Figs. 7.1 and 7.2). The conserved putative GTPase interaction surface located in the central pocket indicates that TCTP proteins share GTPase binding property and GTPase activity regulating function (Cao et al. 2010; Li et al. 2013; Santa Brígida et al. 2014; Gachet et al. 1999) (Figs. 7.1 and 7.2). The N-terminal of TCTP contains a conserved MCL/Bcl-xL binding domain known to promote suppression of apoptosis in mammals (Yang et al. 2005). Like for animal TCTPs, the MCL/Bcl-xL

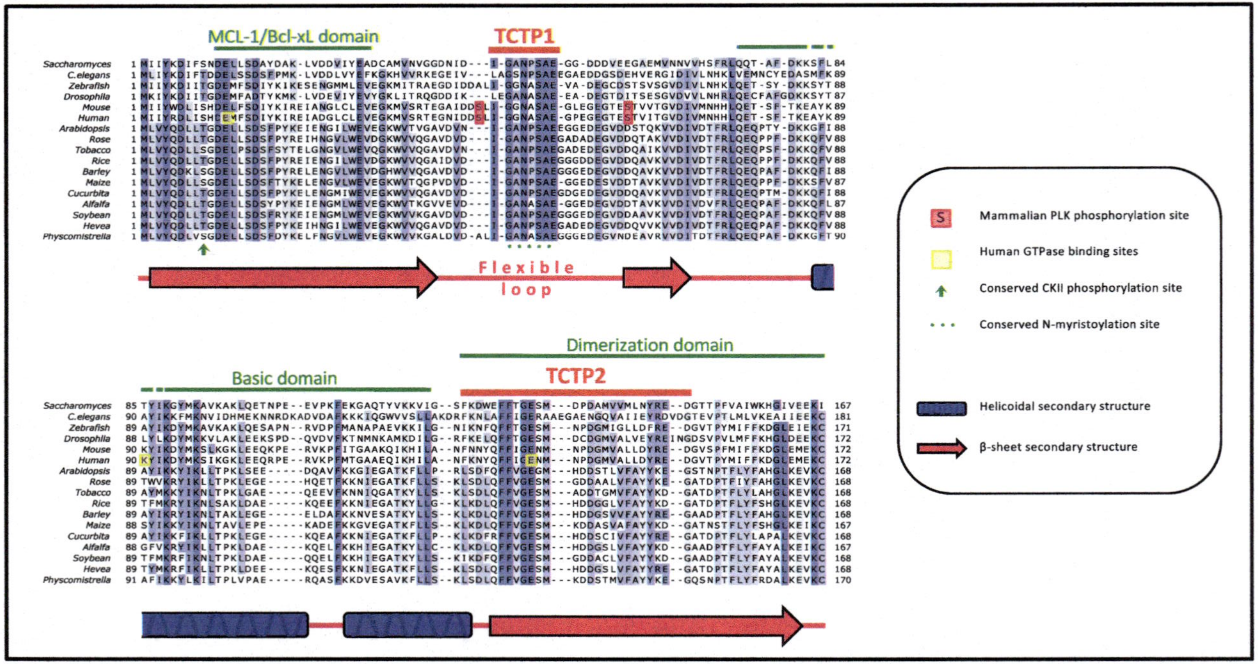

Fig. 7.1 Conservation of TCTP protein across kingdom. *Arabidopsis* AtTCTP protein shares 53.6%, 56%, and 62% amino acid similarity with human, drosophila, and yeast counterparts, respectively. Conserved TCTP signatures and binding domains are represented above amino acid sequences and secondary structure of AtTCTP at the bottom. Sequence alignments were performed using MuscleWS (Edgar 2004) software and edited with Jalview2.0 (Waterhouse et al. 2009). High conserved residues are *colored in blue* following BLOSUM 62 scoring matrix. TCTP sequences were retrieved from UniProtKB: P35691 (*Saccharomyces cerevisiae*), Q93573 (*Caenorhabditis elegans*), Q9DGK4 (Zebrafish), Q9VGS2 (*Drosophila melanogaster*), P63028 (*Mus musculus*), P13693 (human), P31265 (*Arabidopsis thaliana*), KX951492 (Rose), Q9XHL7 (Tobacco, *Nicotiana tabacum*), P35681 (Rice, *Oryza sativa*), Q9M5G3 (Barley, *Hordeum vulgare*), Q8H6A5 (Maize, *Zea mays*), Q2PS27 (*Cucurbita maxima*), P28014 (Alfalfa, *Medicago sativa*), Q9ZSW9 (*Hevea brasiliensis*), A9RT49 (*Physcomitrella patens*)

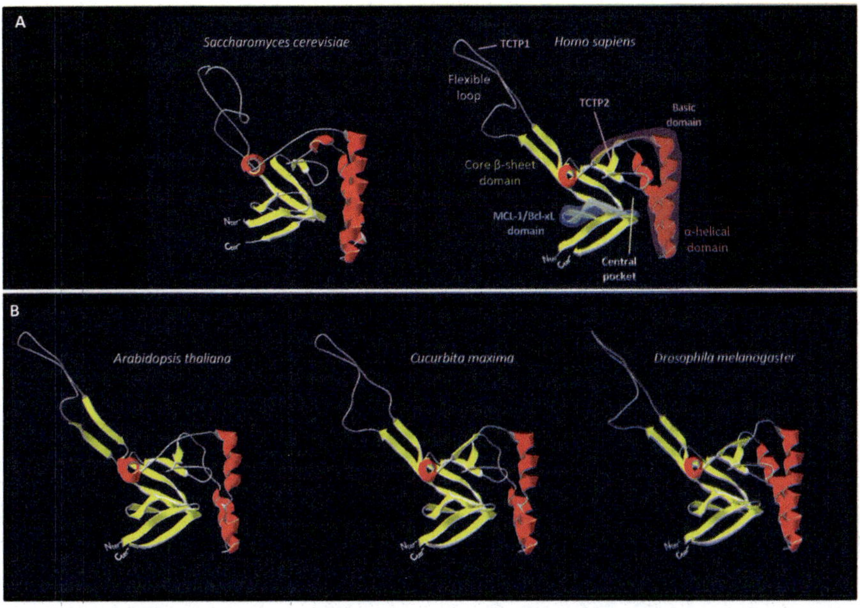

Fig. 7.2 Tridimensional structure of TCTP. (**a**) Computer graphic representation of three-dimensional structure of yeast and human TCTPs using PDB 1H6Q and 2hR9, respectively (Thaw et al. 2001; Susini et al. 2008). Ribbon representation of the alpha-carbon backbone of one TCTP molecule is shown. The N and C termini as well as the position of the various TCTP functional domains are shown on human TCTP structure. (**b**) Structure-based modeling of *Arabidopsis*, *Cucurbita*, and *Drosophila* TCTPs. 3D-structure models were developed using SWISS-MODEL (Biasini et al. 2014) server. Swiss-PDBViewer (Guex and Peitsch 1997) was used to visualized 3D structures. Note that the TCTP core domain is strongly conserved between species. The major alterations between the different TCTP structures lie in the flexible loops and in the length of the second alpha helix

domain of tobacco TCTP was demonstrated to be implicated in cell death suppression (Gupta et al. 2013). Moreover, plant and non-plant TCTPs contain conserved posttranslational modification sites such as the Casein Kinase II (CKII) phosphorylation site and the N-myristoylation site (Brioudes et al. 2010; Thayanithy 2005; Bruckner et al. 2016) (Fig. 7.1). These similarities suggest that plant and non-plant TCTPs likely harbor similar activities and may act in similar regulatory pathways. Despite the high degree of sequence homology between all eukaryotic TCTPs, there are slight differences suggesting some divergent functions of TCTP in plants and animals. For example, the Polo Like Kinase (PLK) phosphorylation site, previously shown to be functional in mouse (Yarm 2002), is conserved only in mammalian TCTPs and is absent in plant TCTPs (Fig. 7.1). The biological significance of these differences remains unclear, but its discovery may help unravel plant specific function(s) (Thayanithy 2005).

In agreement with the conserved TCTP primary and secondary structures, the predicted tri-dimensional structure of plant TCTP is very similar to yeast and human structures (Hinojosa-Moya et al. 2008; Feng et al. 2007; Thaw et al. 2001) (Fig. 7.2). Three distinct structural domains are found in TCTP: a core β-sheet domain, an α-helical domain, and a flexible loop structure. Major differences between plant and animal TCTPs are observed in the flexible loop, while the alpha-hairpin, which includes the basic domain known to be the interface for many interactions in animals, is well conserved (Bommer 2012; Gutierrez-Galeano et al. 2014; Berkowitz et al. 2008; Hinojosa-Moya et al. 2008). In plants, the predictive structures of TCTP support the phylogenetic evidence that they fall into two sub-clades, AtTCTP1-like and CmTCTP-like (*Cucurbita max.*), that differ in the structure of the central "pocket" region and in the flexible loop, suggesting different functions. It should be noted that in plant species harboring a single *TCTP* gene, the sequence is usually *AtTCTP1*-like (Gutierrez-Galeano et al. 2014).

Considering the high degree of amino-acids conservation and the high similarity of predictive tri-dimensional structure among phyla, it is tempting to suggest that plant and animal TCTPs share many of their roles. In agreement with this hypothesis, Brioudes et al. (2010) demonstrated in vivo that *Drosophila dTCTP* could complement cell proliferation defects associated with *AtTCTP* loss-of-function in *Arabidopsis* and vice versa.

In animals, TCTP was reported to interact with several proteins such as the tumor suppressor p53, BAX (Bcl2 Associated X protein), or MDM2 (Mouse Double Minute 2) (Amson et al. 2012). Interestingly, to date no orthologous proteins for these TCTP interactors could be found in plant genomes. It will be very interesting to know if nevertheless these interactions described in animals exist or not in plants. However, it has been reported that plant TCTP is able to interact with mammalian Bax protein (Hoepflinger et al. 2013). One could imagine that despite the obvious absence of primary amino acid similarities of TCTP interacting proteins, other functionally conserved proteins can fulfill their role and secure TCTP signaling in plants. Such information might give precious indications about evolution of TCTP function and interactors across kingdoms.

7.3 TCTP Is Essential for Plant Development

TCTP was demonstrated to have a major role in development and in organ size control in plants as in animals. Final organ shape and size are the result of coordinated cell proliferation, cell expansion, and cell death processes. These three processes have to be tightly controlled and coordinated in order to obtain organs and organisms of species-specific size and shape (Day and Lawrence 2000). As in animals, many reports proposed plant *TCTP* as good candidate to control cell proliferation and cell death. Compared to animals, the role of plant *TCTP* in development is much less understood. All reports in the literature show that plant *TCTP* is essential for the correct development and for the determination of final plant size. Like in animals, knockout of plant *TCTP* leads to embryo lethality (Brioudes et al. 2010). The fact that *TCTP* knockouts are lethal hampered the

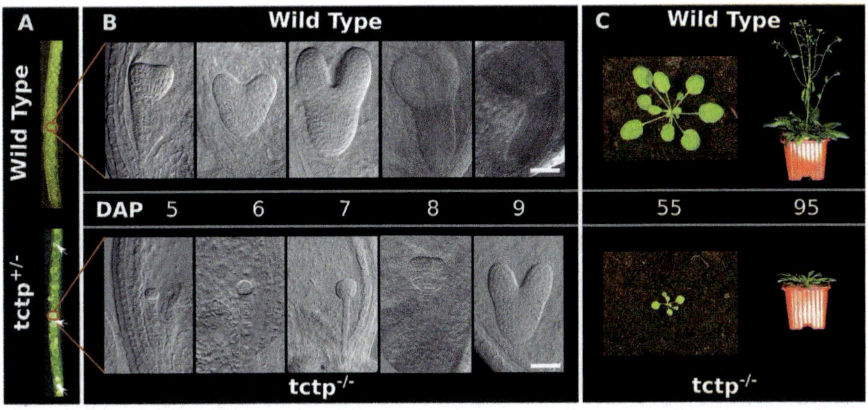

Fig. 7.3 Developmental phenotype associated with TCTP loss of function in *Arabidopsis thaliana*. (**a**) Fruit of *Arabidopsis* plants wild-type or heterozygous for *tctp* mutation (*tctp+/−*). In *tctp+/−* fruit, 25% of the seeds are homozygous for the *tctp* mutation (*tctp−/−*) leading to a white seed phenotype (*white arrows*). (**b**) Wild-type and *tctp−/−* embryos. *tctp−/−* embryos exhibit retarded growth compared to the wild-type. Six days after pollination **(DAP),** wild-type embryos are at heart stage while *tctp−/−* are still at the globular stage. Such delay in development is maintained in the subsequent developmental stages. Scale bar 100 μm. (**c**) Supplementing *tctp−/−* embryos with nutrient in vivo [embryo rescue, for detailed protocol see Brioudes et al. (2010)] allows them to continue their development and became adult plants full knockout for *TCTP* (*tctp−/−*). Note that *tctp−/−* plants are delayed in development compared to the wild-type

studies to address in detail *TCTP* function. To overcome this difficulty, *TCTP* knockdown, by mean of RNA interference approach, was used to explore TCTP roles during development. Although such approach led to significant reduction of *TCTP* expression, full obliteration of *TCTP* expression couldn't be achieved, thus making it difficult to address in detail TCTP function (Brioudes et al. 2010; Cao et al. 2010; Tao et al. 2015; Zhang et al. 2013; Berkowitz et al. 2008; Hsu et al. 2007; Hu et al. 2015).

Brioudes et al. (2010) used an embryo-rescue approach in *Arabidopsis thaliana* to generate the first *tctp−/−* full knockout adult organism in eukaryote (Fig. 7.3c). The authors supplemented *Arabidopsis tctp−/−* knockout embryos with nutrients *in vitro* that allowed their development to adult plants. However, the generated *tctp−/−* knockout plants were delayed in their development and showed severe growth defects, including small organs and plant size, late flowering, and sterility (Fig. 7.3). Nevertheless, the obtained *tctp−/−* knockout plants were very useful and helped dissect the multiple roles of *TCTP* in plant development (Brioudes et al. 2010).

Like in animals, TCTP knockout plants die early during embryo development (Brioudes et al. 2010) and the TCTP down-regulation delayed development and resulted in reduced organ size (Brioudes et al. 2010; Berkowitz et al. 2008). Even though it is generally accepted that *TCTP* is essential for organ and plant development, there are some discrepancies in observations and hypothesis, and a number of conclusions are still under debate. Brioudes et al. (2010) and Berkowitz et al. (2008)

showed a clear negative effect of *TCTP* downregulation on plant development and growth in *A. thaliana*. However, these studies do not agree on the causes of these phenotypes. Berkowitz et al. (2008) proposed that *tctp* lethality was associated with male gametophytic defects leading to failed fertilization and suggested a role of *TCTP* as a general stimulator of cell expansion. Conversely, Brioudes et al. (2010) demonstrated that fertilization took place in *tctp−/−* knockouts, and the lethality was a result of retarded growth of the developing embryos (Fig. 7.3). The fact that Brioudes et al. (2010) were able to rescue the *tctp−/−* knockout embryos to generate the first full knockout adult organism for *tctp* supports the conclusion that the *tctp* lethality is not associated with male gametophytic defect. In agreement with Brioudes et al. (2010), very recently Hafidh et al. (2016) showed that pollen competitiveness between *tctp* and wild type pollen is not different and confirmed that the fertilization between *tctp* pollen and a *tctp* ovule could occur. They also confirmed that the lethality is due to a delay in embryo growth in the early stages of development and thus in agreement with Brioudes et al. (2010) (Fig. 7.3b). Moreover, the authors demonstrated that like animal TCTPs, plant TCTP, even though lacking signal secretion peptide, can be secreted to plant apoplastic spaces *via* exosomes (Hafidh et al. 2016; Amzallag et al. 2004). This feature of TCTP might contribute to the signaling between pollen tube and pistil during fertilization and thus affect pollen tube guidance, ovule targeting, and seed development (Hafidh et al. 2016). Therefore, the data by Brioudes et al. (2010), recently confirmed by Hafidh et al. (2016), support the conclusion that the lethality of *tctp* loss-of-function mutants is a result of a retarded embryo growth that leads to embryo abortion in the developing siliques.

In *Arabidopsis*, the characterization of *tctp−/−* knockout (obtained via embryo rescue), *RNAi-TCTP*, and *TCTP* overexpressing lines through detailed kinematic analysis of leaf growth allowed to demonstrate that AtTCTP controls cell proliferation but not cell expansion (Brioudes et al. 2010). Similar data were recently reported for tobacco, cabbage, and tomato (Cao et al. 2010; Tao et al. 2015; Gupta et al. 2013; Bruckner et al. 2017). In these plants, *TCTP* downregulation leads to delayed plant development and smaller organs compared to wild type. Moreover, flowers were smaller and root growth was reduced (Tao et al. 2015; Bruckner et al. 2017). The authors suggest that at least part of these phenotypes can be explained by a reduction of cell proliferation activity in TCTP-RNAi lines as *TCTP* plays a positive role in plant growth regulation. In agreement with the role of *TCTP* in the control of cell proliferation, *Arabidopsis* AtTCTP protein was shown to accumulate in highly dividing cells (Brioudes et al. 2010). In *Pisum sativum*, mRNA was localized predominantly in dividing cells of root caps and in other rapidly growing tissues as young leaves and stems (Woo and Hawes 1997; Kang et al. 2003). Accordingly, TCTP protein accumulation was correlated with the accumulation of other cell proliferation proteins in the skin of young potato tubers, an actively dividing tissue (Barel and Ginzberg 2008).

To the best of our knowledge, only two studies showed the direct implication of *TCTP* in the control of cell proliferation. In a first study, Brioudes et al. (2010) used synchronized tobacco (*Nicotiana tabacum*) BY-2 cells knockdown for *NtTCTP* to demonstrate that *TCTP* regulates cell cycle progression (Brioudes et al. 2010). In this study, they showed a 4 h delay of cell cycle progression, and such delay affected more specifically G1/S transition (Brioudes et al. 2010). Recently, report by Tao et al. (2015) confirmed these results. Moreover, measurements of leaf size and cell number in *Arabidopsis* and tobacco plants knockdown for *TCTP* showed that delayed leaf growth and smaller leaf size were due to a decrease in the cell number but not in cell size. However, the precise molecular pathway by which TCTP controls cell proliferation in plants is still unknown. Tao et al. (2015) suggested that TCTP could prevent the polyubiquitination of NTHK1 (a Type 2 ethylene receptor) to control cell cycle.

The role of TCTP in controlling cell proliferation and mitotic growth is conserved between plants and animals. Brioudes et al. (2010) demonstrated that *Drosophila dTCTP* could fully complement cell proliferation defects associated with *TCTP* loss-of-function phenotypes in *Arabidopsis* and *vice versa*. However, this study also showed that loss-of-function of *Drosophila dTCTP* also leads to defects in cell expansion, although such defect is not observed in *tctp−/−* knockout plants. In agreement with these data, *AtTCTP* could not complement the cell expansion defect in *Drosophila* mutants. These interspecies complementation experiments highlighted the conserved role of TCTP in controlling cell proliferation and demonstrated that conversely to plant *TCTP*, *Drosophila TCTP* also controls cell expansion.

In animals, TCTP was shown to have an anti-apoptotic role (Susini et al. 2008). This role was also investigated in plants. As for animals, cell death process occurs in plants with genetically and environmentally defined temporal and spatial patterns and is absolutely required for normal plant development (Greenberg 1996). Lliso et al. (2007) showed that TCTP protein accumulation decreased during postharvest aging process in citrus fruits and suggested an anti-apoptotic activity for TCTP in relation with microtubule stabilization. In agreement with these data, Hoepflinger et al. (2013) reported that constitutive expression of *AtTCTP* prevents the apoptotic effect of programmed cell death (PCD)-inducing agent tunicamycin on tobacco leaf disc. The authors then proposed that, as in animals, plant TCTP could act as a cytosolic Ca^{2+} sequester to protect cells against Ca^{2+}-dependent PCD (Graidist et al. 2007).

Cell death process is also induced during an incompatible interaction between plants and pathogens in order to limit pathogen spread and disease development. This process, also termed hypersensitive response (HR), is used as a model to study cell death in plants (Morel and Dangl 1997). In tobacco, it was demonstrated that the downregulation of *NtTCTP* promotes HR and the constitutive transient expression of *NtTCTP* decreases the HR rate, thus another argument to support the anti-apoptotic activity of TCTP in plants (Gupta et al. 2013). In this study, the authors also demonstrated that TCTP inhibits the Reactive Oxygen Species (ROS) production and the MAPK (Mitogen Activating Protein Kinase) cascade observed in HR. Taking together, these published work show that like in animals, plant TCTPs very likely have an anti-apoptotic role. However, the mechanism by which

TCTP prevents PCD is still unknown as counterparts of TCTP interacting mammalian proteins are missing in plants.

7.4 Role of TCTP in Plant Signaling

As discussed above, *TCTP* is absolutely required for plant development and organ size determination. In *Arabidopsis* and tobacco, the downregulation of *TCTP* leads to severe developmental defects that are at least in part due to perturbation of cell proliferation, but many other processes are affected. A number of published work associates TCTP to other cellular functions and signaling molecules/pathways. In the next paragraph, we will provide the up-to-date information on the putative links between TCTP and these cellular functions and signaling molecules/pathways.

7.4.1 Is TCTP a Component of the TOR Pathway?

Hsu et al. (2007) reported that *Drosophila dTCTP* controls organ growth by positively regulating the TARGET OF RAPAMYCIN (TOR) pathway. TOR kinase is part of a signaling complex that controls cell proliferation and growth in animals and in plants, in response to environmental conditions, growth factors (*e.g*., insulin), nutrients, energy, or stress (Robaglia et al. 2004; Deprost et al. 2007; Wullschleger et al. 2006; Oldham et al. 2000; Zhang et al. 2000). In mammals, the TOR pathway (mTOR) controls cell growth by acting as a central regulator of protein synthesis and ribosome biogenesis at the transcriptional and translational levels. This is performed by integrating signals from mitogens and nutrients to downsream signaling pathways (Wullschleger et al. 2006). mTOR activity is positively controlled by the small Ras GTPase, Rheb, that binds directly to mTOR kinase domain in order to activate the mTOR complex in a GTP-dependent manner (Wullschleger et al. 2006). The Tuberous Sclerosis Complex (TSC) negatively regulates TOR pathway by inactivating Rheb through a GTPase-activating protein (GAP) activity.

Based on epistatic analysis, Hsu et al. (2007) showed that in *Drosophila melanogaster*, *dTCTP* acts in parallel to *TSC*, but upstream of *dRheb*. Using GST pull-down and in vivo GDP release assays, they demonstrated that dTCTP binds to nucleotide-free dRheb and specifically displays Guanine nucleotide exchange factor (GEF) activity on it. In this model, dTCTP directly associates with dRheb and is required for its activation in vivo, which in turn positively controls TOR activity; thus, TCTP may have an opposite function to TSC. Mouse TCTP was also reported to function as GEF upstream of S6K (Chen et al. 2007) and human TCTP was demonstrated to activate the mTOR pathway in vivo (Dong et al. 2009). In plants, TOR is highly conserved and its expression positively correlates with growth (Deprost et al. 2007; Wullschleger et al. 2006). Homologous members and substrates of the mammalian mTORC1 complex have been identified in plants, such as

TOR, LST8, S6 kinase, and RAPTOR (Menand et al. 2004; Mahfouz 2006), but the presence of TORC2 components has yet to be proven. In plants, the TOR pathway is conserved and has been shown to be important for plant growth, development, flowering, senescence, and life span by modulating transcription, translation, autophagy, and primary and secondary metabolism (Deprost et al. 2007; Ahn et al. 2011; Ren et al. 2011, 2012; Moreau et al. 2012; Xiong and Sheen 2012; Xiong et al. 2013; Caldana et al. 2013).

There are only indirect indications about the putative link between TCTP and TOR pathway in plants. BiFC (Bi-Fluorscent Complementation) and GST pull-down experiments showed that *Arabidopsis* TCTP is able to bind plant Rab GTPases and also *Drosophila* Rheb, and similarly, *Drosophila* TCTP can bind *Arabidopsis* Rab (Brioudes et al. 2010). However, no data were reported regarding a putative GEF activity of plant TCTP. Similarly, no genetic studies are available to confirm that TCTP acts as a regulator of the TOR pathway. However, knockout or knockdown of *TCTP* or *TOR* in plants exhibit similar phenotypes. For both genes, knockouts are lethal and knockdown display prolonged reduction in cell proliferation and reduced growth (Brioudes et al. 2010; Deprost et al. 2007). Another indication of the putative link between TCTP and TOR in plants is the fact that transcript level of EBP1, a TOR downstream component, is significantly decreased in *TOR* and in *TCTP* knockdown plants, suggesting a common pathway (Berkowitz et al. 2008; Deprost et al. 2007).

It should be noted that in animals, the putative GEF activity of TCTP also remains a matter of debate, and the molecular function of TCTP as an upstream activator of Rheb has been questioned. Rehmann et al. (2008) addressed the putative interaction between TCTP and Rheb and compared the Rheb activation and S6K or S6 phosphorylation status in presence of TCTP and in mammalian cells knockdown for TCTP. Conversely to Hsu et al. (2007), they didn't observe any difference on Rheb and its downstream targets (Rehmann et al. 2008). Similarly, overexpression of *hTCTP* had no significant effect on the phosphorylation state of S6, a substrate of S6K, in stressed human cell lines (Wang et al. 2008). Furthermore, NMR spectroscopy failed to confirm any interaction between TCTP and Rheb (Rehmann et al. 2008). These data do not support the idea that TCTP is an upstream activator of mammalian TOR signaling. It is probable that TCTP acts on Rheb but not as a GEF. Cans et al. (2003) reported that human TCTP was able to interact with a GTP-binding protein (eEF1A) and its GEF (eEF1B) and to act as a guanine nucleotide dissociation inhibitor (GDI). The atomic structure of *Schizosaccharomyces pombe* and human TCTP proteins revealed that TCTP is structurally related to the Mss4/Dss4 protein family (Thaw et al. 2001; Susini et al. 2008). In human, Mss4 has been shown to bind to several exocytic but not endocytic Rab GTPases in vivo and in vivo, and analysis of its catalytic activity towards them confirmed its function as a relatively inefficient GEF, thus supporting the idea that it would most likely function as a guanine nucleotide-free chaperone (GFC) (Nuoffer et al. 1997; Wixler et al. 2011). Moreover, Gnanasekar et al. (2009) identified heat shock protein function and chaperone-like activities of human TCTP and a TCTP homolog from *Schistosoma mansoni*. Given that TCTP is structurally similar to the Mss4/Dss4 family (Thaw et al. 2001) and that it can interact with many cellular proteins involved in cell growth or survival control, it was tempting to propose animal TCTP as a protein chaperone. Although it is still not clear whether TCTP is a GEF, a GDI, or a GFC, the studies above revealed TCTP as a regulator of GTPases activities and associated it with the control of translation. Whether TCTP acts as a GEF or not on Rheb proteins remains to be clarified. There are reported studies that defend both hypotheses. Clearly, more studies are required in order to decipher the relationship between TCTP and TOR pathway in animals as in plants.

7.4.2 Role of TCTP in Hormone Signaling

Very little information exists on the putative roles of *TCTP* in hormonal signaling in plants (Fig. 7.4). Only few studies investigated changes in *TCTP* expression in response to treatment with phytohormones, such as auxin, abscisic acid (ABA), ethylene, or methyl jasmonate (MeJA) (Cao et al. 2010; Li et al. 2013; Kim et al. 2012). Most of the published work simply provided hypothesis based on observations, and sometimes results are contradictory.

In plants, auxin is an essential hormone involved in a wide variety of functions such as cell division, organogenesis, senescence, apical dominance, gravitropism, root growth, etc. (Vanneste and Friml 2009). Berkowitz et al. (2008) showed that *Arabidopsis* TCTP-RNAi plants were less sensitive to increasing concentrations of exogenous auxin compared to the wild-type and proposed an implication of TCTP in the auxin signaling pathway. These data are surprising since auxin signaling mutants often show severe and aberrant phenotypes that are different from that observed in TCTP-RNAi (Estelle and Somerville 1987; Křeček et al. 2009). In cabbage, TCTP-RNAi plants grow slower and accumulate low auxin contents compared to the wild-type (Cao et al. 2010). The authors suggested that such decrease of endogenous auxin content could be responsible of the observed delayed growth. However, this remains highly hypothetical as no direct proof has been provided by the authors to support such hypothesis. Clearly, more studies are required to confirm the putative role of TCTP in auxin signaling pathway.

Several studies reported interaction of TCTP with the abscisic acid (ABA) signaling pathway. ABA is a plant hormone known to be implicated in seed germination and response to drought. It has been suggested in *Arabidopsis* that AtTCTP interacts with ABA signaling pathway to control stomatal closure mechanism in response to drought (Kim et al. 2012). Stomata are differentiated leaf cells that tightly control gas and water exchanges between plant leaves and environment. During the night or in drought condition, plants synthesize ABA to control stomatal closure, avoiding water evaporation (Venkatachalam et al. 2007). Kim et al. (2012) reported that overexpression of *Arabidopsis AtTCTP* confers drought tolerance by rapid ABA-mediated stomatal closure. The authors also proposed that this rapid ABA-mediated stomatal closure is mediated by the interaction of *AtTCTP* with

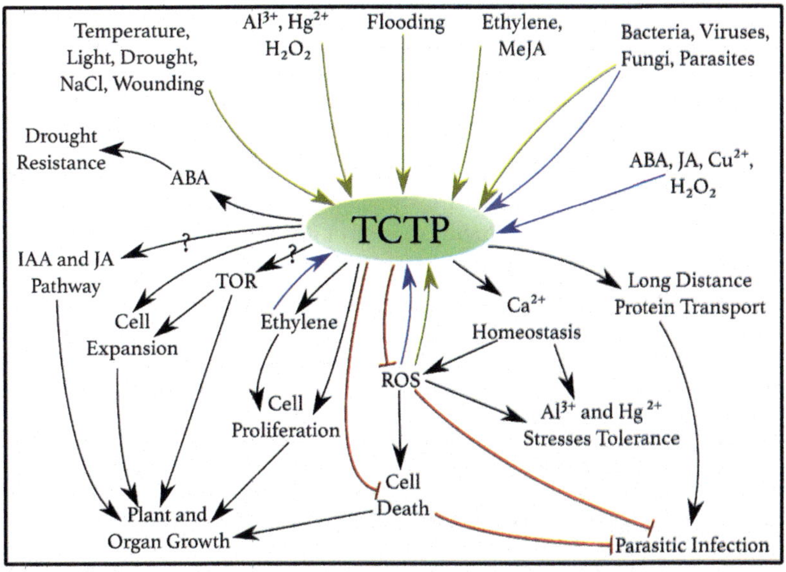

Fig. 7.4 Schematic representation of the multiple stimuli affecting *TCTP* expression (*green arrows*) or TCTP protein accumulation (*blue arrows*) and of the various putative biological processes associated with TCTP function (*red* and *black arrows*) in plants

microtubules during water stress leading to microtubules depolymerization, stomata closure, and drought resistance. Moreover, this interaction between TCTP and tubulin is increased by calcium, a microtubule depolymerization factor. As the expression domain of *TCTP* is increased following ABA-treatments and because putative ABA-responsive elements were identified in the *AtTCTP* promoter, it is possible that there is a feedback loop between ABA and *TCTP* expression that regulates stomatal closure in leaves. Unfortunately, the data by Kim et al. (2012) are based on *TCTP* overexpression experiments and on in vivo data only, and direct link between ABA and TCTP for stomatal closure is not shown. To date, no other study links stomatal behavior in response to ABA and *TCTP*.

In cabbage, knockdown of *TCTP* is associated with higher levels of ABA (Cao et al. 2010) and thus an inverse response to that reported by Kim et al. (2012). In rice, OsTCTP protein accumulation is induced in response to ABA, while in cabbage no difference in *TCTP* expression was observed under ABA stress (Cao et al. 2010; Wang et al. 2015). In Orchardgrass (*Dactylis glomerata*), *TCTP* has been demonstrated to be one of the most stable genes in response to ABA (Huang et al. 2014). All these data suggest that there is a link between the ABA pathway and TCTP, but the inconsistency of the results is evident and therefore more studies are required to consolidate these observations.

Ethylene is an important phytohormone for plant growth and development, and it is associated with fruit ripening. Tao et al. (2015) reported that tobacco NtTCTP protein accumulation increases in response to ethylene treatments. NtTCTP then interacts with the ethylene receptor NTHK1, and such interaction is enhanced by ethylene and protects NTHK1 from degradation by the 26S proteasome and promotes seedling growth. Some studies also report a change in *TCTP* expression upon Jasmonic acid (JA) treatment. JA is a lipid-derived plant hormone that regulates a wide range of processes, such as biotic and abiotic plant stress responses (insects, pathogens, wounding…) and developmental processes (root growth, senescence…) (Abe et al. 2008; Devoto and Turner 2003; Browse 2005). In *Hevea brasiliensis*, *HbTCTP* expression is strongly reduced in response to MeJA treatment (Li et al. 2013). In cotton, overexpression of *TCTP* leads to important decrease of expression of genes in the JA pathway (Zhang et al. 2014). Conversely, Wang et al. (2015) reported that OsTCTP protein accumulation is not affected by MeJA in *Oryza sativa*. These results are contradictory, and further experiments are required to address more precisely the putative role of *TCTP* in JA signaling.

To summarize, we have little knowledge about the link between *TCTP* and phytohormone signaling, and clearly, more studies are needed to determine the role of *TCTP* in phytohormone signaling during plant growth and development.

7.4.3 Response to Abiotic and Biotic Stresses

Conversely to animals, plants have sessile lifestyle and they have to cope with and adapt to their environment. In order to grow, plants need light, water, and nutrients. The availability of those elements directly impact their development, morphology, and size. While some reports show that the expression of *Arabidopsis AtTCTP* and tomato *SlTCTP* mRNAs is highly stable in different growth conditions (Brioudes et al. 2010; Coker and Davies 2003), several other studies reported variations of *TCTP* mRNA expression and/or protein accumulation in response to environmental stresses (Fig. 7.4). In plants, *TCTP* expression was reported to vary in response to salt stress, high temperature, drought, and pathogen attacks (Fig. 7.4). In this section, we will provide and discuss the up-to-date knowledge on the putative roles of TCTP in response to environmental signals and stresses.

Different studies suggest that TCTP may have a role in tolerance to drought and heat stresses. In *Jatropha curcas* and in cabbage, the expression of *TCTP* was shown to be induced following heat shock (Cao et al. 2010; Qin et al. 2011). In *Hevea brasiliensis*, *HbTCTP1* transcripts were rapidly increased after drought treatment and then returned to normal level (Li et al. 2013). However, an inverse behavior was observed in grapevine, where *TCTP* expression was repressed under water deficit stress (Vincent et al. 2007). Kim et al. (2012) showed that in *Arabidopsis*, plants overexpressing *AtTCTP* were more tolerant to drought due to reduced water loss. As described above, AtTCTP interacts with microtubules in guard cells, which leads to microtubule depolymerization and rapid ABA-mediated stomatal closure and reduced water loss (Kim et al. 2012). The stomatal closure induced by ABA is one of the major adaptive responses to drought stress (Zhang et al. 2006). The binding of plant TCTP to tubulins seems calcium dependent. Calcium is a known factor in microtubule depolymerization (Kim et al. 2012; O'Brien et al. 1997; Yu et al. 2001). It was reported that animal TCTP also associates transiently with microtubules during cell cycle and binds calcium (Gachet et al. 1999; Kim et al. 2000). Visibly, plant and animal TCTPs bind to and associate with microtubules and calcium to perform different roles, although more data are required to address how conserved these roles are between plants and animals and what are the underlying molecular mechanisms.

Altogether, it seems that heat shock and drought rapidly induce *TCTP* expression, and high level of TCTP might be required to resist such stress. However, the reported data remain as observation, and no data on the molecular and genetic mechanisms that might be involved are available.

In many plant species, *TCTP* mRNA accumulation was also found to be stimulated by a number of other abiotic stress signals such as aluminum, mercury, Cu^{2+}, H_2O_2, high salt, ethrel, wounding, and MeJA (Sage-Ono et al. 1998; Li et al. 2013; Santa Brígida et al. 2014; Wang et al. 2012, 2015; Ermolayev 2003) (Fig. 7.4). *TCTP* expression was shown to be induced by aluminum treatment in an aluminum-tolerant soybean cultivar, but not in an aluminum-sensitive cultivar, suggesting a putative role of TCTP in aluminum homeostasis maintanance (Ermolayev 2003). More recently, it was reported that rice *OsTCTP* expression was induced in response to H_2O_2 and to high levels of mercury in soil and that *OsTCTP* expression enhances tolerance to mercury in rice *via* a decrease of the Hg-induced reactive oxygen species (ROS) (Wang et al. 2012, 2015). In rubber tree (*Hevea brasiliensis),* transient changes in *HbTCTP1* transcript levels were also observed after diverse abiotic treatments as high salt, ethrel, and H_2O_2 (Li et al. 2013). Moreover, *TCTP* expression is reduced in this species during wounding and Tapping Panel Dryness (TPD) syndrome, which is the result of repeated wounding (Li et al. 2013; Venkatachalam et al. 2007). This *TCTP* down-regulation is preceded by ROS outburst that induces hypersensitive response (HR) indicating that programmed cell death due to repeated wounding might be the cause of TPD (Li et al. 2013).

Although these data remain as observations with no clear data to support the role of *TCTP* in response to stresses, many of these abiotic stresses are known to lead to the production of ROS. In animals as in plants, ROS are natural by-products of the normal metabolism and have important roles in cell signaling and homeostasis (Kardeh et al. 2014; Russell and Cotter 2015; Li and Yu 2015; del Rio 2015; Choudhury et al. 2013). ROS levels are known to increase dramatically in response to various environmental stresses, which can result in significant damage to cells and tissues (Møller et al. 2007; Xia et al. 2015). In animals, ROS also play a role in tissue protection and defense against cancer by controlling cell proliferation through apoptosis induction (Kardeh et al. 2014).

In plants, the majority of published data suggest that ROS either induce *TCTP* expression and promote stress tolerance and adaptation or inhibit *TCTP* expression in parallel with PCD induction. Some authors also suggested that TCTP could control the adaptive response to abiotic stresses *via* its Ca^{2+} bindings ability (Li et al. 2013; Gupta et al. 2013; Hoepflinger et al. 2013). However, part of the data remains contradictory. To reconcile these contradictory published data, we can imagine that despite the amino acid sequence conservation, TCTP proteins from different species may have different roles in response to abiotic stresses. Further investigations are required to address the discrepancies in the literature and more importantly to shed light to the molecular mechanisms underlying the role of TCTP in abiotic stress response.

As mentioned above, TCTP is an anti-apoptotic protein associated with plant hypersensitive response (HR) during incompatible interaction with many pathogens. HR is one of the best-characterized defense responses in plants that occurs during incompatible host–pathogen interaction. It is characterized by rapid, localized PCD to restrict the spread of pathogens and protect the plant. Because HR is one of the major defense mechanisms against pathogen attacks (Morel and Dangl 1997), TCTP is assumed to play a role in plants defense. Wide-genome as well as proteomics analysis show that both *TCTP* transcript and protein were differentially expressed and accumulated in response to attacks by viruses, bacteria, and fungi.

Analysis of proteomic changes in *Arabidopsis* infected by *Pseudomonas syringae* demonstrated that infection with an avirulent strain or induction of HR by bacteria correlates with *TCTP* downregulation, and conversely, in the case of established bacterial infection, *TCTP* expression is upregulated (Jones et al. 2006; Fabro et al. 2008). Successful infection by *Agrobacterium tumefaciens* that requires suppression of plant defense mechanism, also correlates with increased *TCTP* expression (Veena et al. 2003). In agreement with these reports, *NbTCTP* silencing in tobacco accelerates HR following infection by various bacteria, while *TCTP* overexpression diminished HR response to bacterial infection (Gupta et al. 2013). *NtTCTP* seems to negatively regulate HR cell death by acting on the MAPK-regulated cell death pathway. Moreover, expression of the BAX protein in *NtTCTP* downregulated plants accelerated the HR phenotype. BAX is a mammalian pro-apoptotic protein, which in mammals antagonizes TCTP effect. In this study, the authors clearly demonstrate the anti-apoptotic effect of plant TCTP (Gupta et al. 2013).

TCTP has been proposed to play a major role during plant infection by potyvirus. In a first study, *TCTP* was identified among the most strongly induced genes after potyvirus infection in tomato (Alfenas-Zerbini et al. 2009). In a more recent study, it was demonstrated that TCTP is an important host factor for an efficient infection by potyviruses in tomato and *Nicotiana benthamiana* plants. Plants silenced for *TCTP* exhibited reduced accumulation of PepYMV early during infection (Bruckner et al. 2017). Furthermore, TCTP was observed in both nuclei and cytoplasm of non-infected cells, while only in the cytoplasm of infected cells, which suggest that the virus may alter TCTP localization to promote viral infection. However, the mechanism by which TCTP promote successful infection by viruses is not yet clear. In animals, the highly structured *TCTP* mRNA was shown to activate the dsRNA-dependent protein kinase PKR (Bommer et al. 2002), thus suggesting antiviral responses. In plants, dsRNA binding is used by plant viruses to suppress RNA silencing (Merai et al. 2006). It is possible that viruses induce *TCTP* expression to deactivate dsRNA-induced plant defense response in order to allow efficient plant infection. Alternatively, TCTP could be associated with viral movement and selective loading of various viral components at specific cell boundaries during virus infection (Zhu 2002; Voinnet et al. 1998; Itaya et al. 2002). In pumpkin, *CmTCTP* was shown to associate with phloem RNA binding proteins CmPP16-1 and CmPP16-2 (Aoki et al. 2005), and this complex was demonstrated to move selectively in pumpkin phloem. In *Ricinus communis,* TCTP protein was also found in the vascular system phloem sap that carries organic nutrients or photosynthate (Barnes et al. 2004). The authors suggest that TCTP might be involved in the selective transport and/or unload of macromolecules such as viral components by the phloem, for long-distance movement in the sieve tube.

The role of *TCTP* seems to be not solely restricted to response to bacteria and viruses. In *Arabidopsis*, infection with obligate biotroph fungi induces *TCTP* expression, and conversely *TCTP* expression is negatively regulated by the SA and JA-mediated host defense pathway (Fabro et al. 2008). Obligate biotroph fungi must establish compatible interactions with their hosts to survive, and TCTP could be necessary for a successful fungal infection (Fabro et al. 2008). However, it is still unclear how these parasites are able to avoid plant defense activation (Panstruga 2003).

Similar to animal parasitic nematodes, TCTP from two plant parasitic nematodes (*Meloidogyne incognita* and *M. enterolobii*) was found among the proteins secreted during host plant invasion and in agreement with such observation, nematode infection was promoted by overexpressing *Meloidogyne MeTCTP in planta* (Bellafiore et al. 2008; Zhuo et al. 2017). In *M. incognita*, TCTP is localized in the sub-ventral gland while in *M. enterolobii* it was found in the dorsal gland, indicating that the site of TCTP secretion might be species dependent. Similar to observations during bacterial infection, nematode TCTP could suppress BAX-induced programmed cell death. However, the authors did not show direct evidence of MeTCTP secretion into host organism and did not quantify the changes in host *TCTP* expression/accumulation upon nematode infection (Zhuo et al. 2017). All together these published data indicate that nematode TCTP might be a plant-parasitic effector excreted into the host plants to promote parasitism. However, how nematode TCTP accomplishes this role remains unclear. Analysis of other proteins secreted by *M. incognita* indicates that nematode infection can interfere with plant ROS signaling to suppress cell death. Analysis of *M. incognita* secretome demonstrated the presence of detoxification enzymes that may be able to degrade ROSs (Bellafiore et al. 2008). In cotton plant, Zhang et al. (2014) show that TCTP was downregulated upon aphid attack and that cotton *TCTP* overexpression in *Arabidopsis* leads to a reduction of the symptoms induced by the parasite.

The published data suggest that TCTP likely plays a role in response to various pathogens. It is possible that TCTP is acting to circumvent the plant defense in order to establish compatible pathogen infection probably through diverse processes, such as preventing cell death and/or allowing the selective transport of macromolecules.

7.5 Conclusion

In plants as in animals, TCTP is involved and likely plays major roles in many cellular processes (Fig. 7.4), such as cell proliferation and cell death control or response to abiotic and biotic stresses. The precise molecular mechanisms associated with many of these processes in plants are still unknown. In animals, TCTP is involved in diverse cancers and is thought to be one of the best therapeutic targets to fight against the disease (Acunzo et al. 2014). In plants, tumorogenesis occurs but rarely induces physiological disorder (Doonan and Sablowski 2010). It is clear in many published work that *TCTP* has conserved functions in plants and in animals. However, to the best of our knowledge, only one report addressed the TCTP functional conservation between plants and animals at the molecular level (Brioudes et al. 2010). Future directions must take into account animal and plant model organisms to investigate at the molecular, genetic and biochemical levels the mechanisms by which TCTP act in these various processes. Such work may help in understanding some development disorders associated with TCTP misexpression in mammals and in plants.

Acknowledgements This work was funded by the "Biologie et Amélioration des Plantes" Department of the French "Institut National de la Recherche Agronomique", by The "Ecole Normale Supérieure de Lyon", by the Claude Bernard University at Lyon (UCBL), and by the CIFRE program of the ANRT.

References

Abe H et al (2008) Function of jasmonate in response and tolerance of Arabidopsis to thrip feeding. Plant Cell Physiol 49:68–80

Acunzo J, Baylot V, So A, Rocchi P (2014) TCTP as therapeutic target in cancers. Cancer Treat Rev 40:760–769

Ahn CS, Han J-A, Lee H-S, Lee S, Pai H-S (2011) The PP2A regulatory subunit Tap46, a component of the TOR signaling pathway, modulates growth and metabolism in plants. Plant Cell 23:185–209

Alfenas-Zerbini P et al (2009) Genome-wide analysis of differentially expressed genes during the early stages of tomato infection by a potyvirus. Mol Plant-Microbe Interact 22:352–361

Amson R et al (2012) Reciprocal repression between P53 and TCTP. Nat Med 18:91–99

Amzallag N et al (2004) TSAP6 facilitates the secretion of translationally controlled tumor protein/histamine-releasing factor via a nonclassical pathway. J Biol Chem 279:46104–46112

Anastasiou E, Lenhard M (2007) Growing up to one's standard. Curr Opin Plant Biol 10:63–69

Aoki K et al (2005) Destination-selective long-distance movement of phloem proteins. Plant Cell 17:1801–1814

Arya R, White K (2015) Cell death in development: signaling pathways and core mechanisms. Semin Cell Dev Biol 39:12–19

Barel G, Ginzberg I (2008) Potato skin proteome is enriched with plant defence components. J Exp Bot 59:3347–3357

Barnes A et al (2004) Determining protein identity from sieve element sap in Ricinus communis L. by quadrupole time of flight (Q-TOF) mass spectrometry. J Exp Bot 55:1473–1481

Barreau C, Paillard L, Osborne HB (2005) AU-rich elements and associated factors: are there unifying principles? Nucleic Acids Res 33:7138–7150

Bellafiore S et al (2008) Direct identification of the Meloidogyne incognita secretome reveals proteins with host cell reprogramming potential. PLoS Pathog 4:e1000192

Berkowitz O, Jost R, Pollmann S, Masle J (2008) Characterization of TCTP, the translationally controlled tumor protein, from Arabidopsis thaliana. Plant Cell Online 20:3430–3447

Biasini M et al (2014) SWISS-MODEL: modelling protein tertiary and quaternary structure using evolutionary information. Nucleic Acids Res 42:W252–W258

Bögre L, Magyar Z, López-Juez E (2008) New clues to organ size control in plants. Genome Biol 9:226

Bommer UA (2012) Cellular function and regulation of the translationally controlled tumour protein TCTP. Open Allergy J 5:19–32

Bommer UA, Thiele BJ (2004) The translationally controlled tumour protein (TCTP). Int J Biochem Cell Biol 36:379–385

Bommer UA et al (2002) The mRNA of the translationally controlled tumor protein P23/TCTP is a highly structured RNA, which activates the dsRNA-dependent protein kinase PKR. RNA 8:478–496

Brioudes F, Thierry A-M, Chambrier P, Mollereau B, Bendahmane M (2010) Translationally controlled tumor protein is a conserved mitotic growth integrator in animals and plants. Proc Natl Acad Sci 107:16384–16389

Browse J (2005) Jasmonate: an oxylipin signal with many roles in plants. Vitam Horm 72:431–456

Bruckner FP et al (2017) Translationally controlled tumour protein (TCTP) from tomato and Nicotiana benthamiana is necessary for successful infection by a potyvirus: TCTP is a host factor for potyvirus infection. Mol Plant Pathol 18:672–683

Busov VB, Brunner AM, Strauss SH (2008) Genes for control of plant stature and form. New Phytol 177:589–607

Caldana C et al (2013) Systemic analysis of inducible target of rapamycin mutants reveal a general metabolic switch controlling growth in Arabidopsis thaliana. Plant J 73:897–909

Cans C et al (2003) Translationally controlled tumor protein acts as a guanine nucleotide dissociation inhibitor on the translation elongation factor eEF1A. Proc Natl Acad Sci USA 100:13892–13897

Cao B, Lu Y, Chen G, Lei J (2010) Functional characterization of the translationally controlled tumor protein (TCTP) gene associated with growth and defense response in cabbage. Plant Cell Tissue Organ Cult 103:217–226

Chen SH et al (2007) A knockout mouse approach reveals that TCTP functions as an essential factor for cell proliferation and survival in a tissueor cell type–specific manner. Mol Biol Cell 18:2525–2532

Chen W et al (2013) Tumor protein translationally controlled 1 is a p53 target gene that promotes cell survival. Cell Cycle 12:2321–2328

Choudhury S, Panda P, Sahoo L, Panda SK (2013) Reactive oxygen species signaling in plants under abiotic stress. Plant Signal Behav 8:e23681

Coker JS, Davies E (2003) Selection of candidate housekeeping controls in tomato plants using EST data. Biotechniques 35:740–749

Cook M, Tyers M (2007) Size control goes global. Curr Opin Biotechnol 18:341–350

Crickmore MA, Mann RS (2008) The control of size in animals: insights from selector genes. Bioessays 30:843–853

Day SJ, Lawrence PA (2000) Measuring dimensions: the regulation of size and shape. Development 127:2977–2987

del Rio LA (2015) ROS and RNS in plant physiology: an overview. J Exp Bot 66:2827–2837

Deprost D et al (2007) The Arabidopsis TOR kinase links plant growth, yield, stress resistance and mRNA translation. EMBO Rep 8:864–870

Devoto A, Turner JG (2003) Regulation of jasmonate-mediated plant responses in Arabidopsis. Ann Bot 92:329–337

Dong X, Yang B, Li Y, Zhong C, Ding J (2009) Molecular basis of the acceleration of the GDP-GTP exchange of human ras homolog enriched in brain by human translationally controlled tumor protein. J Biol Chem 284:23754–23764

Doonan JH, Sablowski R (2010) Walls around tumours—why plants do not develop cancer. Nat Rev Cancer 10:794–802

Edgar RC (2004) MUSCLE: multiple sequence alignment with high accuracy and high throughput. Nucleic Acids Res 32:1792–1797

Ermolayev V (2003) Comparison of Al-induced gene expression in sensitive and tolerant soybean cultivars. J Exp Bot 54:2745–2756

Estelle MA, Somerville C (1987) Auxin-resistant mutants of Arabidopsis thaliana with an altered morphology. Mol Gen Genet 206:200–206

Fabro G et al (2008) Genome-wide expression profiling Arabidopsis at the stage of Golovinomyces cichoracearum haustorium formation. Plant Physiol 146:1421–1439

Feng Y, Liu D, Yao H, Wang J (2007) Solution structure and mapping of a very weak calcium-binding site of human translationally controlled tumor protein by NMR. Arch Biochem Biophys 467:48–57

Gachet Y et al (1999) The growth-related, translationally controlled protein P23 has properties of a tubulin binding protein and associates transiently with microtubules during the cell cycle. J Cell Sci 112:1257–1271

Gnanasekar M, Thirugnanam S, Zheng G, Chen A, Ramaswamy K (2009) Gene silencing of translationally controlled tumor protein (TCTP) by siRNA inhibits cell growth and induces apoptosis of human prostate cancer cells. Int J Oncol 34:1241–1246

Graidist P et al (2007) Fortilin binds Ca2+ and blocks Ca2+-dependent apoptosis in vivo. Biochem J 408:181–191

Greenberg JT (1996) Programmed cell death: a way of life for plants. Proc Natl Acad Sci 93:12094–12097

Guex N, Peitsch MC (1997) SWISS-MODEL and the Swiss-Pdb Viewer: an environment for comparative protein modeling. Electrophoresis 18:2714–2723

Gupta M et al (2013) A translationally controlled tumor protein negatively regulates the hypersensitive response in Nicotiana benthamiana. Plant Cell Physiol 54:1403–1414

Gutierrez-Galeano DF, Toscano-Morales R, Calderon-Perez B, Xoconostle-Cazares B, Ruiz-Medrano R (2014) Structural divergence of plant TCTPs. Front Plant Sci 5:361

Hafidh S et al (2016) Quantitative proteomics of the tobacco pollen tube secretome identifies novel pollen tube guidance proteins important for fertilization. Genome Biol 17:81

Han Y-J, Kim Y-M, Hwang O-J, Kim J-I (2015) Characterization of a small constitutive promoter from Arabidopsis translationally controlled tumor protein (AtTCTP) gene for plant transformation. Plant Cell Rep 34:265–275

Hinojosa-Moya J et al (2008) Phylogenetic and structural analysis of translationally controlled tumor proteins. J Mol Evol 66:472–483

Hinojosa-Moya JJ et al (2013) Characterization of the pumpkin translationally-controlled tumor protein CmTCTP. Plant Signal Behav 8:e26477

Hoepflinger MC, Reitsamer J, Geretschlaeger AM, Mehlmer N, Tenhaken R (2013) The effect of translationally controlled tumour protein (TCTP) on programmed cell death in plants. BMC Plant Biol 13:135

Hsu Y-C, Chern JJ, Cai Y, Liu M, Choi K-W (2007) Drosophila TCTP is essential for growth and proliferation through regulation of dRheb GTPase. Nature 445:785–788

Hu C et al (2015) Suppression of intestinal immunity through silencing of TCTP by RNAi in transgenic silkworm, Bombyx mori. Gene 574:82–87

Huang L et al (2014) Reference gene selection for quantitative real-time reverse-transcriptase PCR in orchardgrass subjected to various abiotic stresses. Gene 553:158–165

Itaya A, Matsuda Y, Gonzales RA, Nelson RS, Ding B (2002) Potato spindle tuber viroid strains of different pathogenicity induces and suppresses expression of common and unique genes in infected tomato. Mol Plant-Microbe Interact 15:990–999

Johnson K, Lenhard M (2011) Genetic control of plant organ growth. New Phytol 191:319–333
Jones AME, Thomas V, Bennett MH, Mansfield J, Grant M (2006) Modifications to the Arabidopsis defense proteome occur prior to significant transcriptional change in response to inoculation with Pseudomonas syringae. Plant Physiol 142:1603–1620
Kang JG, Yun J, Chung KS, Song PS, Park CM (2003) Promoter system of plant translationally controlled tumor protein gene. Patent US6518484 B2
Kardeh S, Ashkani-Esfahani S, Alizadeh AM (2014) Paradoxical action of reactive oxygen species in creation and therapy of cancer. Eur J Pharmacol 735:150–168
Kim M, Jung Y, Lee K, Kim C (2000) Identification of the calcium binding sites in translationally controlled tumor protein. Arch Pharm Res 23:633–636
Kim Y-M et al (2012) Overexpression of Arabidopsis translationally controlled tumor protein gene AtTCTP enhances drought tolerance with rapid ABA-induced stomatal closure. Mol Cells 33:617–626
Křeček P et al (2009) The PIN-FORMED (PIN) protein family of auxin transporters. Genome Biol 10:249
Krizek BA (2009) Making bigger plants: key regulators of final organ size. Curr Opin Plant Biol 12:17–22
Lau OS, Deng XW (2010) Plant hormone signaling lightens up: integrators of light and hormones. Curr Opin Plant Biol 13:571–577
Leivar P, Monte E (2014) PIFs: systems integrators in plant development. Plant Cell 26:56–78
Li L, Yu A-Q (2015) The functional role of peroxiredoxin 3 in reactive oxygen species, apoptosis, and chemoresistance of cancer cells. J Cancer Res Clin Oncol 141:2071–2077
Li D, Deng Z, Liu X, Qin B (2013) Molecular cloning, expression profiles and characterization of a novel translationally controlled tumor protein in rubber tree (Hevea brasiliensis). J Plant Physiol 170:497–504
Lliso I, Tadeo FR, Phinney BS, Wilkerson CG, Talón M (2007) Protein changes in the albedo of citrus fruits on postharvesting storage. J Agric Food Chem 55:9047–9053
Lloyd AC (2013) The regulation of cell size. Cell 154:1194–1205
Lopez AP, Franco AR (2006) Cloning and expression of cDNA encoding translationally controlled tumor protein from strawberry fruits. Biol Plant 50:447–449
Mahfouz MM (2006) Arabidopsis target of rapamycin interacts with RAPTOR, which regulates the activity of S6 kinase in response to osmotic stress signals. Plant Cell 18:477–490
Masura SS, Ahmad Parveez GK, Eng Ti LL (2011) Isolation and characterization of an oil palm constitutive promoter derived from a translationally control tumor protein (TCTP) gene. Plant Physiol Biochem 49:701–708
Menand B, Meyer C, Robaglia C (2004) Plant growth and the TOR pathway. Curr Top Microbiol Immunol 279:97–113
Merai Z et al (2006) Double-stranded RNA binding may be a general plant RNA viral strategy to suppress RNA silencing. J Virol 80:5747–5756
Meyuhas O, Kahan T (2015) The race to decipher the top secrets of TOP mRNAs. Biochim Biophys Acta 1849:801–811
Møller IM, Jensen PE, Hansson A (2007) Oxidative modifications to cellular components in plants. Annu Rev Plant Biol 58:459–481
Moreau M et al (2012) Mutations in the Arabidopsis homolog of LST8/GβL, a partner of the target of rapamycin kinase, impair plant growth, flowering, and metabolic adaptation to long days. Plant Cell 24:463–481
Morel JB, Dangl JL (1997) The hypersensitive response and the induction of cell death in plants. Cell Death Differ 4:671–683
Nakkaew A, Chotigeat W, Phongdara A (2010) Molecular cloning and expression of EgTCTP, encoding a calcium binding protein, enhances the growth of callus in oil palm (Elaeis guineensis, Jacq). Sonklanakarin. J Sci Technol 32:561–569
Narsai R et al (2007) Genome-wide analysis of mRNA decay rates and their determinants in Arabidopsis thaliana. Plant Cell 19:3418–3436

Nibau C, Wu H, Cheung AY (2006) RAC/ROP GTPases: 'hubs' for signal integration and diversification in plants. Trends Plant Sci 11:309–315

Niklas KJ (2015) A phyletic perspective on cell growth. Cold Spring Harb Perspect Biol 7: a019158

Nuoffer C, Wu SK, Dascher C, Balch WE (1997) Mss4 does not function as an exchange factor for Rab in endoplasmic reticulum to Golgi transport. Mol Biol Cell 8:1305–1316

O'Brien ET, Salmon ED, Erickson HP (1997) How calcium causes microtubule depolymerization. Cell Motil Cytoskeleton 36:125–135

Ohme-Takagi M, Taylor CB, Newman TC, Green PJ (1993) The effect of sequences with high AU content on mRNA stability in tobacco. Proc Natl Acad Sci USA 90:11811–11815

Oldham S, Montagne J, Radimerski T, Thomas G, Hafen E (2000) Genetic and biochemical characterization of dTOR, the Drosophila homolog of the target of rapamycin. Genes Dev 14:2689–2694

Pan D (2007) Hippo signaling in organ size control. Genes Dev 21:886–897

Panstruga R (2003) Establishing compatibility between plants and obligate biotrophic pathogens. Curr Opin Plant Biol 6:320–326

Pavy N et al (2005) Generation, annotation, analysis and database integration of 16,500 white spruce EST clusters. BMC Genomics 6:144

Pay A, Heberle-Bors E, Hirt H (1992) An alfalfa cDNA encodes a protein with homology to translationally controlled human tumor protein. Plant Mol Biol 19:501–503

Peleg Z, Blumwald E (2011) Hormone balance and abiotic stress tolerance in crop plants. Curr Opin Plant Biol 14:290–295

Penzo-Méndez AI, Stanger BZ (2015) Organ-size regulation in mammals. Cold Spring Harb Perspect Biol 7:a019240

Qin X et al (2011) Molecular cloning, characterization and expression of cDNA encoding translationally controlled tumor protein (TCTP) from Jatropha curcas L. Mol Biol Rep 38:3107–3112

Rehmann H et al (2008) Biochemical characterisation of TCTP questions its function as a guanine nucleotide exchange factor for Rheb. FEBS Lett 582:3005–3010

Ren M et al (2011) Target of rapamycin regulates development and ribosomal RNA expression through kinase domain in Arabidopsis. Plant Physiol 155:1367–1382

Ren M et al (2012) Target of rapamycin signaling regulates metabolism, growth, and life span in Arabidopsis. Plant Cell 24:4850–4874

Rexin D, Meyer C, Robaglia C, Veit B (2015) TOR signalling in plants. Biochem J 470:1–14

Rho SB et al (2011) Anti-apoptotic protein TCTP controls the stability of the tumor suppressor p53. FEBS Lett 585:29–35

Robaglia C et al (2004) Plant growth: the translational connection. Biochem Soc Trans 32:581–584

Russell EG, Cotter TG (2015) New insight into the role of reactive oxygen species (ROS) in cellular signal-transduction processes. Int Rev Cell Mol Biol 319:221–254

Sage-Ono K, Ono M, Harada H, Kamada H (1998) Dark-induced accumulation of mRNA for a homolog of translationally controlled tumor protein (TCTP) in Pharbitis. Plant Cell Physiol 39:357–360

Santa Brígida AB et al (2014) Molecular cloning and characterization of a cassava translationally controlled tumor protein gene potentially related to salt stress response. Mol Biol Rep 41:1787–1797

Susini L et al (2008) TCTP protects from apoptotic cell death by antagonizing bax function. Cell Death Differ 15:1211–1220

Szécsi J et al (2006) BIGPETALp, a bHLH transcription factor is involved in the control of Arabidopsis petal size. EMBO J 25:3912–3920

Tao JJ et al (2015) Tobacco translationally controlled tumor protein interacts with ethylene receptor tobacco Histidine Kinase1 and enhances plant growth through promotion of cell proliferation. Plant Physiol 169:96–114

Thaw P et al (2001) Structure of TCTP reveals unexpected relationship with guanine nucleotide-free chaperones. Nat Struct Mol Biol 8:701–704
Thayanithy V (2005) Evolution and expression of translationally controlled tumour protein (TCTP) of fish. Comp Biochem Physiol B Biochem Mol Biol 142:8–17
Thiele H, Berger M, Skalweit A, Thiele BJ (2000) Expression of the gene and processed pseudogenes encoding the human and rabbit translationally controlled tumour protein (TCTP). Eur J Biochem 267:5473–5481
Van Hautegem T, Waters AJ, Goodrich J, Nowack MK (2015) Only in dying, life: programmed cell death during plant development. Trends Plant Sci 20:102–113
Vanneste S, Friml J (2009) Auxin: a trigger for change in plant development. Cell 136:1005–1016
Veena JH, Doerge RW, Gelvin SB (2003) Transfer of T-DNA and Vir proteins to plant cells by Agrobacterium tumefaciens induces expression of host genes involved in mediating transformation and suppresses host defense gene expression. Plant J Cell Mol Biol 35:219–236
Venkatachalam P, Thulaseedharan A, Raghothama K (2007) Identification of expression profiles of tapping panel dryness (TPD) associated genes from the latex of rubber tree (Hevea brasiliensis Muell. Arg.) Planta 226:499–515
Vincent D et al (2007) Proteomic analysis reveals differences between Vitis vinifera L. cv. Chardonnay and cv. Cabernet Sauvignon and their responses to water deficit and salinity. J Exp Bot 58:1873–1892
Voinnet O, Vain P, Angell S, Baulcombe DC (1998) Systemic spread of sequence-specific transgene RNA degradation in plants is initiated by localized introduction of ectopic promoterless DNA. Cell 95:177–187
Wang X et al (2008) Re-evaluating the roles of proposed modulators of mammalian target of Rapamycin Complex 1 (mTORC1) signaling. J Biol Chem 283:30482–30492
Wang F, Shang Y, Yang L, Zhu C (2012) Comparative proteomic study and functional analysis of translationally controlled tumor protein in rice roots under Hg2+ stress. J Environ Sci 24:2149–2158
Wang ZQ, Li GZ, Gong QQ, Li GX, Zheng SJ (2015) OsTCTP, encoding a translationally controlled tumor protein, plays an important role in mercury tolerance in rice. BMC Plant Biol 15:123
Waterhouse AM, Procter JB, Martin DMA, Clamp M, Barton GJ (2009) Jalview Version 2—a multiple sequence alignment editor and analysis workbench. Bioinformatics 25:1189–1191
Wixler V et al (2011) Identification and characterisation of novel Mss4-binding Rab GTPases. Biol Chem 392:239–248
Woo H-H, Hawes MC (1997) Cloning of genes whose expression is correlated with mitosis and localized in dividing cells in root caps of Pisum sativum L. Plant Mol Biol 35:1045–1051
Wullschleger S, Loewith R, Hall MN (2006) TOR signaling in growth and metabolism. Cell 124:471–484
Xia XJ et al (2015) Interplay between reactive oxygen species and hormones in the control of plant development and stress tolerance. J Exp Bot 66:2839–2856
Xiong Y, Sheen J (2012) Rapamycin and glucose-target of rapamycin (TOR) protein signaling in plants. J Biol Chem 287:2836–2842
Xiong Y et al (2013) Glucose–TOR signalling reprograms the transcriptome and activates meristems. Nature 496:181–186
Yang Y et al (2005) An N-terminal region of translationally controlled tumor protein is required for its antiapoptotic activity. Oncogene 24:4778–4788
Yarm FR (2002) Plk phosphorylation regulates the microtubule-stabilizing protein TCTP. Mol Cell Biol 22:6209–6221
Yu R, Huang RF, Wang XC, Yuan M (2001) Microtubule dynamics are involved in stomatal movement of Vicia faba L. Protoplasma 216:113–118
Zhang H, Stallock JP, Ng JC, Reinhard C, Neufeld TP (2000) Regulation of cellular growth by the Drosophila target of rapamycin dTOR. Genes Dev 14:2712–2724

Zhang J, Jia W, Yang J, Ismail AM (2006) Role of ABA in integrating plant responses to drought and salt stresses. Field Crop Res 97:111–119

Zhang L, Li W, Han S, Yang W, Qi L (2013) cDNA cloning, genomic organization and expression analysis during somatic embryogenesis of the translationally controlled tumor protein (TCTP) gene from Japanese larch (Larix leptolepis). Gene 529:150–158

Zhang JM et al (2014) Cotton TCTP1 gene encoding a translationally controlled tumor protein participates in plant response and tolerance to aphids. Plant Cell Tiss Org Cult 117:145–156

Zhu Y (2002) Movement of potato spindle tuber viroid reveals regulatory points of phloem-mediated RNA traffic. Plant Physiol 130:138–146

Zhuo K et al (2017) A novel Meloidogyne enterolobii effector MeTCTP promotes parasitism by suppressing programmed cell death in host plants. Mol Plant Pathol 18:45–54

Chapter 8
Function of Translationally Controlled Tumor Protein in Organ Growth: Lessons from Drosophila Studies

Kwang-Wook Choi, Sung-Tae Hong, and Thao Phuong Le

Abstract Regulation of cell growth and proliferation is crucial for development and function of organs in all animals. Genetic defects in growth control can lead to developmental disorders and cancers. Translationally controlled tumor protein (TCTP) is a family of evolutionarily conserved proteins implicated in cancer. Recent studies have revealed multiple roles of TCTP in diverse cellular events, but TCTP functions in vivo are poorly understood in vertebrate systems. We have used *Drosophila melanogaster*, the fruit fly, as a model organism for genetic dissection of Tctp function. Our studies have shown that Tctp is essential for organ development by regulating growth signaling. Furthermore, it is required for genome stability by promoting DNA repair and chromatin remodeling in the nucleus. Thus, Tctp acts as a multifaceted cytosolic and nuclear factor for regulating organ growth and genome stability. In this chapter, we describe an overview of our findings on Tctp functions in *Drosophila* and discuss their implications in cancer.

8.1 Introduction

TCTP family proteins are widely expressed in eukaryotes. It was first identified in mouse tumor cells in the early growth phase (Bohm et al. 1989; Chitpatima et al. 1988). TCTP is abundantly expressed in rat and human testes, a tissue that undergoes intense mitotic activity (Guillaume et al. 2001). In yeast, TCTP is expressed throughout the cell cycle (Chung et al. 2000), but is greatly upregulated during exponential growth (Norbeck and Blomberg 1997), whereas it is repressed in conditions of growth arrest

K.-W. Choi (✉) • T.P. Le
Department of Biological Sciences, Korea Advanced Institute of Science and Technology, Daejeon 305-701, South Korea
e-mail: kchoi100@kaist.ac.kr

S.-T. Hong
Department of Anatomy & Cell Biology, College of Medicine, Chungnam National University, Daejeon 35015, South Korea

A. Telerman, R. Amson (eds.), *TCTP/tpt1 - Remodeling Signaling from Stem Cell to Disease*, Results and Problems in Cell Differentiation 64,
DOI 10.1007/978-3-319-67591-6_8

(Bonnet et al. 2000). Furthermore, extensive analysis of differential gene expression in the tumor cells and revertants revealed that TCTP is upregulated in cancer cells derived from different organs and is among the most downregulated genes during tumor reversion, the process of "quitting the malignant phenotype" (Amson et al. 2012, 2013; Bommer and Thiele 2004; Tuynder et al. 2002, 2004). These studies suggest important roles of TCTP in growth regulation, tumorigenesis, and its reversion process.

In addition, TCTPs have been implicated in several other functions, including histamine release (Nielsen et al. 1998), microtubule association (Gachet et al. 1999; Jeon et al. 2016), and Ca^{++} regulation (Kim et al. 2000; Sanchez et al. 1997). An important clue to the molecular function of TCTP was provided from its structure. TCTP family proteins turned out to be related to a guanine nucleotide exchange factor (GEF) for Rab proteins (Dong et al. 2009; Thaw et al. 2001), suggesting that TCTPs might be involved in intracellular vesicle trafficking (Predic et al. 2002). Many of these TCTP functions are based on biochemical interactions and functional assays in culture cells but not in animals. Thus, whether TCTP plays such roles in tissues and organs of animals in vivo is an important question to be addressed.

Drosophila is an ideal system for studying gene functions in vivo. We became interested in Tctp initially by chance (hereafter, *Drosophila TCTP* gene and protein are labeled "*Tctp*" and "Tctp", respectively, according to the FlyBase nomenclature). Our work since then has revealed important functions of Tctp in the regulation of growth and genome stability, providing new insights into the roles of mammalian TCTP genes. Firstly, we will begin with a brief background of our Tctp work. Secondly, we will discuss the function of Tctp in TOR signaling for organ growth. Thirdly, we will present a nuclear function of Tctp for DNA repair and genome stability. Lastly, we will discuss possible implications of our works on mammalian TCTP functions and cancer.

8.2 Identification of *Drosophila* Tctp Function in Organ Growth

Nothing was known about the *Tctp* gene in *Drosophila* when we first noticed its putative role in tissue growth. At the time, we had been studying how the adult compound eye develops from the eye imaginal disc, an epithelial primordium for the eye. In the early stages of eye development, establishment of the dorso-ventral (DV) axes of the eye disc is critically important for its growth and patterning (Singh et al. 2005). While searching for genes involved in the growth of the eye disc, we found a strong genetic interaction between a mutation affecting eye growth and a deficiency chromosome uncovering the *Tctp* locus.

This genetic enhancement of the eye growth phenotype suggested a role of Tctp in organ development. Because there was no known mutation in the *Tctp* gene, we first generated *Tctp* mutants by imprecise excision of a P-element inserted in an intron of the

Tctp gene (Hsu et al. 2007). Genetic tests suggested one of these mutations to be a null allele. This mutant was lethal as homozygotes, resulting in death during early larval stages. Some *Tctp* null homozygotes go through embryogenesis probably due to the presence of wild-type maternal Tctp protein deposited from heterozygous mothers, but they all die during early larval stages. Thus, it is clear that Tctp plays important functions for larval stages during which primordia for adult organs undergo active development.

To understand the function of Tctp in organ development, we mainly used the eye and the wing, two appendages that have been studied extensively (Baker 2007; Cohen and Di Nardo 1993). These organs develop from the eye and the wing imaginal disc, respectively. Although development of these organs involves several conserved signaling mechanisms, the ways these signaling pathways are used for axial patterning and differentiation of the eye and the wing are quite different (Baker 2007). Nonetheless, tissue-specific Tctp knockdown using RNA interference (RNAi) in these developing organs by using Gal4–UAS system (Brand and Perrimon 1993) results in similar growth defects. Because organ growth depends not only on cell proliferation but also on the cell size, it is important to determine whether Tctp is involved in the control of cell size, cell number, or both. Since each cell of adult wing has a single hair protruding from the cell surface, the hair density provides an approximate measure of cell size and number. Using the hair density, we showed that both cell number and size are reduced when Tctp is partially depleted by RNAi (Hsu et al. 2007), indicating the requirements of Tctp for cell growth as well as proliferation.

To support the results of Tctp RNAi phenotypes in organ growth, *Tctp* null mutant was also examined using genetic mosaic clones. Because *Tctp* null mutant flies die during early stage of development, we utilized methods of making genetic mosaic animals in which patches of homozygous *Tctp* mutant clones can be generated in heterozygous flies by mitotic recombination (Xu and Rubin 1993) (Fig. 8.1a). The clone size depends on the number of cell division after the first mitotic recombination event. When *Tctp* mutant clones were examined in wing discs soon after mitotic recombination was induced at first instar larval stage, both +/+ wild-type twin spots and $Tctp^{-}/Tctp^{-}$ mutant clones were very small but showed similar sizes (Fig. 8.1b). On the contrary, when mosaic wings were examined during late third instar stage, $Tctp^{-}/Tctp^{-}$ clones remained small or were eliminated while +/+ twin spot clones had grown much faster, eventually competing out $Tctp^{-}/Tctp^{-}$ mutant cells (Fig. 8.1c). This clonal analysis demonstrated that Tctp is essential for growth of tissues in wing discs. Another interesting point is that *Tctp* mutant cells cannot compete well with adjacent wild-type cells, eventually being lost, thus indicating the importance of Tctp in cell survival.

8.3 Role of Tctp in TOR Signaling

Clonal analysis of *Tctp* null mutation demonstrated that Tctp is required for organ growth. We tested whether the growth defects of *Tctp* mutant cells are due to abnormalities in cell proliferation and/or cell survival. A genetic technique called

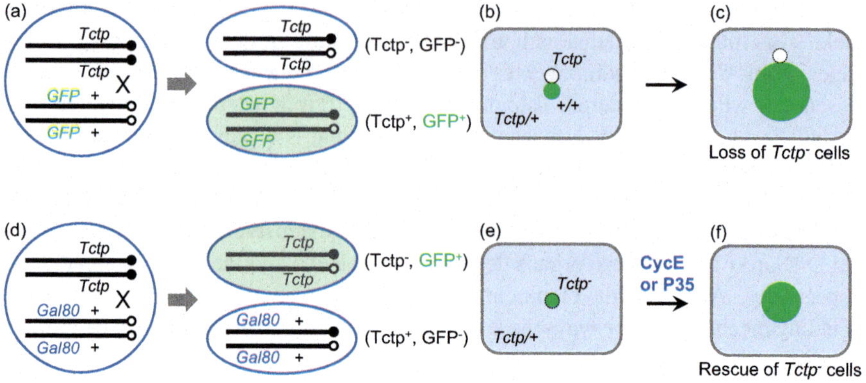

Fig. 8.1 Suppression of *Tctp* mutant defects by CycE and P35. (**a–c**) Growth defects of *Tctp* mutant clones in wing disc. (**a**) *Tctp*$^-$ mutant clones are generated by mitotic recombination in *Tctp*$^-$/+ heterozygote cells. *Tctp*$^-$ clones and wild-type twin spots (+/+) are marked by the absence or presence of GFP expression. (**b**) At 24 h after clone induction, both *Tctp*$^-$ and +/+ twin spot clones are small but similar in size. (**c**) By 60 h after clone induction, +/+ clones grow large but most *Tctp*$^-$ clones are eliminated. (**d–f**) Partial suppression of *Tctp*$^-$ mutant defects by CycE or P35. (**d**) Clones are generated by the MARCM method. The presence of *Gal80* in *Tctp* $^-$/+ heterozygous wing cells represses Gal4-dependent GFP expression. The presence of *Tub-Gal4* and *UAS-GFP* are not shown for simplicity. After recombination, *Tctp* mutant cells can express GFP due to loss of *Gal80* whereas +/+ twin spot cannot. (e) *Tctp*$^-$ clones are very small or not formed. (**f**) Expression of CycE or P35 in mutant clones partially suppresses growth defects, resulting in the formation of *Tctp*$^-$ clones

MARCM (Mosaic Analysis with a Recessive Cell Marker) (Lee and Luo 2001) provides a powerful in vivo tool for expressing a gene within mutant clones. Using this technique, we overexpressed either Cyclin E (CycE) or P35 caspase inhibitor in *Tctp* mutant clones. Notably, small sizes of *Tctp* null mutant clones were enlarged by expressing cyclin E or P35 in the MARCM mutant clones (Fig. 8.1d–f), indicating that loss of Tctp not only affects cell proliferation but also impairs cell survival (Hsu et al. 2007).

Genetic analysis of *Tctp* mutation and knockdown indicates that Tctp is involved in the regulation of both cell proliferation and cell size. Because the Target of Rapamycin (TOR) signaling regulates both cell growth and proliferation (Laplante and Sabatini 2012b), it was conceivable that Tctp might function in the TOR pathway to control organ growth. A central component of this pathway is TOR protein kinase that phosphorylates S6 kinase (S6k) and Thor (*Drosophila* 4EBP) to promote cell growth (Miron et al. 2001; Oldham et al. 2000). TOR kinase is activated by parallel inputs from growth factor signaling and nutrient conditions (Jewell et al. 2013). Upon insulin receptor (InR) signaling, the GTPase Activating Protein (GAP) activity of the Tuberous sclerosis complex (Tsc1/Tsc2) is inhibited, thus activating Rheb (Ras-related human protein enriched in brain) GTPase and its associated effector TOR (Dong and Pan 2004; Garami et al. 2003; Saucedo et al. 2003; Stocker et al. 2003; Zhang et al. 2000, 2003). Indeed, there was striking genetic interaction between *Tctp* and the TOR pathway genes, including upstream genes like *Tsc1/2*, *Rheb*, and *InR* as well as a downstream gene *S6k*. For instance, tissue growth induced by overexpression of InR or Rheb can be suppressed by Tctp knockdown (Hsu et al. 2007). Genetic relationships between Tctp and these TOR components suggest that Tctp acts at a step close to Rheb.

In addition to the genetic interaction of Tctp with Rheb, we noted that yeast TCTP/DSS4 (mammalian MSS4) was initially identified as a genetic suppressor of *sec4* gene encoding a Rab GTPase (Burton et al. 1993; Moya et al. 1993). Since TCTP is structurally similar to the Mss4 GTPase regulator (Thaw et al. 2001), Tctp might function in TOR signaling by interacting with Rheb GTPase. Biochemical evidence indicated that Tctp can physically interact with Rheb, and it has a guanine nucleotide exchange (GEF) activity for Rheb. The glutamate residue at the 12th position of TCTP was implicated in binding to Sec4 Rab GTPase (Thaw et al. 2001). We showed that the E to V mutation (Tctp^{E12V}) abolished the GEF activity. This residue is also critical for the in vivo function of Tctp because, unlike wild-type Tctp, Tctp^{E12V} fails to rescue the growth defects of Tctp RNAi (Hsu et al. 2007). Taken together, multiple pieces of evidence suggest that Tctp acts through Rheb in order to activate TOR signaling for organ growth (Fig. 8.2a).

The proposed Tctp function as a GEF toward Rheb had been questioned by Wang et al. (2008) and Rehmann et al. (2008) based on the observations that mammalian TCTP could not bind to Rheb and did not reproducibly affect mTORC1 signaling. The cause of these discrepancies on the function of TCTP is not yet clear but is likely to be due to differences in cell cultures and assay conditions. In fact, Dong et al. (2009) demonstrated that human Tctp not only binds to Rheb but also accelerates GDP release from hRheb. Additionally, they showed that hTCTP can prolong the activation of mTOR signaling in amino acid-depleted cells whereas hTCTP^{E12V} mutant form cannot. hTCTP also acts upstream to Rheb for the activation of S6k phosphorylation. All of these results were consistent with the Tctp–Rheb relationships shown in *Drosophila*. In addition, analysis of the structure model of the hRheb–hTCTP complex showed that hTCTP binding to hRheb opens the nucleotide binding site to facilitate the dissociation of GDP. Moreover, key residues involved in the hTCTP–hRheb interaction were experimentally validated (Dong et al. 2009), supporting the function of TCTP as a GEF for Rheb.

The binding between TCTP and Rheb has also been shown in *Arabidopsis* (Brioudes et al. 2010), indicating that this interaction is conserved in invertebrates, vertebrates, and plant systems. We have shown that human TCTP can fully rescue the growth defects in Tctp-depleted *Drosophila* organs (Hsu et al. 2007). Remarkably, *Drosophila* Tctp can also restore the defects in *Arabidopsis TCTP* mutants (Brioudes et al. 2010). These studies suggest strong structural and functional conservation of TCTP family genes among plants, invertebrate animals, and humans.

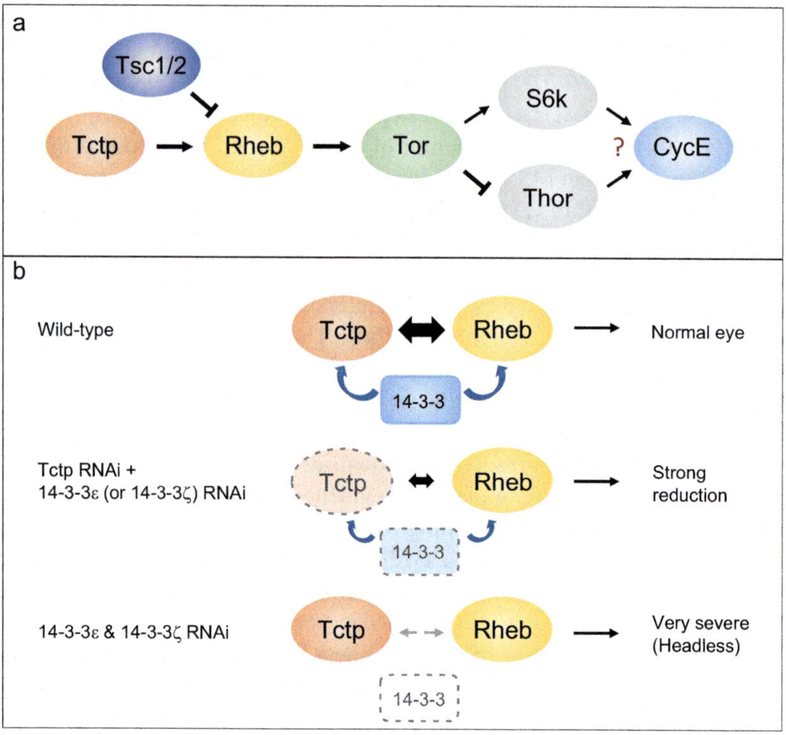

Fig. 8.2 14-3-3 promotes Tctp–Rheb interaction. (**a**) Tor kinase in TORC1 is activated by Rheb GTPase. Tor phosphorylates S6k and Thor/4EBP. Phosphorylated S6k promotes protein synthesis for cell growth. Phosphorylated Thor/4EBP cannot inhibit the function of eIF4, thereby increasing protein synthesis. Increased translation leads to expression of cell cycle regulators. Rheb activity is inhibited by TSC1/2 while facilitated by Tctp. (**b**) In wild-type condition, 14-3-3 isoforms directly interact with Tctp and Rheb, resulting in normal eye size. Tctp RNAi causes a reduction in the eye size. Knockdown of either 14-3-3ε or 14-3-3ζ has no effect, but it strongly enhances the Tctp RNAi eye phenotype. Knockdown of both forms of 14-3-3 abolishes the Tctp–Rheb interaction and disrupts eye disc development causing the headless phenotype. It is unknown whether homo- or hetero-dimerization of 14-3-3 isoform directly links Tctp and Rheb

8.4 Regulation of Tctp Function by 14-3-3

Our study described above suggests that Tctp facilitates the Rheb function in vivo for TOR signaling. Because growth signaling must be precisely controlled for normal development, it is an intriguing question how Tctp function is regulated during organogenesis. In an effort to identify factors that modulate the function of Tctp, we have performed a genetic screen using the Gal4-UAS system (Brand and Perrimon 1993) for targeted knockdown of gene(s). This screen was based on the fact that Tctp RNAi in the eye disc using *eyeless (ey)-Gal4* driver results in a reduction of the eye size. We screened a library of UAS-dsRNA (RNAi for short) lines to identify specific RNAi lines that either enhance or suppress the Tctp RNAi eye phenotype.

From this screen, we found many RNAi lines that modify the Tctp RNAi eye phenotype. Tctp modifier genes identified in this screen belong to diverse categories, including the ones that are involved in growth signaling, cell death, cytoskeleton, and transcription. Interestingly, 14-3-3 RNAi was found as an enhancer of Tctp RNAi. 14-3-3 is a family of highly conserved proteins expressed in all eukaryotic cells. 14-3-3 proteins play important roles as adaptors in multiple signaling pathways (Morrison 2009). 14-3-3 genes have also been implicated in oncogenesis as well as tumor suppression (Aghazadeh and Papadopoulos 2016; Aitken et al. 2002; Zhao et al. 2011). Mammals have several isoforms of 14-3-3 that have redundant functions as well as cell-type specific roles (Aghazadeh and Papadopoulos 2016; Obsilova et al. 2008). *Drosophila* has two genes, 14-3-3ε and 14-3-3ζ (Skoulakis and Davis 1996). Although silencing of either isoform of 14-3-3 in the eye disc does not affect the eye growth, it synergistically enhances the effects of Tctp RNAi, resulting in much smaller eyes than the size of Tctp RNAi eyes (Fig. 8.2b, Le et al. 2016). Strong genetic interaction between Tctp and 14-3-3 isoforms raised a possibility that 14-3-3 might be involved in regulating the Tctp function in TOR signaling.

Previous studies in mammalian cells have also found that 14-3-3 proteins participate in TORC1 signaling by interacting with TSC2 and PRAS40. Both TSC2 and PRAS40 are negative regulators of Rheb and TOR kinase, respectively (Fonseca et al. 2007; Inoki et al. 2002; Li et al. 2002). Upon phosphorylation of TSC2 by Akt, 14-3-3 binds to phosphorylated TSC2 to inhibit the GAP function of TSC2. PRAS40 directly binds to TOR kinase to inhibit its kinase activity. Binding of 14-3-3 to PRAS40 leads to its dissociation from the TORC1 complex, thereby activating the TORC1 activity (Jewell et al. 2013; Morrison 2009). Hence, 14-3-3 proteins promote TOR signaling by inhibiting these negative regulators of TOR signaling. The serine and threonine phosphorylation sites of Akt (S924 and T1518) are conserved in *Drosophila* and mammalian TSC2 proteins. Interestingly, however, a mutated TSC2 with both substitutions of S924 and T1518 with unphosphorylatable alanine is fully functional to inhibit TOR signaling in *Drosophila* (Dong and Pan 2004). Thus, this phosphorylation-dependent 14-3-3 binding is probably not essential for the TSC2 function in *Drosophila*. Moreover, PRAS40 in *Drosophila* regulates fertility but is not required for growth of the fly, despite its importance in TOR signaling in mammalian cells (Pallares-Cartes et al. 2012). It has not been tested in vivo whether loss of 14-3-3 impairs TOR signaling in mammals. Thus, further studies are necessary to determine the importance of 14-3-3 interaction with TSC2 and PRAS40 in vivo.

Due to the fact that TSC2 phosphorylation is dispensable in *Drosophila*, we figured that genetic interaction between 14-3-3 and Tctp is probably independent of 14-3-3 binding to TSC2. Instead, we found that both 14-3-3 isoforms could directly interact with Tctp (Le et al. 2016), raising the possibility that 14-3-3 might promote TOR signaling through a new mechanism by binding to Tctp. Furthermore, 14-3-3 isoforms can physically interact with Rheb protein as well. Because Tctp and Rheb act together, these proteins seem to function together with 14-3-3 isoforms. One possibility is that 14-3-3 isoforms might be involved in facilitating the interaction between Tctp and Rheb. In testing this possibility, we found that knockdown of both 14-3-3 isoforms abolished the interaction between Tctp and Rheb to an undetectable level in co-immunoprecipitation assays using *Drosophila* S2 cells. However, depletion of either 14-3-3ε or 14-3-3ζ isoform did not significantly affect the Tctp–Rheb interaction, which is correlated with the observation that knockdown of a single 14-3-3 isoform causes noticeable defects in neither eye nor wing. These results indicate that 14-3-3 isoforms are critical for the interaction between Tctp and Rheb, although the two isoforms share redundant roles.

To test the functional redundancy of the 14-3-3 isoforms, we examined the phenotypes of depleting either one or both of the 14-3-3 isoforms in specific tissues. In striking contrast to the single knockdown of either isoform of 14-3-3, knockdown of both 14-3-3 isoforms using *ey-Gal4* results in pupal lethality. *Ey-Gal4* drives the expression of Gal4 in the primordia for both eye and head during the early stage of development. Examination of dead pupae showed relatively normal body parts, but there was specific loss of entire head and eye tissues to which Gal4 expression was targeted (Fig. 8.2b). Knockdown of both 14-3-3 isoforms in developing wing discs also resulted in severe loss of cell proliferation and induction of specific cell death in the targeted wing area (Le et al. 2016). These tests clearly indicate that two isoforms of 14-3-3 can function redundantly, thereby compensating the loss of one isoform. However, based on the findings that 14-3-3ε or 14-3-3ζ null mutations are semi- or fully lethal as homozygotes, respectively (Acevedo et al. 2007), we presume that these two 14-3-3 isoforms cannot be entirely redundant. Thus, 14-3-3ε and 14-3-3ζ isoforms seem to be partially redundant in certain conditions or tissues, while they also have unique functions. Nevertheless, knockdown of both 14-3-3 isoforms critically impairs normal growth of imaginal discs.

The levels of pS6k and pThor are convenient readouts of TOR signaling, as they are phosphorylation targets of TOR kinase. Interestingly, assays in S2 cells have shown that single knockdown of either of the 14-3-3 isoforms causes considerable reduction of pS6k and pThor (4EBP) levels, indicating that each 14-3-3 isoform is required for full TOR signaling (Le et al. 2016). Thus, TOR signaling in S2 cells might be more sensitive to a reduction of 14-3-3 isoforms than it is for developing organ tissues, although the molecular basis for this difference is unknown. Another important question is how 14-3-3 isoforms promote the interaction between Tctp and Rheb. 14-3-3 proteins are known to function as a homo- or a hetero-dimer (Acevedo et al. 2007; Yaffe 2002). Thus, it is plausible that dimerization of 14-3-3 proteins bound to Tctp and Rheb might bring them together to facilitate their direct interaction. Alternatively, 14-3-3 may be involved in modifying the structure and/or subcellular localization of Tctp and Rheb so that they can form a complex to activate Rheb activity. Additional studies are necessary to pinpoint the precise function of 14-3-3 in the formation of Tctp–Rheb complex.

8.5 Effects of 14-3-3 Isoforms and Tctp on Cyclin E

CycE, as a key regulator of the G1-S transition in cell cycle, plays an essential role in cell proliferation. TOR signaling regulates not only cell growth through activation of S6k and 4EBP but also regulates proliferation by promoting cell cycle. Consistent with genetic and physical interaction of 14-3-3 and Tctp, loss of function clones of *14-3-3ε* or *14-3-3ζ* null mutation results in a partial reduction of CycE level in the eye imaginal disc (Le et al. 2016). However, the partial CycE reduction by depleting one 14-3-3 isoform is insufficient to disrupt proliferation or differentiation of retinal cells of eye disc. We have shown that reduced organ size caused by *Tctp* mutation or RNAi can be rescued by CycE overexpression (Hsu et al. 2007). Therefore, it is likely that defects in organ growth caused by double knockdown of Tctp and one of 14-3-3 isoforms might be due to loss of CycE. As expected, organ growth defects caused by reducing both Tctp and 14-3-3 (or knockdown of both 14-3-3 isoforms) were strongly suppressed by CycE (Le et al. 2016).

In *Drosophila*, TOR kinase activates cell proliferation by regulating the level of CycE (Zhang et al. 2000), although it is not clearly defined how activated TOR leads to CycE expression. Phosphorylation of S6k and 4EBP by activated mTOR signaling results in increased protein synthesis, leading to the promotion of cell cycle (Jewell et al. 2013; Laplante and Sabatini 2012b). In mammalian cells, the eIF4E pathway activated by phosphorylation of 4EBP results in increased translation of mRNAs for cell cycle regulators such as CycD1 (Hashemolhosseini et al. 1998; Laplante and Sabatini 2012a; Rosenwald et al. 1995), thus allowing more CycE–Cdk2 complex to promote cell proliferation. *Drosophila* CycD is not only involved in cell cycle progression but also cell growth through an independent pathway (Datar et al. 2006). It is currently unknown whether the regulation of TOR-dependent CycE level in *Drosophila* is mediated by CycD. Interestingly, suppression of human 14-3-3ε inhibits proliferation of cancer cells and tumor growth. 14-3-3ε inhibition suppresses CycE expression while inducing the cell cycle inhibitor $p27^{kip1}$ at the G1 stage, consistent with the antitumor effect of 14-3-3ε inhibition (Gong et al. 2014). This study supports the role of *Drosophila* 14-3-3 in cell cycle progression during organ development. It would be interesting to see whether the effects of human14-3-3ε inhibition are at least in part mediated by TCTP/Rheb-dependent TOR signaling.

8.6 Tctp Function in DNA Damage Control

TCTP is known to be expressed in the cytosol and the nucleus of normal as well as cancer cells. Although the roles of TCTP in the nucleus have not been extensively studied, it has been reported that TCTP can function as a transcriptional regulator to induce *oct4* genes for the maintenance of stem cell fate (Cheng et al. 2012; Koziol et al. 2007). These findings suggest that TCTP can have important nuclear functions distinct from its known functions in the cytosol.

In *Drosophila*, first clue to the potential nuclear function of Tctp was provided by the finding that Tctp directly interacts with Ataxia Telangiectasia Mutated (ATM). Previously, we have used a phage display method to identify specific peptide ligands that bind to Tctp. Screening of aptamer libraries helped identify Tctp-binding peptide sequences. ATM was one of the proteins that contain Tctp-binding aptamer sequences and was confirmed to bind Tctp (Hong and Choi 2013). ATM is a serine–threonine protein kinase pivotal for repairing DNA damages caused by double strand break (DSB). Importantly, Tctp was found to form nuclear foci colocalized with ATM upon γ-ray irradiation, which is consistent with its function in DNA damage response. Mammalian ATM forms a protein complex with other factors such as Mre11 and Rad50 for DNA repair. *Drosophila* Tctp was also found to be associated with Mre11 and Rad50, suggesting that Tctp is a member of the functional DNA repair complex. In fact, *Tctp* mutants were defective in various DNA damage responses, showing frequent aberrant chromosomes, abnormal G2/M checkpoint response, and higher cell death upon ionizing irradiation. Consistent with the physical interaction between Tctp and ATM, defective DNA damage responses in *atm* mutants including abnormal chromosomes were strongly enhanced by reducing the level of Tctp. Likewise, growth defects by Tctp RNAi were strongly enhanced by reduced dosages of *atm* and other known DNA damage control genes. Together with the physical interaction between Tctp and ATM, these genetic interactions support the direct involvement of Tctp in DNA damage control in vivo.

Upon sensing DSBs, ATM kinase phosphorylates H2Av (H2AX in mammals), a variant form of histone 2A. The generation of phosphorylated H2Av (γH2Av) is an initial step for recruiting DNA repair proteins. Thus, γH2Av is a biomarker for DSBs and the sites of damage repair foci (Kuo and Yang 2008). Larval salivary gland cells undergo endocycling without mitosis and have a high level of γH2Av. In *Tctp* mutants, levels of γH2Av in salivary glands were diminished compared with the wild-type level. The lower level of γH2Av was rescued by adding a wild-type *Tctp* gene in the mutant background. Furthermore, the level of γH2Av induction upon γ-irradiation was also significantly lower in *Tctp* mutant wing discs than wild-type discs. These results provide in vivo evidence that Tctp is required for repairing both endogenous and exogenous DNA damages.

An important question is how the interaction between Tctp and ATM leads to DSB repair. Since *Tctp* mutations result in a reduction of the γH2Av level, Tctp might be involved in the promotion of ATM kinase activity. In vitro assays showed that addition of Tctp increased the ATM kinase activity toward the H2Av substrate in a dose-dependent manner. Thus, Tctp is directly involved in enhancing the ATM kinase activity, facilitating the DSB repair process. It is worth noting that defective Tctp with E12V substitution could not activate ATM kinase activity, indicating that the E12 residue is critical for the function of Tctp in TOR signaling as well as for the interaction with ATM. Remarkably, human TCTP is also associated with ATM and several proteins involved in DSB repair such as Ku70/80, DNA-binding

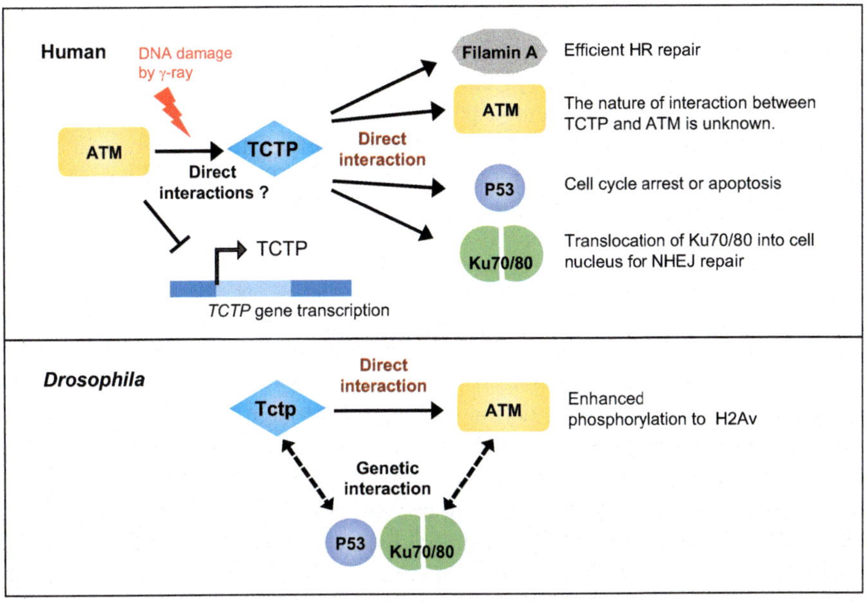

Fig. 8.3 Tctp–ATM interaction for DNA repair in mammals and fly. Relationships between TCTP and ATM in DNA damage response in humans and fly. In humans, ATM inhibits TCTP expression in nonirradiated cells. DNA damage by γ-ray irradiation leads to an increase in the TCTP level through ATM. TCTP forms a protein complex with Filamin A, P53, and Ku70/80 for DNA repair. It is unknown whether TCTP directly binds to ATM. In *Drosophila*, γ-irradiation has little effect on the Tctp level. Tctp directly binds to ATM and promotes its kinase activity. Tctp shows genetic interaction with P53 and Ku70/78, but it is unknown whether this interaction is direct

subunits of DNA-dependent protein kinase. Knockdown of TCTP impairs its ability to repair DSBs in irradiated human cells (Zhang et al. 2012). Thus, *Drosophila* and human TCTP proteins show strong conservation in their nuclear functions for DNA damage response (Fig. 8.3).

We have mentioned earlier that growth defects of *Tctp* mutant cell clones can be partially suppressed by providing CycE as well as the P35 cell death inhibitor, suggesting that Tctp is also required for preventing cell death. This is an important point because mammalian TCTP proteins are anti-apoptotic (Telerman and Amson 2009). One of the mechanisms of TCTP's anti-apoptotic function is related to its ability to bind Bcl-2 family proteins Bcl-xL and MCL1 that inhibit the pro-apoptotic activity of Bax, located in the outer mitochondrial membrane (Liu et al. 2005; Susini et al. 2008). A recent study has shown that TCTP contains a BH3-like domain that recognizes the BH3 domain of Bcl-xL to activate the anti-apoptotic function of Bcl-xL (Thebault et al. 2016). Interestingly, *Drosophila* Tctp also has a putative BH3 domain with conserved hydrophobic residues. In addition to the TCTP interaction with anti-apoptotic mitochondrial proteins, mammalian TCTPs are known to inhibit p53 tumor suppressor-induced cell death upon DNA damage (Amson et al. 2011; Kloc et al. 2012). TCTP not only represses transcription of p53 but it also destabilizes p53 in order to inhibit apoptosis (Rho et al. 2011). We have shown that reduced eye caused by Tctp RNAi can be suppressed by overexpression of dominant-negative p53 (Hong and Choi 2013). This suggests that cell death resulting from Tctp knockdown is due in part to p53 upregulation or activation and that the relationship between p53 and TCTP in mammalian systems might also be conserved in *Drosophila*.

8.7 Tctp in Chromatin Remodeling and Genome Stability

A closer examination of the nuclear localization of Tctp in the salivary gland indicates that it is associated with most interband regions of polytene chromosomes. Furthermore, from a yeast two-hybrid screen, Tctp was found to interact with Brahma (Brm), the *Drosophila* homolog of yeast Swi/SNF chromatin remodeling factor (Hong and Choi 2016). These findings raised the possibility that Tctp may have a regulatory role at the level of chromatin. A series of genetic and biochemical tests revealed that Tctp is required for inhibiting the ATPase activity of Brm and therefore antagonizing the Brm function in developing organs. Consistent with the role of the Brm remodeler complex in transcriptional gene regulation, loss of Tctp increases RNA polymerase II activity, enhancing transcription in a number of genes.

Interestingly, a reduction in the Tctp level leads to dramatic increases in the transcriptional expression of retrotransposons inserted in the pericentromeric regions, situated near the centromere and highly modified by heterochromatin marks (Hong and Choi 2016). Silencing of retrotransposons by heterochromatin marks is important for the maintenance of genome stability (Larson et al. 2012; Peng and Karpen 2007, 2009; Shi et al. 2008). In addition to its role for inhibiting transposon expression, Tctp is necessary to maintain the stability of *rDNA* genes and other repeated sequences. Methylation of histone 3 at lysine 9 (H3K9me2/3) by SU(VAR)3–9 histone methyl transferase (HMT) is a critical mark for HP1a-dependent heterochromatin formation and gene silencing. Our data demonstrate that Tctp is required for transcription of *su(var)3–9*, hence affecting the levels of H3K9 methylation and HP1a protein (Fig. 8.4).

Position effect variegation (PEV) is a genetic phenomenon caused by repression of the genes abnormally transposed near heterochromatic regions. *Tctp* mutations suppress various PEV phenotypes, consistent with its role in promoting the gene silencing effects of heterochromatin. Tctp and Brm also show the opposing relationship in their effects on PEV. These observations support that the antagonistic relationship between Tctp and Brm contributes to the regulation of the chromatin boundary between euchromatin and heterochromatin (Hong and Choi 2016).

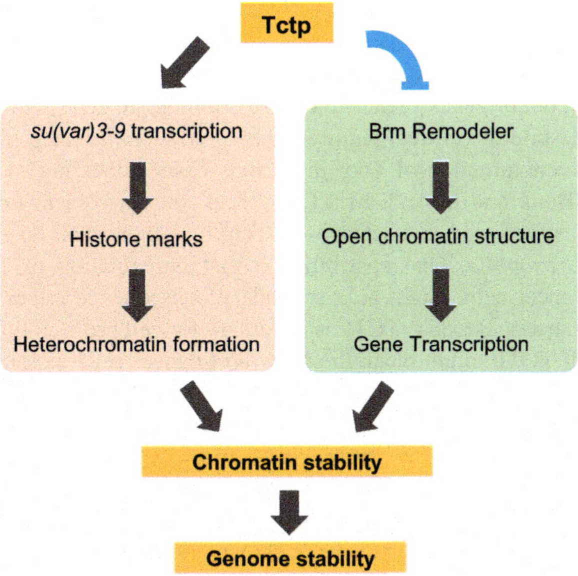

Fig. 8.4 A model for Tctp function in chromatin remodeling and genome stability. Tctp keeps genome stability through two different compaction mechanisms: (1) Heterochromatin formation by inducing transcription of *su(var)3–9*, thus increasing H3K9 methylation at the initial step of chromatin compaction and (2) maintenance of the proper level of chromatin opening by negatively modulating the Brm chromatin remodeler

8.8 Concluding Remarks

Tctp is an evolutionarily conserved protein. Most of structural features of mammalian TCTP family proteins are shared in invertebrate systems like *Drosophila*. TCTP family proteins also seem to be conserved in their in vivo functions, based on the successful complementation of mutant phenotypes by TCTP transgenes from other species. Despite the structural and functional conservation, however, there are several differences between the *Drosophila* Tctp and vertebrate TCTP proteins.

Firstly, TCTP is known as a histamine releasing factor (HRF) that is secreted to promote immune responses (MacDonald et al. 1995). TCTP homologs are secreted from the parasite *Plasmodium falciparum* that causes malaria and found in the plasma of infected hosts (MacDonald et al. 2001). Evidence suggests that TCTP is secreted via exosomal secretory pathway (Amzallag et al. 2004; Lespagnol et al. 2008). It is an intriguing question whether secretion is a general property of TCTP family proteins. If *Drosophila* Tctp can be secreted to act at a distance, phenotypes of *Tctp* mutant clones in genetically mosaic tissues might be rescued by Tctp protein secreted from the adjacent wild-type cells. Thus far, we have not noticed any obvious sign of non-cell autonomous function of Tctp in developing *Drosophila* tissues. It has been shown that truncation of the N terminal sequence and dimerization of TCTP is necessary for its cytokine-like extracellular activity (Kim et al. 2009). A critical cysteine residue of human TCTP involved in the dimerization is also conserved in *Drosophila* Tctp. Therefore, although the function of Tctp in organ growth seems to be cell-autonomous, we cannot exclude the possibility that Tctp can act as a secreted factor in *Drosophila*.

Secondly, one of the most unexpected findings from our *Drosophila* Tctp studies is the apparent lack of gain-of-function Tctp phenotypes. Human TCTP is upregulated in various cancer cells (Tuynder et al. 2002) and has been implicated in tumorigenesis (Bae et al. 2015; Jung et al. 2011; Niforou et al. 2008). However, despite the critical requirements of *Drosophila* Tctp for organ growth, ectopic or overexpression of Tctp in various tissues does not induce tumorous overgrowth (Hong and Choi 2013). Thus, Tctp functions seem to be permissive rather than instructive. It is unclear why Tctp overexpression does not induce overgrowth in *Drosophila*. One possibility is that upregulation of TCTP observed in various cancer cells might be a secondary consequence rather than the primary cause of cancers. Because TCTP is necessary for cell proliferation and survival, knockdown of TCTP might block the growth of cancer cells, possibly explaining the phenomenon of tumor reversion. Alternatively, although *Drosophila* Tctp overexpression may not be sufficient to induce tumorous growth in normal tissues, it might trigger overgrowth in the presence of an additional factor(s) or under certain cellular conditions. If that is the case, systematic searches for such factors would be critical to understand the mechanism for TCTP-induced tumorigenicity. Studies with transgenic TCTP mice have shown that overexpression of TCTP causes hypertension but with normal appearance at any age (Kim et al. 2008). Therefore, mammalian TCTP may also require additional factors to induce cancer, and it is an interesting possibility that upregulation of TCTP might predispose normal cells to be transformed when combined with an additional factor.

Thirdly, the role of Tctp in DNA damage repair seems to be conserved, but important differences have also been noticed (Fig. 8.3). Both *Drosophila* Tctp and human TCTP are involved in two distinct mechanisms of DNA repair: homologous recombination (HR) and nonhomologous end-joining (NHEJ). In human cells, depletion of ATM in nonirradiated cells results in an increase of TCTP level. Further, low dose γ-rays upregulate TCTP to protect against DNA damage, and this upregulation depends on ATM (Zhang et al. 2012). Thus, ATM seems to be involved in regulating the level of TCTP expression and/or stability in nonirradiated cells or cells exposed to low dose γ-irradiation. In contrast, the level of *Drosophila* Tctp is not influenced by irradiation or ATM function (Hong and Choi 2013). Instead, Tctp directly promotes the ATM kinase activity to phosphorylate H2Av. Therefore, loss of Tctp results in a reduction of γH2Av level. This is also in contrast to TCTP-depleted human fibroblast cells in which the number of γH2AX nuclear foci and the level of γH2AX remains high. Human fibroblast cells might have a mechanism to compensate the loss of TCTP to maintain the γH2AX levels. It is also possible that upregulated human TCTP by ATM might be able to activate ATM kinase activity as in the case for *Drosophila*. Future studies are necessary to understand the basis for these variations in the functional relationship between TCTP and ATM in different organisms.

There is profound evidence that many cancers are associated with genome instability. Because loss of TCTP function impairs DNA repair, it may lead to genome instability causing cancer. Therefore, not only upregulation of TCTP but also its loss may be associated with cancer through distinct mechanisms.

Furthermore, *Drosophila* Tctp is critical for genome stability by regulating global gene expression and chromatin modification. It is an important question to be addressed in the future whether mammalian TCTP might play similar roles in the regulation of chromatin modification and genome stability.

Acknowledgements We thank Kyungok Cho and Jean Jung for comments on the manuscript. This work was supported by NRF-2014K1A1A2042982, NRF-2017R1A2B3007516 (KWC), 2016R1D1A1B03932093 and 35B-2011-1-C00033 (STH) through the National Research Foundation of Korea funded by the Korean Ministry of Education Science & Technology.

References

Acevedo SF, Tsigkari KK, Grammenoudi S, Skoulakis EM (2007) In vivo functional specificity and homeostasis of Drosophila 14-3-3 proteins. Genetics 177:239–253

Aghazadeh Y, Papadopoulos V (2016) The role of the 14-3-3 protein family in health, disease, and drug development. Drug Discov Today 21:278–287

Aitken A, Baxter H, Dubois T, Clokie S, Mackie S, Mitchell K, Peden A, Zemlickova E (2002) Specificity of 14-3-3 isoform dimer interactions and phosphorylation. Biochem Soc Trans 30:351–360

Amson R, Kubiak JZ, Van Montagu M, Telerman A (2011) Could TCTP contribute to Armin Braun's paradigm of tumor reversion in plants? Cell Cycle 10:1

Amson R, Pece S, Lespagnol A, Vyas R, Mazzarol G, Tosoni D, Colaluca I, Viale G, Rodrigues-Ferreira S, Wynendaele J et al (2012) Reciprocal repression between P53 and TCTP. Nat Med 18:91–99

Amson R, Pece S, Marine JC, Di Fiore PP, Telerman A (2013) TPT1/TCTP-regulated pathways in phenotypic reprogramming. Trends Cell Biol 23:37–46

Amzallag N, Passer BJ, Allanic D, Segura E, Thery C, Goud B, Amson R, Telerman A (2004) TSAP6 facilitates the secretion of translationally controlled tumor protein/histamine-releasing factor via a nonclassical pathway. J Biol Chem 279:46104–46112

Bae SY, Kim HJ, Lee KJ, Lee K (2015) Translationally controlled tumor protein induces epithelial to mesenchymal transition and promotes cell migration, invasion and metastasis. Sci Rep 5:8061

Baker NE (2007) Patterning signals and proliferation in Drosophila imaginal discs. Curr Opin Genet Dev 17:287–293

Bohm H, Benndorf R, Gaestel M, Gross B, Nurnberg P, Kraft R, Otto A, Bielka H (1989) The growth-related protein P23 of the Ehrlich ascites tumor: translational control, cloning and primary structure. Biochem Int 19:277–286

Bommer UA, Thiele BJ (2004) The translationally controlled tumour protein (TCTP). Int J Biochem Cell Biol 36:379–385

Bonnet C, Perret E, Dumont X, Picard A, Caput D, Lenaers G (2000) Identification and transcription control of fission yeast genes repressed by an ammonium starvation growth arrest. Yeast 16:23–33

Brand AH, Perrimon N (1993) Targeted gene expression as a means of altering cell fates and generating dominant phenotypes. Development 118:401–415

Brioudes F, Thierry AM, Chambrier P, Mollereau B, Bendahmane M (2010) Translationally controlled tumor protein is a conserved mitotic growth integrator in animals and plants. Proc Natl Acad Sci U S A 107:16384–16389

Burton J, Roberts D, Montaldi M, Novick P, De Camilli P (1993) A mammalian guanine-nucleotide-releasing protein enhances function of yeast secretory protein Sec4. Nature 361:464–467
Cheng X, Li J, Deng J, Li Z, Meng S, Wang H (2012) Translationally controlled tumor protein (TCTP) downregulates Oct4 expression in mouse pluripotent cells. BMB Rep 45:20–25
Chitpatima ST, Makrides S, Bandyopadhyay R, Brawerman G (1988) Nucleotide sequence of a major messenger RNA for a 21 kilodalton polypeptide that is under translational control in mouse tumor cells. Nucleic Acids Res 16:2350
Chung S, Kim M, Choi W, Chung J, Lee K (2000) Expression of translationally controlled tumor protein mRNA in human colon cancer. Cancer Lett 156:185–190
Cohen SM, Di Nardo S (1993) Wingless: from embryo to adult. Trends Genet 9:189–192
Datar SA, Galloni M, de la Cruz A, Marti M, Edgar BA, Frei C (2006) Mammalian cyclin D1/Cdk4 complexes induce cell growth in Drosophila. Cell Cycle 5:647–652
Dong J, Pan D (2004) Tsc2 is not a critical target of Akt during normal Drosophila development. Genes Dev 18:2479–2484
Dong X, Yang B, Li Y, Zhong C, Ding J (2009) Molecular basis of the acceleration of the GDP-GTP exchange of human ras homolog enriched in brain by human translationally controlled tumor protein. J Biol Chem 284:23754–23764
Fonseca BD, Smith EM, Lee VH, MacKintosh C, Proud CG (2007) PRAS40 is a target for mammalian target of rapamycin complex 1 and is required for signaling downstream of this complex. J Biol Chem 282:24514–24524
Gachet Y, Tournier S, Lee M, Lazaris-Karatzas A, Poulton T, Bommer UA (1999) The growth-related, translationally controlled protein P23 has properties of a tubulin binding protein and associates transiently with microtubules during the cell cycle. J Cell Sci 112(Pt 8):1257–1271
Garami A, Zwartkruis FJ, Nobukuni T, Joaquin M, Roccio M, Stocker H, Kozma SC, Hafen E, Bos JL, Thomas G (2003) Insulin activation of Rheb, a mediator of mTOR/S6K/4E-BP signaling, is inhibited by TSC1 and 2. Mol Cell 11:1457–1466
Gong X, Yan L, Gu H, Mu Y, Tong G, Zhang G (2014) 14-3-3epsilon functions as an oncogene in SGC7901 gastric cancer cells through involvement of cyclin E and p27kip1. Mol Med Rep 10:3145–3150
Guillaume E, Pineau C, Evrard B, Dupaix A, Moertz E, Sanchez JC, Hochstrasser DF, Jegou B (2001) Cellular distribution of translationally controlled tumor protein in rat and human testes. Proteomics 1:880–889
Hashemolhosseini S, Nagamine Y, Morley SJ, Desrivieres S, Mercep L, Ferrari S (1998) Rapamycin inhibition of the G1 to S transition is mediated by effects on cyclin D1 mRNA and protein stability. J Biol Chem 273:14424–14429
Hong ST, Choi KW (2013) TCTP directly regulates ATM activity to control genome stability and organ development in Drosophila melanogaster. Nat Commun 4:2986
Hong ST, Choi KW (2016) Antagonistic roles of Drosophila Tctp and Brahma in chromatin remodelling and stabilizing repeated sequences. Nat Commun 7:12988
Hsu YC, Chern JJ, Cai Y, Liu M, Choi KW (2007) Drosophila TCTP is essential for growth and proliferation through regulation of dRheb GTPase. Nature 445:785–788
Inoki K, Li Y, Zhu T, Wu J, Guan KL (2002) TSC2 is phosphorylated and inhibited by Akt and suppresses mTOR signalling. Nat Cell Biol 4:648–657
Jeon HJ, You SY, Park YS, Chang JW, Kim JS, Oh JS (2016) TCTP regulates spindle microtubule dynamics by stabilizing polar microtubules during mouse oocyte meiosis. Biochim Biophys Acta 1863:630–637
Jewell JL, Russell RC, Guan KL (2013) Amino acid signalling upstream of mTOR. Nat Rev Mol Cell Biol 14:133–139
Jung J, Kim HY, Kim M, Sohn K, Kim M, Lee K (2011) Translationally controlled tumor protein induces human breast epithelial cell transformation through the activation of Src. Oncogene 30:2264–2274

Kim M, Jung Y, Lee K, Kim C (2000) Identification of the calcium binding sites in translationally controlled tumor protein. Arch Pharm Res 23:633–636

Kim MJ, Kwon JS, Suh SH, Suh JK, Jung J, Lee SN, Kim YH, Cho MC, Oh GT, Lee K (2008) Transgenic overexpression of translationally controlled tumor protein induces systemic hypertension via repression of Na+,K+-ATPase. J Mol Cell Cardiol 44:151–159

Kim M, Min HJ, Won HY, Park H, Lee JC, Park HW, Chung J, Hwang ES, Lee K (2009) Dimerization of translationally controlled tumor protein is essential for its cytokine-like activity. PLoS One 4:e6464

Kloc M, Tejpal N, Sidhu J, Ganachari M, Flores-Villanueva P, Jennings NB, Sood AK, Kubiak JZ, Ghobrial RM (2012) Inverse relationship between TCTP/RhoA and p53/cyclin a/actin expression in ovarian cancer cells. Folia Histochem Cytobiol 50:358–367

Koziol MJ, Garrett N, Gurdon JB (2007) Tpt1 activates transcription of oct4 and nanog in transplanted somatic nuclei. Curr Biol 17:801–807

Kuo LJ, Yang LX (2008) Gamma-H2AX - a novel biomarker for DNA double-strand breaks. In Vivo 22:305–309

Laplante M, Sabatini DM (2012a) mTOR signaling. Cold Spring Harb Perspect Biol 4

Laplante M, Sabatini DM (2012b) mTOR signaling in growth control and disease. Cell 149:274–293

Larson K, Yan SJ, Tsurumi A, Liu J, Zhou J, Gaur K, Guo D, Eickbush TH, Li WX (2012) Heterochromatin formation promotes longevity and represses ribosomal RNA synthesis. PLoS Genet 8:e1002473

Le TP, Vuong LT, Kim AR, Hsu YC, Choi KW (2016) 14-3-3 proteins regulate Tctp-Rheb interaction for organ growth in Drosophila. Nat Commun 7:11501

Lee T, Luo L (2001) Mosaic analysis with a repressible cell marker (MARCM) for Drosophila neural development. Trends Neurosci 24:251–254

Lespagnol A, Duflaut D, Beekman C, Blanc L, Fiucci G, Marine JC, Vidal M, Amson R, Telerman A (2008) Exosome secretion, including the DNA damage-induced p53-dependent secretory pathway, is severely compromised in TSAP6/Steap3-null mice. Cell Death Differ 15:1723–1733

Li Y, Inoki K, Yeung R, Guan KL (2002) Regulation of TSC2 by 14-3-3 binding. J Biol Chem 277:44593–44596

Liu H, Peng HW, Cheng YS, Yuan HS, Yang-Yen HF (2005) Stabilization and enhancement of the antiapoptotic activity of mcl-1 by TCTP. Mol Cell Biol 25:3117–3126

MacDonald SM, Rafnar T, Langdon J, Lichtenstein LM (1995) Molecular identification of an IgE-dependent histamine-releasing factor. Science 269:688–690

MacDonald SM, Bhisutthibhan J, Shapiro TA, Rogerson SJ, Taylor TE, Tembo M, Langdon JM, Meshnick SR (2001) Immune mimicry in malaria: Plasmodium falciparum secretes a functional histamine-releasing factor homolog in vitro and in vivo. Proc Natl Acad Sci U S A 98:10829–10832

Miron M, Verdu J, Lachance PE, Birnbaum MJ, Lasko PF, Sonenberg N (2001) The translational inhibitor 4E-BP is an effector of PI(3)K/Akt signalling and cell growth in Drosophila. Nat Cell Biol 3:596–601

Morrison DK (2009) The 14-3-3 proteins: integrators of diverse signaling cues that impact cell fate and cancer development. Trends Cell Biol 19:16–23

Moya M, Roberts D, Novick P (1993) DSS4-1 is a dominant suppressor of sec4-8 that encodes a nucleotide exchange protein that aids Sec4p function. Nature 361:460–463

Nielsen HV, Johnsen AH, Sanchez JC, Hochstrasser DF, Schiotz PO (1998) Identification of a basophil leukocyte interleukin-3-regulated protein that is identical to IgE-dependent histamine-releasing factor. Allergy 53:642–652

Niforou KM, Anagnostopoulos AK, Vougas K, Kittas C, Gorgoulis VG, Tsangaris GT (2008) The proteome profile of the human osteosarcoma U2OS cell line. Cancer Genomics Proteomics 5:63–78

Norbeck J, Blomberg A (1997) Two-dimensional electrophoretic separation of yeast proteins using a non-linear wide range (pH 3-10) immobilized pH gradient in the first dimension; reproducibility and evidence for isoelectric focusing of alkaline (pI > 7) proteins. Yeast 13:1519–1534
Obsilova V, Silhan J, Boura E, Teisinger J, Obsil T (2008) 14-3-3 proteins: a family of versatile molecular regulators. Physiol Res 57(Suppl 3):S11–S21
Oldham S, Montagne J, Radimerski T, Thomas G, Hafen E (2000) Genetic and biochemical characterization of dTOR, the Drosophila homolog of the target of rapamycin. Genes Dev 14:2689–2694
Pallares-Cartes C, Cakan-Akdogan G, Teleman AA (2012) Tissue-specific coupling between insulin/IGF and TORC1 signaling via PRAS40 in drosophila. Dev Cell 22:172–182
Peng JC, Karpen GH (2007) H3K9 methylation and RNA interference regulate nucleolar organization and repeated DNA stability. Nat Cell Biol 9:25–35
Peng JC, Karpen GH (2009) Heterochromatic genome stability requires regulators of histone H3 K9 methylation. PLoS Genet 5:e1000435
Predic J, Soskic V, Bradley D, Godovac-Zimmermann J (2002) Monitoring of gene expression by functional proteomics: response of human lung fibroblast cells to stimulation by endothelin-1. Biochemistry 41:1070–1078
Rehmann H, Bruning M, Berghaus C, Schwarten M, Kohler K, Stocker H, Stoll R, Zwartkruis FJ, Wittinghofer A (2008) Biochemical characterisation of TCTP questions its function as a guanine nucleotide exchange factor for Rheb. FEBS Lett 582:3005–3010
Rho SB, Lee JH, Park MS, Byun HJ, Kang S, Seo SS, Kim JY, Park SY (2011) Anti-apoptotic protein TCTP controls the stability of the tumor suppressor p53. FEBS Lett 585:29–35
Rosenwald IB, Kaspar R, Rousseau D, Gehrke L, Leboulch P, Chen JJ, Schmidt EV, Sonenberg N, London IM (1995) Eukaryotic translation initiation factor 4E regulates expression of cyclin D1 at transcriptional and post-transcriptional levels. J Biol Chem 270:21176–21180
Sanchez JC, Schaller D, Ravier F, Golaz O, Jaccoud S, Belet M, Wilkins MR, James R, Deshusses J, Hochstrasser D (1997) Translationally controlled tumor protein: a protein identified in several nontumoral cells including erythrocytes. Electrophoresis 18:150–155
Saucedo LJ, Gao XS, Chiarelli DA, Li L, Pan D, Edgar BA (2003) Rheb promotes cell growth as a component of the insulin/TOR signalling network. Nat Cell Biol 5:566–571
Shi S, Larson K, Guo D, Lim SJ, Dutta P, Yan SJ, Li WX (2008) Drosophila STAT is required for directly maintaining HP1 localization and heterochromatin stability. Nat Cell Biol 10:489–496
Singh A, Lim J, Choi KW (2005) Dorsoventral boundary for organizing growth and planar polarity in the Drosophila eye. In: Mlodzik M (ed) Advances in developmental biology: planar cell polarization during development. Elsevier, Amsterdam, p 59, 172p
Skoulakis EM, Davis RL (1996) Olfactory learning deficits in mutants for leonardo, a Drosophila gene encoding a 14-3-3 protein. Neuron 17:931–944
Stocker H, Radimerski T, Schindelholz B, Wittwer F, Belawat P, Daram P, Breuer S, Thomas G, Hafen E (2003) Rheb is an essential regulator of S6K in controlling cell growth in Drosophila. Nat Cell Biol 5:559–566
Susini L, Besse S, Duflaut D, Lespagnol A, Beekman C, Fiucci G, Atkinson AR, Busso D, Poussin P, Marine JC et al (2008) TCTP protects from apoptotic cell death by antagonizing bax function. Cell Death Differ 15:1211–1220
Telerman A, Amson R (2009) The molecular programme of tumour reversion: the steps beyond malignant transformation. Nat Rev Cancer 9:206–216
Thaw P, Baxter NJ, Hounslow AM, Price C, Waltho JP, Craven CJ (2001) Structure of TCTP reveals unexpected relationship with guanine nucleotide-free chaperones. Nat Struct Biol 8:701–704
Thebault S, Agez M, Chi X, Stojko J, Cura V, Telerman SB, Maillet L, Gautier F, Billas-Massobrio I, Birck C et al (2016) TCTP contains a BH3-like domain, which instead of inhibiting, activates Bcl-xL. Sci Rep 6:19725

Tuynder M, Susini L, Prieur S, Besse S, Fiucci G, Amson R, Telerman A (2002) Biological models and genes of tumor reversion: cellular reprogramming through tpt1/TCTP and SIAH-1. Proc Natl Acad Sci U S A 99:14976–14981

Tuynder M, Fiucci G, Prieur S, Lespagnol A, Geant A, Beaucourt S, Duflaut D, Besse S, Susini L, Cavarelli J et al (2004) Translationally controlled tumor protein is a target of tumor reversion. Proc Natl Acad Sci U S A 101:15364–15369

Wang X, Fonseca BD, Tang H, Liu R, Elia A, Clemens MJ, Bommer UA, Proud CG (2008) Re-evaluating the roles of proposed modulators of mammalian target of rapamycin complex 1 (mTORC1) signaling. J Biol Chem 283:30482–30492

Xu T, Rubin GM (1993) Analysis of genetic mosaics in developing and adult Drosophila tissues. Development 117:1223–1237

Yaffe MB (2002) How do 14-3-3 proteins work?-- gatekeeper phosphorylation and the molecular anvil hypothesis. FEBS Lett 513:53–57

Zhang H, Stallock JP, Ng JC, Reinhard C, Neufeld TP (2000) Regulation of cellular growth by the Drosophila target of rapamycin dTOR. Genes Dev 14:2712–2724

Zhang Y, Gao XS, Saucedo LJ, Ru BG, Edgar BA, Pan DJ (2003) Rheb is a direct target of the tuberous sclerosis tumour suppressor proteins. Nat Cell Biol 5:578–581

Zhang J, de Toledo SM, Pandey BN, Guo G, Pain D, Li H, Azzam EI (2012) Role of the translationally controlled tumor protein in DNA damage sensing and repair. Proc Natl Acad Sci U S A 109:E926–E933

Zhao J, Meyerkord CL, Du Y, Khuri FR, Fu H (2011) 14-3-3 proteins as potential therapeutic targets. Semin Cell Dev Biol 22:705–712

Chapter 9
Translationally Controlled Tumor Protein (TCTP/HRF) in Animal Venoms

Andrea Senff-Ribeiro

Abstract Proteins from TCTP/HRF family were identified as venom toxins of spiders from different genus. We have found a TCTP toxin in the venom gland of *Loxosceles intermedia*, a venomous spider very common in South Brazil. TCTP from *L. intermedia*, named LiTCTP, was cloned, produced in a heterologous prokaryotic system, and the recombinant toxin was biochemically characterized. Our results point that LiTCTP is involved in the inflammatory events of Loxocelism, the clinical signs triggered after *Loxosceles* sp. bite, which include intense inflammatory reaction at the bite site followed by local necrosis. TCTP toxins were also identified in spiders from different genus. There are very few articles about TCTP toxins in other venomous animals in the literature, although a NCBI database search on the protein sequences reveals TCTP on snake's venom glands transcriptomic and genomic studies. Studies on TCTP as a venom toxin are very few and its biological role as a venom component in prey capture is still unknown.

9.1 Introduction

Proteins from TCTP/HRF family have already been described in the gland secretion of arthropods, venom, and saliva (Gremski et al. 2014). TCTP/HRF activates multiple human cells including basophils, eosinophils, T cells, and B cells, which participate in the allergic response (MacDonald 2012). Therefore, extracellular functions of HRF/TCTP may exacerbate the allergic and inflammatory cascade observed in venomous accidents. These TCTP venom toxins were identified in venom glands of spiders from different genus (Sade et al. 2012; Kimura et al. 2012; Zobel-Thropp et al. 2014). We have studied a TCTP from *Loxosceles intermedia* venom. *Loxosceles* spiders are encountered in all continents and more than 100 different species have been reported. *Loxosceles* sp. spiders are venomous animals whose bites trigger a set of clinical signs called Loxocelism. Victims present

A. Senff-Ribeiro (✉)
Department of Cell Biology, Federal University of Paraná, Curitiba, Paraná, Brazil
e-mail: senffribeiro@gmail.com; senffribeiro@ufpr.br

A. Telerman, R. Amson (eds.), *TCTP/tpt1 - Remodeling Signaling from Stem Cell to Disease*, Results and Problems in Cell Differentiation 64,
DOI 10.1007/978-3-319-67591-6_9

intense inflammatory reaction followed by local necrosis with gravitational lesion spreading at the bite site and occasional systemic manifestations (Vetter 2008). The TCTP toxin from *L. intermedia* venom (LiTCTP) is involved in the inflammatory events of Loxoscelism. LiTCTP directly causes edema and increases vascular permeability in mice (Sade et al. 2012). In some cases, *Loxosceles* spider bites can cause hypersensitivity or even allergic reactions (Rattmann et al. 2008).

TCTP protein was also described in the venom of spider species: Gtx-TCTP was identified in *Grammostola rosea* tarantula venom gland (Kimura et al. 2012), SthTCTP in *Scytodes* spiders (Zobel-Thropp et al. 2014). But these toxins were not biochemically or biologically characterized.

Although research articles on TCTP from spiders are very few, we can find several protein sequences related to TCTP/HRF on NCBI database; there are also TCTP sequences described in snakes, but no related protein in scorpions.

9.2 *Loxosceles intermedia* TCTP

Loxosceles (brown spiders) is one of the medically important spider's genuses, which also include Lactrodectus (widow spiders), Phoneutria (armed spiders), and Hadronyche spp. (funnel-web spiders). Five species of Loxosceles are responsible for most cases of human envenomation (*L. rufescens, L. laeta, L. intermedia, L. gaucho*, and *L. reclusa*) (da Silva et al. 2004). The classical clinical symptoms of the dermonecrotic lesion caused by *Loxosceles* spider bites includes erythema, itching, and pain and are classified as cutaneous Loxoscelism (more than 70% of the cases) (Gremski et al. 2014). Accidents can also cause hypersensibility or even allergic reactions, symptoms which could be associated with histaminergic events such as an increase in vascular permeability and vasodilatation. Loxosceles venoms have a complex molecular composition, enriched with proteins that range from 3 to 40 kDa in molecular mass. The venom is composed of several different toxins and the mechanisms by which the venom exerts its effects are still under investigation, although studies have shown that venom components act synergistically (Gremski et al. 2014).

A TCTP/HRF protein was identified in the cDNA library of the *L. intermedia* venom gland. The complete cDNA sequence of LiTCTP comprises 536 bp and has an open reading frame that encodes a deduced 172-amino acid protein. The calculated molecular mass of the mature protein for LiTCTP was 22.3 kDa and the predicted pI 4.7 (Sade et al. 2012). We performed the cloning, heterologous expression, purification, and functional characterization on this novel member of the TCTP family from the *L. intermedia* venom gland (Sade et al. 2012). TCTP from *Loxosceles intermedia* (LiTCTP) was cloned and expressed as a heterologous protein in an *E. coli* expression system as a fusion protein with a 6× His-tag at the N-terminus. Purification of LiTCTP was performed by two steps chromatography, immobilized metal–ion affinity chromatography (Ni-NTA) agarose and ion exchange chromatography (DEAE-sepharose).

Transcriptome analysis of *L. intermedia* venom gland revealed that TCTP coding transcripts represent 0.4% of the encoded toxins (Gremski et al. 2010). LiTCTP recombinant protein and native *L. intermedia* venom toxins showed immunological cross-reactivity by immunoblot and ELISA assays (Sade et al. 2012). Recently, we showed there are also TCTP-related proteins in the venoms of *L. laeta* and *L. gaucho* by the detection of LiTCTP-related epitopes on toxins from these venoms (Buch et al. 2015).

LiTCTP is most homologous to TCTP from the ixodid ticks *Ixodes scapularis* and *Amblyomma americanum* (69% of sequence identity) and the tick *Dermacentor variabilis* (68% of identity). Phylogenetic analysis demonstrates that the *L. intermedia* TCTP protein is closely related to histamine releasing factors of ixodid ticks.

Most of the symptoms observed during Loxoscelism can be triggered by the phospholipase-D toxins, the most characterized and studied class of toxins from Loxosceles venoms. However, recombinant isoforms of phospholipase-D were not able to induce paw edemas of the same intensity as observed in whole venom tests. The edema at the bite site is a well-described symptom of cutaneous Loxoscelism (Ribeiro et al. 2007; Paludo et al. 2009). As TCTP is histamine releasing factor, we investigated the edematogenic effect of recombinant LiTCTP in a mouse model. LiTCTP induced subcutaneous paw edema in vivo in a time- and concentration-dependent manner. The LiTCTP edematogenic effect started rapidly (5 min) decreasing until a minimum thickness was reached after 240 min (Sade et al. 2012).

The effects of LiTCTP in vascular permeability were evaluated by observing vascular leakage of Evans Blue dye into the extravascular compartment of the skin in mice injected with the toxin (Sade et al. 2012). The dye leakage pattern varied between the vascular effect of the venom and the recombinant toxin, venom showed several extravasation points near the injection site, and LiTCTP presented a more diffuse profile. It has already been shown that *L. intermedia* venom can degranulate mast cells and release mediators such as histamine, which increase vascular permeability and induce vascular relaxation (Paludo et al. 2009). LiTCTP seems to be an earlier and quicker component of edema formation compared with the inflammatory response in mouse paws triggered by whole venom (Sade et al. 2012). In the context of Loxoscelism, extracellular functions of HRF/TCTP could exacerbate the inflammatory cascades and allergic response by the activation of immune cells involved in these process (including basophils, eosinophils, T cells, and B cells) contributing to the clinical signs observed following *Loxosceles* spider's bites.

The TCTP biological role as a Loxosceles venom component in prey capture is still unknown. In the case of the *Loxosceles* spiders, TCTP and other constituents of whole venom are secreted via holocrine secretion by venom gland as revealed by ultrastructural studies of the venom gland (dos Santos et al. 2000). Maybe TCTP is present in the Loxosceles venom because it is a cellular component of venom gland cells, as other cytoplasmatic proteins which were also described in the transcriptome study (Gremski et al. 2010).

9.3 Other TCTPs Found in Animal Venoms

Other studies in spider toxinology also described TCTP toxins. A TCTP sequence was identified in the cDNAs encoding toxin-like peptides from the venom gland of the Chilean common tarantula *Grammostola rosea* (Kimura et al. 2012). This spider belongs to the Theraphosidae family of spiders, which are large and often hairy arachnids with a widespread distribution throughout the tropics and subtropics (Escoubas and Rash 2004). Unlike *Loxosceles*, most of these animals are not involved in venomous accidents with humans (Escoubas and Rash 2004). After ESTs techniques applied to the cDNA library, Gtx-TCTP was revealed in *Grammostola rosea* tarantula venom gland (Kimura et al. 2012). GTx-TCTP transcript was expressed in both the venom gland and the pereopodal muscle. Real-time PCR showed that GTx-TCTP transcript in the pereopodal muscle was one-13th of that in the venom gland. Authors assume that GTx-TCTP acts as both growth-related cytosolic protein and secretory protein and that further investigation is needed to elucidate the bifunctional features of GTx-TCTP (Kimura et al. 2012).

TCTP is among the proteins expressed in the venom gland of *Scytodes thoracica*. This genus has a worldwide distribution and contains nearly 200 species. In *Scytodes* transcriptome and proteomic study, three cDNAs were identified as SthTCTP (*S. thoracica* TCTP). The molecular mass of SthTCTP amino acid sequence is predicted to be 19.3 kDa with a pI of 4.63. Phylogenetic analysis and several significant hits ($e \leq 10^{-5}$) from NCBI support a close relationship between SthTCTP and LiTCTP (86% of identity). *Scytodes* is a close relative of sicariids, including *Loxosceles* and *Sicarius* spiders whose venoms are toxic to mammals (Vetter 2008).

These are the only data on spiders TCTP available in scientific literature, but a simple search on protein sequences related to TCTP/HRF on NCBI database finds 10 TCTP sequences from spiders: Scytodidae *(3)*, Theraphosidae *(3)*, Sicariidae *(2)*, Lycosidae *(1)*, Theridiidae *(1)* families. Figure 9.1 shows the alignment of TCTP sequences from spiders and human. There is a TCTP partial sequence from *Lycosa singoriensis* venom gland on the database, although this sequence is not specifically mentioned as a toxin-like or a cellular transcript in the related manuscript (Zhang et al. 2010). Concerning spiders from *Theridiidae* family, *Latrodectus hesperus* (western black widow), it is not clear if the partial sequence was identified in venom glands or other part of the spider. From *Theraphoside* family, besides the sequence from *Grammostola rosea*, there is also a partial TCTP sequence from the spider *Haplopelma schmidti* (*Selenocosmia huwena*) on the database but the results of the study are still unpublished. The database also points to 16 TCTPs sequences from mites and ticks, other arthropods.

There is no data on the literature about TCTP in other venomous animals, but a search on the protein sequences from NCBI database (http://www.ncbi.nlm.nih.gov/protein) reveals 10 TCTP sequences on snakes: from *Viperidae (5)*, *Elapidae (3)*, *Colubridae (1)*, and *Pythonidae (1)* families. Figure 9.2 shows the alignment of TCTP sequences from snakes. Venoms from *Viperidae* and *Elapidae* families

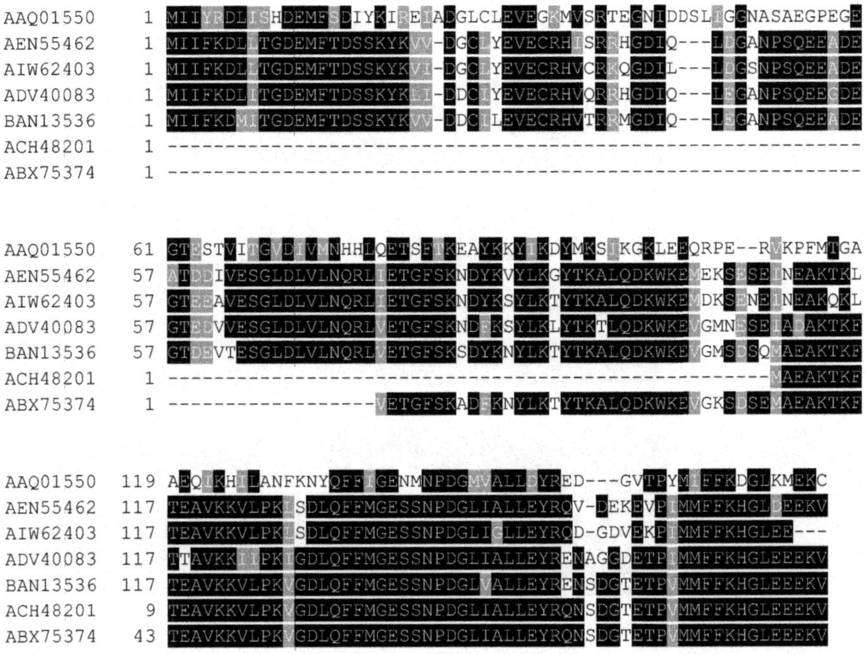

Fig. 9.1 Multiple alignment of TCTP protein sequences from different species of spiders compared to human and *Loxosceles intermedia* TCTP. Sequence alignment was performed using the Clustal Omega program and formatted with the BOXSHADE program (version 3.21). Fully conserved positions are shaded in *black* and conservative substitutions are in *gray*. GenBank sequences: *Homo sapiens* (AAQ01550.1); *Loxosceles intermedia* (AEN55462.1); *Scytodes thoracica* (AIW62403.1); *Grammostola rosea* (BAN13536.1); *Haplopelma schmidti* (ACH48201.1); *Latrodectus Hesperus* (ADV40083.1); *Lycosa singoriensis* (ABX75374.1)

present direct and negative impacts on human health. *Crotalus adamanteus* and *Crotalus horridus* (*Viperidae* family) TCTPs are on the database. *C. adamanteus*, the largest member of the genus, is a pit viper native to the southeastern United States whose TCTP was identified in its venom gland transcriptome. TCTP was also found in the venom gland transcriptome from a specific population of *C. horridus*, which presents potent and lethal venom and is found in northern Florida (USA). Interestingly this venom does not have the hemorrhagic effects typical of rattlesnake bites. *Micrurus fulvius* (eastern coral snake) and Ophiophagus are New World coral snakes (*Elapidae* family). TCTPs identified in the venom gland transcriptome from these venomous snakes are on the database. Although, these TCTP sequences which are in the database are not mentioned in the reference manuscripts (Rokyta et al. 2013; Rokyta et al. 2012; Margres et al. 2013). From *Colubridae* family, *Boiga irregularis* has a TCTP sequence on the database, although it is not mentioned in the reference manuscript (McGivern et al. 2014). The annotation of TCTP from *Python bivittatus* (Burmese python) was derived from a genomic sequence using gene prediction method. The search for TCTP

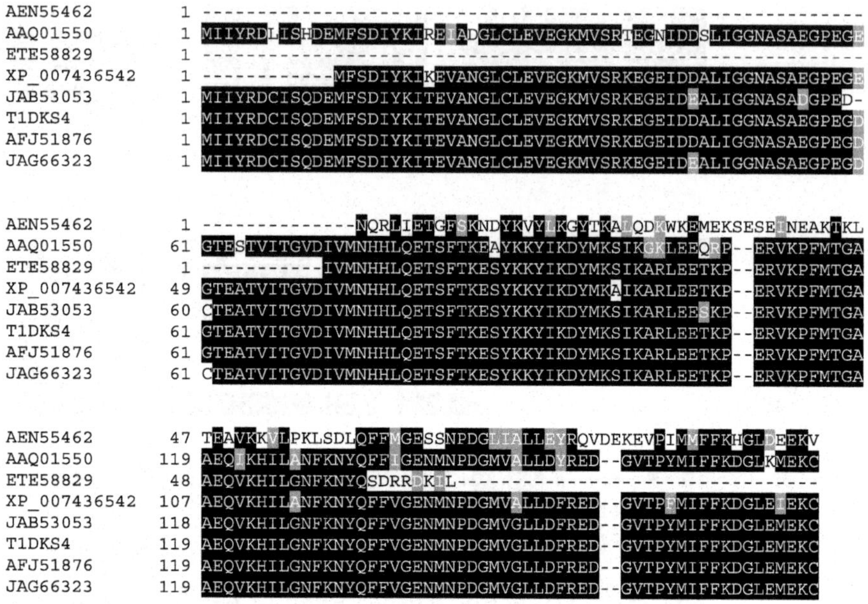

Fig. 9.2 Multiple alignment of TCTP protein sequences from different species of snakes compared to human and *Loxosceles intermedia* TCTP. Sequence alignment was performed using the Clustal Omega program and formatted with the BOXSHADE program (version 3.21). Fully conserved positions are shaded in *black* and conservative substitutions are in *gray*. *Loxosceles intermedia* TCTP (AEN55462.1); *Homo sapiens* TCTP (AAQ01550.1); *Ophiophagus hannah* (ETE58829.1); *Python bivittatus* (XP_007436542.1); *Micrurus fulvius* (JAB53053.1); *Crotalus horridus* (T1DKS4.1); *Crotalus adamanteus* (AFJ51876.1); *Boiga irregularis* (JAG66323.1)

sequences on the scorpions' database results in no matches; however, there is at least 6700 sequences from scorpions on the PUBMED protein database.

9.4 Perspectives

Worldwide over 44,900 species of spiders have been identified, but only a handful of venoms from these species have been characterized with molecular techniques (Platinick 2014). TCTP has already been described in gland secretions of many other arthropods, mainly parasites, such as ixodid ticks (Mulenga and Azad 2005) and the lamprey *Lampetra japonica* (Sun et al. 2008). Thus, it is likely that TCTP's activity as a histamine releasing factor plays a role in inflammation and infection processes of venomous accidents and parasitic conditions pathophysiology, but its primary role as a venom component still needs be elucidated.

References

Buch DR, Souza FN, Meissner GO, Morgon AM, Gremski LH, Ferrer VP, Trevisan-Silva D, Matsubara FH, Boia-Ferreira M, Sade YB, Chaves-Moreira D, Gremski W, Veiga SS, Chaim OM, Senff-Ribeiro A (2015) Brown spider (*Loxosceles* genus) venom toxins: evaluation of biological conservation by immune cross-reactivity. Toxicon 108:154–166

da Silva PH, da Silveira RB, Appel MH, Mangili OC, Gremski W, Veiga SS (2004) Brown spiders and Loxoscelism. Toxicon 44:693–709

dos Santos VL, Franco CR, Viggiano RL, da Silveira RB, Cantao MP, Mangili OC, Veiga SS, Gremski W (2000) Structural and ultrastructural description of the venom gland of *Loxosceles intermedia* (brown spider). Toxicon 38:265–285

Escoubas P, Rash L (2004) Tarantulas: eight-legged pharmacists and combinatorial chemists. Toxicon 43(5):555–574

Gremski LH, Da Silveira RB, Chaim OM, Probst CM, Ferrer VP, Nowatzki J, Weinschutz HC, Madeira HM, Gremski W, Nader HB, Senff-Ribeiro A, Veiga SS (2010) A novel expression profile of the *Loxosceles intermedia* spider venomous gland revealed by transcriptome analysis. Mol BioSyst 6(12):2403–2416

Gremski LH, Trevisan-Silva D, Ferrer VP, Matsubara FH, Meissner GO, Wille AC, Vuitika L, Dias-Lopes C, Ullah A, De Moraes FR, Chávez-Olórtegui C, Barbaro KC, Murakami MT, Arni RK, Senff-Ribeiro A, Chaim OM, Veiga SS (2014) Recent advances in the understanding of brown spider venoms: from the biology of spiders to the molecular mechanisms of toxins. Toxicon 83:91–120

Kimura T, Ono S, Kubo T (2012) Molecular cloning and sequence analysis of the cDNAs encoding toxin-like peptides from the venom glands of tarantula *Grammostola rosea*. Int J Peptides 2012:731293. https://doi.org/10.1155/2012/731293

MacDonald SM (2012) Potential role of histamine releasing factor (HRF) as a therapeutic target for treating asthma and allergy. J Asthma Allergy 5:51–59

Margres MJ, Aronow K, Loyacano J, Rokyta DR (2013) The venom-gland transcriptome of the eastern coral snake (*Micrurus fulvius*) reveals high venom complexity in the intragenomic evolution of venoms. BMC Genomics 14:531. https://doi.org/10.1186/1471-2164-14-531

McGivern JJ, Wray KP, Margres MJ, Couch ME, Mackessy SP, Rokyta DR (2014) RNA-seq and high-definition mass spectrometry reveal the complex and divergent venoms of two rear-fanged colubrid snakes. BMC Genomics 15:1061. https://doi.org/10.1186/1471-2164-15-1061

Mulenga A, Azad AF (2005) The molecular and biological analysis of ixodid ticks histamine release factors. Exp Appl Acarol 37:215–229

Paludo KS, Biscaia SM, Chaim OM, Otuki MF, Naliwaiko K, Dombrowski PA, Franco CR, Veiga SS (2009) Inflammatory events induced by brown spider venom and its recombinant dermonecrotic toxin: a pharmacological investigation. Comp Biochem Physiol C Toxicol Pharmacol 149(3):323–333

Platinick NI (2014) The world spider catalog. American Museum of Natural History, New York, NY. Version 14.5

Rattmann YD, Pereira CR, Cury Y, Gremski W, Marques MC, da Silva-Santos JE (2008) Vascular permeability and vasodilation induced by the *Loxosceles intermedia* venom in rats: involvement of mast cell degranulation, histamine and 5-HT receptors. Toxicon 51(3):363–372

Ribeiro RO, Chaim OM, da Silveira RB, Gremski LH, Sade YB, Paludo KS, Senff-Ribeiro A, de Moura J, Chávez-Olórtegui C, Gremski W, Nader HB, Veiga SS (2007) Biological and structural comparison of recombinant phospholipase D toxins from *Loxosceles intermedia* (brown spider) venom. Toxicon 50(8):1162–1174

Rokyta DR, Lemmon AR, Margres MJ, Aronow K (2012) The venom-gland transcriptome of the eastern diamondback rattlesnake (*Crotalus adamanteus*). BMC Genomics 13:312. https://doi.org/10.1186/1471-2164-13-312

Rokyta DR, Wray KP, Margres MJ (2013) The genesis of an exceptionally lethal venom in the timber rattlesnake (*Crotalus horridus*) revealed through comparative venom-gland transcriptomics. BMC Genomics 14:394. https://doi.org/10.1186/1471-2164-14-394

Sade YB, Bóia-Ferreira M, Gremski LH, Da Silveira RB, Gremski W, Senff-Ribeiro A, Chaim OM, Veiga SS (2012) Molecular cloning, heterologous expression and functional characterization of a novel translationally-controlled tumor protein (TCTP) family member from *Loxosceles intermedia* (brown spider) venom. Int J Biochem Cell Biol 44(1):170–177

Sun J, Wu Y, Wang J, Ma F, Liu X, Li Q (2008) Novel translationally controlled tumor protein homologue in the buccal gland secretion of *Lampetra japonica*. Biochimie 90:1760–1768

Vetter RS (2008) Spiders of the genus Loxosceles (*Araneae, Sicariidae*): a review of biological, medical and psychological aspects regarding envenomations. J Arachnol 36:150–163

Zhang Y, Chen J, Tang X, Wang F, Jiang L, Xiong X, Wang M, Rong M, Liu Z, Liang S (2010) Transcriptome analysis of the venom glands of the Chinese wolf spider *Lycosa singoriensis*. Zoology (Jena) 113(1):10–18. https://doi.org/10.1016/j.zool.2009.04.001

Zobel-Thropp PA, Correa SM, Garb JE, Binford GJ (2014) Spit and venom from *Scytodes* spiders: a diverse and distinct cocktail. J Proteome Res 13:817–835. https://doi.org/10.1021/pr400875s

Chapter 10
Tctp in Neuronal Circuitry Assembly

Cláudio Gouveia Roque and Christine E. Holt

Abstract Although *tctp* expression in many areas of the human brain was reported more than 15 years ago, little was known about how it functions in neurons. The early notion that Tctp is primarily expressed in mitotic cells, together with reports suggesting a relative low abundance in the brain, has perhaps potentiated this almost complete disregard for the study of Tctp in the context of neuron biology. However, recent evidence has challenged this view, as a number of independent genome-wide profiling studies identified *tctp* mRNA among the most enriched in the axonal compartment across diverse neuronal populations, including embryonic retinal ganglion cells. Considering the emerging parallels between axon guidance and cancer cell invasion, the axonal expression of cancer-associated *tctp* was suggestive of it holding an unexplored role in the wiring of neuronal circuits. Our study revealed that Tctp is necessary for the accurate and timely development of axon projections during the formation of vertebrate retinal circuits via its association with the survival machinery of the axon. Globally, the findings indicate that compromised pro-survival signaling in Tctp-deficient axons results in mitochondrial dysfunction and a subsequent decrease in axonal mitochondrial density. These effects likely translate into a metabolic state inadequate to support the normal guidance and extension processes of a developing axon.

C.G. Roque (✉)
The Taub Institute for Research on Alzheimer's Disease and the Aging Brain, Columbia University, New York, NY, USA
e-mail: cag2230@cumc.columbia.edu

C.E. Holt
Department of Physiology, Development and Neuroscience, University of Cambridge, Cambridge, England, UK

A. Telerman, R. Amson (eds.), *TCTP/tpt1 - Remodeling Signaling from Stem Cell to Disease*, Results and Problems in Cell Differentiation 64,
DOI 10.1007/978-3-319-67591-6_10

10.1 Introduction

10.1.1 Features of Axon Development

The arrangement of retinal neurons in the brain reflects that of the light-sensitive cells in the retina and, ultimately, the visual world. During embryonic development, independent of the birthplace in the retina, retinal ganglion cell (RGC) axons extend in the direction of the optic nerve head, where they collect to exit the eye and form the optic nerve. In vertebrates, once past the midline optic chiasm in the ventral diencephalon, retinal ganglion cell axons grow to the optic tectum, their most prominent synaptic target in the midbrain, and arborize in a topographic array that, in essence, copies the spatial map in the retina onto the brain. Likewise, several other neuronal projections are concurrently established in the embryonic brain; so, how do axons succeed in finding their way?

Observations in vivo of developing axonal projections have discovered that their growth is highly directed, with axons navigating along a prescribed trajectory en route to their respective synaptic targets and making very few errors of navigation in the process (Crossland et al. 1974; Lance-Jones and Landmesser 1981; Holt and Harris 1983; Raper et al. 1983). This remarkable pathfinding fidelity depends on successive spatial signals—guidance cues—presented in the embryonic landscape and integrated by the growth cone, a sensory, and motile structure at the tip of developing axons. Axon trajectories are thus seemingly divided into shorter segments in such a way that the effort of navigating towards a distant target is reduced to the simpler task of reaching consecutive intermediate points.

Four evolutionary conserved families of signaling molecules that function as instructive—chemotactic—guidance cues are classically described for their widespread roles in axon guidance: netrins, semaphorins, slits, and ephrins. In addition to these *chemical* signals, growth cones are also instructed by cell–cell and cell–matrix *physical* adhesions that provide not only an effective roadmap for navigation but also an essential platform for the protrusive behavior of the growth cone (Lowery and Van Vactor 2009; O'Donnell et al. 2009). These contacts can be mediated by members of the integrin, cadherin, and, most prominently, immunoglobulin (IgG) superfamilies. Importantly, the actions of these various signals are not mutually exclusive, but rather coordinately act to ensure that axon navigation ensues unerringly. Indeed, the involvement of various instructive signals even along a short trajectory considerably diminishes the likelihood of guidance errors and promotes the necessary fidelity in the establishment of neuronal connections.

10.1.2 Axonal mRNA Localization: One in Thousands

The synapse, the structure formed by two communicating neurons, underlies one of the most striking features of nerve cells: the extreme cellular and molecular polarization of axons and dendrites (Barnes and Polleux 2009). Necessarily asymmetric in function, these two compartments receive an independent assortment of organelles, membrane components, and molecules from the cell body (Hirokawa and Takemura 2005). Subcellular RNA localization has emerged as a particularly prevalent and cost-efficient mechanism of outsourcing genomic information in these highly polarized cells where the site of transcription can be far removed from the final destination of the protein (Jung et al. 2012, 2014). Mechanistically, specific transcripts can be precisely localized to subcellular compartments using the "address" information harbored in their untranslated regions (UTRs), which function as *cis*-acting platforms for regulatory RNA-binding proteins (RBPs) and small noncoding RNAs (Andreassi and Riccio 2009). Subsequently, local, 'on-site' synthesis confers both spatial and temporal precision, as the new protein is present only where and when a biological demand for it exists (Fig. 10.1). It is this mRNA-based mechanism that, for example, allows the growth cone to enjoy a certain degree of functional autonomy in its guidance process (Jung et al. 2012).

Still, it was not until the recent appreciation of the complexity of the axonal transcriptome—several independent genome-wide screens have identified thousands of mRNAs localizing in the axonal compartment of embryonic and adult neuronal cells (Taylor et al. 2009; Andreassi et al. 2010; Zivraj et al. 2010; Gumy et al. 2011)—that the functional significance of this cellular mechanism was fully grasped. Indeed, the prevailing view at the turn of the century was that axons did not synthesize proteins but that instead the entire proteome in this compartment was maintained by a constant provision of proteins synthesized in the cell body and transported along the axon (Jung et al. 2012; Lasek et al. 1973). Significant evidence linking local mRNA translation to many aspects of axonal biology has since overwhelmingly dismissed this notion. It is now known that axonal mRNA translation regulates not only growth cone guidance decisions (Campbell and Holt 2001; Brittis et al. 2002; Wu et al. 2005; Leung et al. 2006, 2013), where its involvement was originally studied, but also axon elongation (van Kesteren et al. 2006; Hengst et al. 2009), axon maintenance and degeneration (Andreassi et al. 2010; Cox et al. 2008; Yoon et al. 2012; Baleriola et al. 2014), as well as nerve injury and axon regeneration responses (Verma et al. 2005; Ben-Yaakov et al. 2012; Perry et al. 2012), among other processes (Jung et al. 2012).

In turn, the impact of local protein synthesis in such diverse cellular mechanisms underlines a crucial aspect of axonal RNA localization: the dynamic nature of the local transcriptome. Indeed, even within the same population of neurons, comparative profiling of two different developmental stages has revealed that axons contain 'age'-specific mRNA pools. For example, mRNAs encoding synaptic vesicle proteins, which intuition suggests being irrelevant during the pathfinding stages, are only found in target-arrived axons (Zivraj et al. 2010). It is noteworthy, however, that all of the axonal populations analyzed to date appear to have a common core of transcripts, such as those encoding mitochondrial and ribosomal proteins (Jung et al. 2012), suggesting that these molecules are implicated in 'everyday' axon upkeep. This also seems to be the case with Tctp: its transcript is ranked among the most enriched in the axonal compartment of diverse embryonic

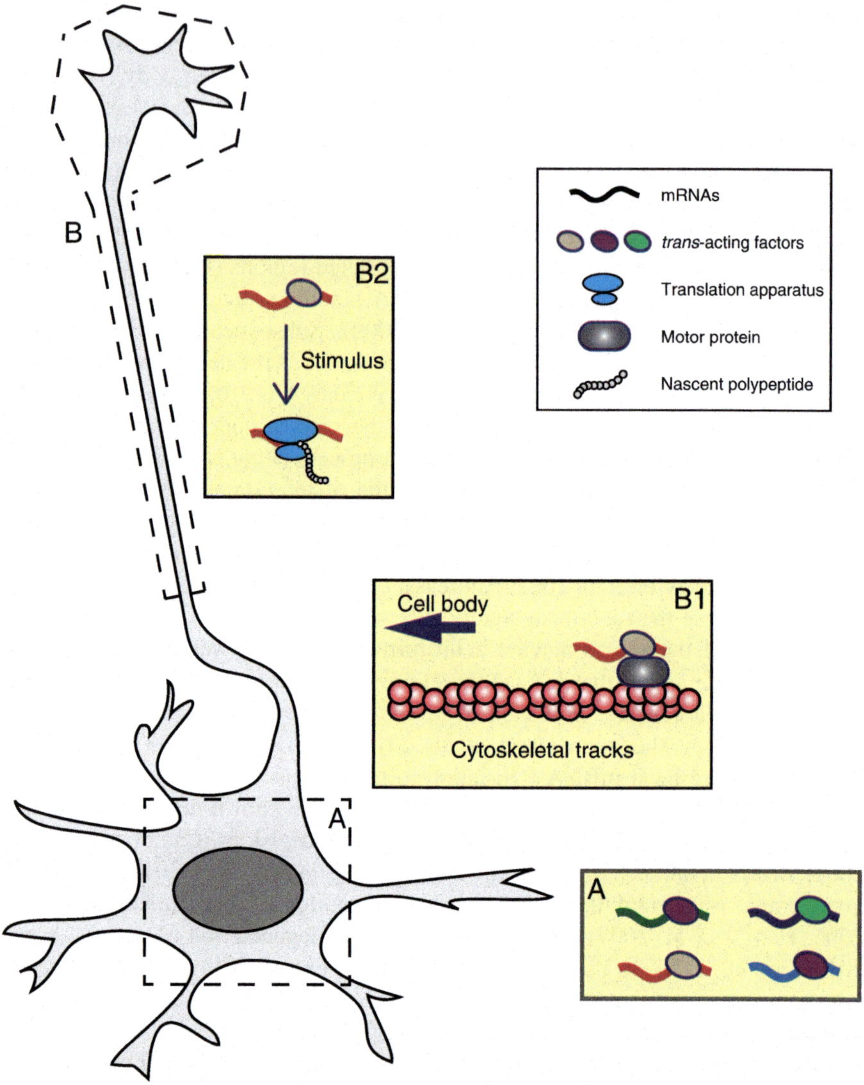

Fig. 10.1 Axonal mRNA localization and local protein synthesis. (*A*) Subcellular targeting of specific mRNAs depends on the recognition of *localization barcodes* by nuclear and cytoplasmic *trans*-acting factors (TAFs), which collectively associate as part of higher-order messenger ribonucleoprotein (mRNP) complexes. Most of the axon-targeting elements that have been identified are situated in the 3′UTR of the mRNAs and are decoded by various TAFs operating synchronously. Some functionally related transcripts share similar axon-targeting motifs and are regulated by common sets of TAFs, a property that allows these messages to be translated simultaneously with temporal and spatial precision. (*B1*) Upon recruiting additional adapter proteins (not depicted), mRNPs are shipped along cytoskeletal tracts by motor-driven active transport mechanisms towards their subcellular destination. Notably, mRNAs are maintained in a translationally dormant state during the assembly and transport phases. (*B2*) By modulating the activation of mTORC1 signaling and, in parallel, eliciting changes in the binding affinity of specific TAFs, various local stimuli, including guidance cues, can bring about concerted alterations in gene expression programs

Table 10.1 *tctp* is a highly enriched axonal mRNA. Although initial evidence put forward the idea that *tctp* was a modestly expressed gene in the brain (Thiele et al. 2000), recent genome-wide RNA profiling efforts have identified its mRNA among the most abundant axonal transcripts in both embryonic and adult neuronal populations.

Neuronal type	Embryonic	Adult	Methodology	Rank	Reference
Dorsal root ganglia (Rat)	✓	✓	Microarray	31st/69th	Gumy et al. (2011)
Cortical neurons (Rat)	Not investigated	✓ (aged in culture)	Microarray	11th	Taylor et al. (2009)
Sympathetic neurons (Rat)	✓ (perinatal)	Not investigated	SAGE analysis	5th	Andreassi et al. (2010)
Retinal ganglion cells (Frog)[a]	✓[b]	Not investigated	Microarray	83rd/72nd	Zivraj et al. (2010)
Retinal ganglion cells (Mouse)[a]	✓	Not investigated	Microarray	23rd	Zivraj et al. (2010)
Retinal ganglion cells (Frog)	✓	Not investigated	RT-qPCR	n.a.[c]	Roque et al. (2016)

[a]mRNAs isolated specifically from the growth cone – the protrusive, sensory structure located at the tip of developing axons
[b]Present in 'pathfinding' and 'target-arrived' axons
[c]10-fold higher than *actb*, a classical axon-enriched transcript

and adult neuronal populations (Table 10.1), which indicates that Tctp has a constitutive axonal function.

10.1.3 Axon Guidance and Cancer: Shared Features

The parallels between the processes of axon guidance and cancer cell invasion hinted that Tctp, a protein associated with malignancy, could play a particularly important role during the wiring of neuronal circuits (Box 10.1). Indeed, from the continuous changes in motility and adhesion, or the crosstalk with the surrounding environment, the challenges faced by a metastatic cell echo those overcome by a pathfinding growth cone as it navigates through the developing brain.

Box 10.1 Neuronal Connectivity and Cancer Metastasis: Historical Parallels?
Historically, the neuroscience field debated two explanatory hypotheses regarding the wiring of the nervous system. The "resonance theory" explained the developmental patterning of the central nerve tracts on a purely mechanical basis, by schemes of initially nonselective growth that, based on the validity of the connection formed, were later maintained or eliminated

(continued)

Box 10.1 (continued)

(Meyer 1998; Weiss 1936). A second framework proposed that selective chemical or electrical forces guided neuronal connections and found initial support in the experiments of John Langley in the late ninetieth century (Langley 1895). The extensive studies of Roger Sperry on how regenerating frog retinal ganglion cell axons are arranged when reinnervating their target categorically proved the latter hypothesis (Sperry 1943, 1944, 1963). In his most dramatic experiment, Sperry rotated the eye 180° on its dorsoventral axis after severing the optic nerve and noted that it lead to the animal having inverted vision; that is, the axons were originating from reversed positions in the eye yet managing to find their appropriate synaptic connections in the brain. He concluded that "the cells and fibers of the brain and cord must carry some kind of individual identification tags, presumably cytochemical in nature, by which they are distinguished one from another almost, in many regions, to the level of the single neuron" (Sperry 1963), a molecular view of the structuring of the nervous system which remains largely unchallenged to date (Zipursky and Sanes 2010).

This idea resonates with the seminal work of Stephen Paget, an English surgeon who published in 1889 what has come to be known as the "seed and soil" hypothesis, for it embodies an idea quite akin to that implied in Sperry's chemoaffinity postulate. Paget noted, in the process of analysis of more than 900 autopsy records, that tumor metastasis contains an organ-specific, nonrandom character: "The evidence seems to me irresistible that in cancer of the breast the bones suffer in a special way (...) Some bones suffer more than others; the disease has its seats of election" (Paget 1889). From these observations, he inferred that metastases depend on certain cancer cells—the "seeds"—having a specific affinity for the environment of certain organs—the "soil"—correctly concluding, with sound resemblance to modern day theories of neural circuitry assembly, that only when both "seed" and "soil" were compatible would metastasis form (Fidler 2003; Valastyan and Weinberg 2011).

Curiously, a short incursion into the history of the classical axon guidance molecules reveals an association with cancer pathology dating back to their discovery, suggesting that common signaling pathways operate in both contexts. The *Epha1* gene, for example, was cloned from a carcinoma cell line in 1987 in a screen for novel tyrosine kinase receptors with oncogenic potential, and the first Ephrin ligand was also described by a group working in the context of cancer (Pasquale 2010). Likewise, the *Dcc* gene, the prototypical Netrin-1 receptor, was originally identified as a tumor suppressor in advanced stages of colorectal carcinoma (hence its designation, deleted in colorectal cancer) (Mehlen et al. 2011). Only in the mid-1990s did their association with axon guidance mechanisms begin to be established (Tessier-Lavigne and Goodman 1996; Serafini et al. 1996).

However, the links between these processes are perhaps best illustrated by the recent characterization of frequent mutations and copy number variations in classical axon guidance genes in tumors derived from pancreatic ductal adenocarcinoma and liver fluke-associated cholangiocarcinoma patients (Biankin et al. 2012; Ong et al. 2012) or the ongoing cancer clinical trials targeting axon guidance molecules. Finally, it is also relevant to note that *tctp* is not the only cancer-associated transcript localizing in developing axons; in fact, the "cancer" gene ontology (GO) term is among the most significantly enriched in these axonal transcriptomes (Zivraj et al. 2010; Gumy et al. 2011), underlying the cellular and molecular commonalities that can be drawn between both contexts.

10.1.4 Axonal Mitochondria

Since, as a general rule, neuronal cells cannot be replaced throughout the individual's lifetime (Dekkers et al. 2013), the preservation of functional neural circuits must necessarily rely on effective protective mechanisms of its components. Classical conjectures supported the view that the process of axonal degeneration ensued from deficient sustenance from the cell body (e.g., as a result of cell body death) (Pease and Segal 2014). However, it is now well established that the axonal degenerative cascade can be actively promoted by in situ death pathways and is counteracted by locally acting and, to some extent, axon-specific pro-survival mechanisms (Pease and Segal 2014). Moreover, adequate metabolic provision—and hence mitochondria—is pivotal to axonal function, as the demand for energy, metabolites, and calcium buffering is particularly elevated at axons terminals (e.g., to support synaptic transmission) (Friedman and Nunnari 2014). Indeed, many mitochondrial dysfunctions trigger neurodegenerative disorders with prominent axonal phenotypes (Nunnari and Suomalainen 2012; Delettre et al. 2000; Alexander et al. 2000), suggesting that axons are particularly vulnerable to compromised mitochondrial output. Similarly, a growing axon is dependent on adequate mitochondrial operation, as it requires the continuous provision of energy for its extension in the embryonic brain. It follows that neurons must preserve a damage-prone mitochondrial network to maintain functionality and integrity.

10.2 TCTP in Neuronal Circuitry Assembly

Given that the identification of *tctp* as a potential candidate of study stemmed from genome-wide profiling screens, we initially sought to validate that its transcripts localize to retinal ganglion cell axons and growth cones at a time when the *Xenopus laevis* retinotectal projection is developing.[1] In situ hybridization showed robust *tctp* signal in the optic fiber layer and in the optic nerve head axon-only structures

[1]The retinotectal projection is formed by the nerve fibers of retinal ganglion cells, which connect the retina to the optic tectum.

through which retinal ganglion cell axons navigate to exit the eye. Additionally, in eye explants, *tctp* mRNA signal could be detected in the growth cone of retinal ganglion cell axons. In concordance with *tctp* mRNA axonal localization, Tctp protein was similarly detected in these retinal ganglion cell structures. Ample mRNA and protein signals were also found in the inner and outer plexiform layers, suggestive of localization in the neurites of other retinal neurons, as well as in the photoreceptor layer, populated by light-sensitive neurons, and the ciliary marginal zone, a neurogenic niche in the retina. Significantly, our initial investigations also showed that *tctp* expression is nearly tenfold higher than *actb* in retinal ganglion cell axons as measured by quantitative PCR,[2] confirming *tctp* as a highly enriched axonal transcript.

Further analyses revealed that Tctp is implicated in the development of the retinotectal projection (Fig. 10.2). Specifically, Tctp depletion using antisense morpholino oligonucleotides results in splayed projections that fail to innervate the optic tectum at the normal developmental time window (Fig. 10.2a). These effects are not a consequence of extracellularly acting Tctp, as normal retinal ganglion cell axons develop unerringly through a Tctp-deficient optic tract pathway (Fig. 10.2b). Moreover, in vivo time-lapse imaging of developing Tctp-depleted retinal axons revealed that their rate of extension was about half of that observed in controls, excluding the possibility that the axonal phenotypes observed are a result of an underlying delay in eye development.

We began our characterization of Tctp axonal mode of action by focusing on mitochondria. This line of investigation unexpectedly arose while examining the histology of Tctp-depleted retinas for signs of delayed development. Curiously, although the gross stratification of the retina was unaffected, we noted obvious signs of degeneration in the photoreceptor layer of Tctp morphants. The subsequent finding that Tctp expression in these cells is confined to the mitochondria-rich inner segments, together with reports documenting Tctp as part of the mitochondrial proteome (Fountoulakis et al. 2002; Rezaul et al. 2005), suggested a potential link between Tctp and mitochondrial function.[3] These indeed proved to be insightful observations, as Tctp morphant retinas show reduced total ATP levels. Following on this result, we measured a ~20% decline in the membrane potential of mitochondria from Tctp-depleted axons,[4] as well as a significant decrease in the number of axonal mitochondria. Importantly, this decrease in axonal mitochondrial density was not accompanied by changes in overall mitochondrial biogenesis or mass, arguing for a phenotype with predominantly axonal repercussions. Indeed, examination of mitochondrial transport dynamics in axons showed that a higher

[2] *actb* is a well-characterized axon-enriched mRNA (Jung et al. 2012; Leung et al. 2006).

[3] It is relevant to note that photoreceptor degeneration is frequently characterized by bioenergetic decline. For example, mitochondrial dysfunction is reported in age-related macular degeneration.

[4] The mitochondrial membrane potential ($\Delta\Psi_m$) is a parameter directly related to the ability of cells to generate ATP by oxidative phosphorylation and thus serves as cardinal indicator of mitochondrial function.

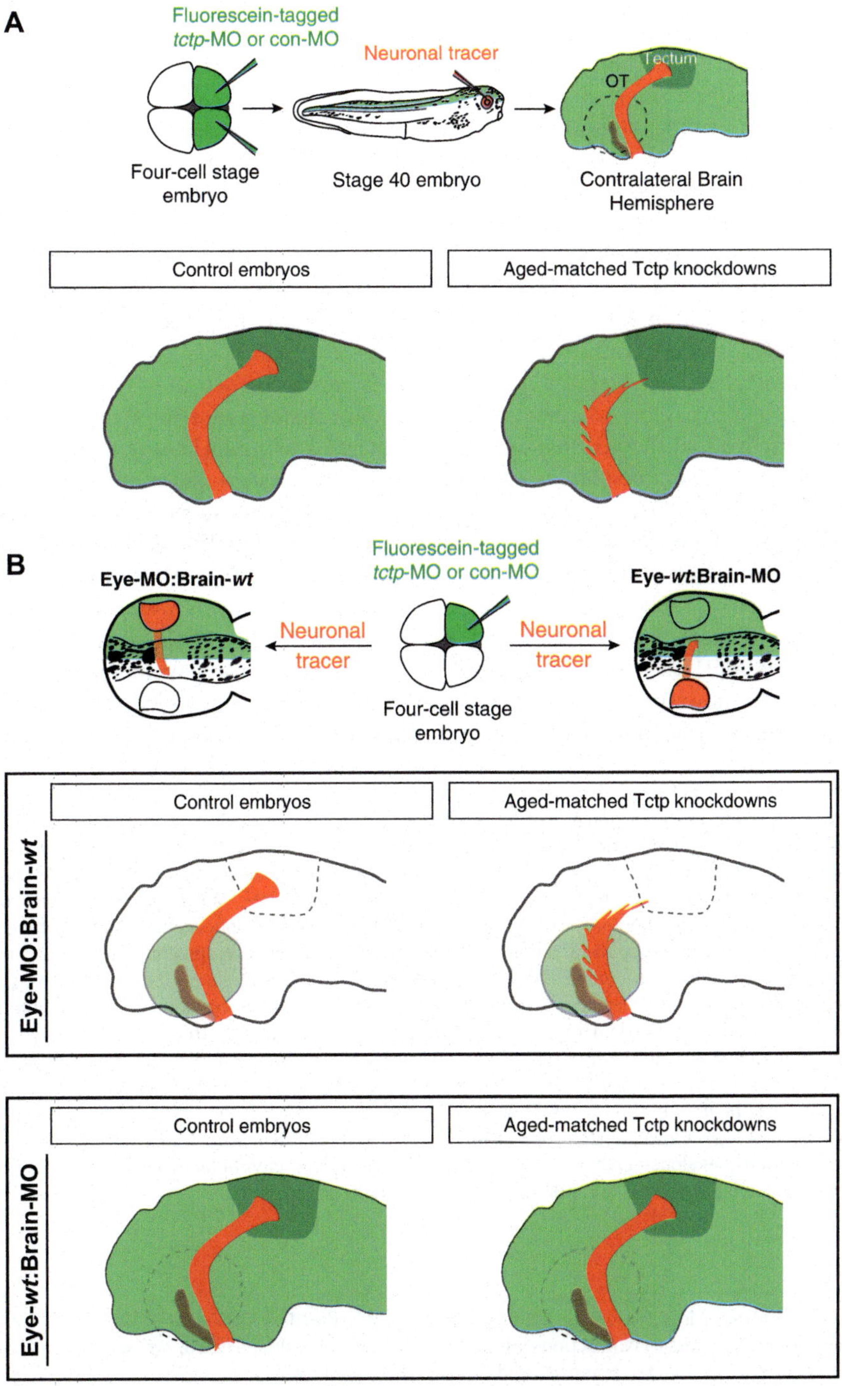

Fig. 10.2 Tctp is required for axon development in the embryonic visual system. (**a**) *tctp* knockdown in vivo was achieved using an antisense oligonucleotide morpholino (MO) delivered into both dorsal

proportion of these organelles move towards the cell body in axons deficient in Tctp than in controls, in line with previous reports showing that dysfunctional mitochondria are selectively "shipped" to the cell body for repair and/or degradation (Miller and Sheetz 2004; Sheng and Cai 2012).

How does mitochondrial dysfunction develop from Tctp deficiency? An interesting possibility stemmed from reports linking Tctp to the B-cell lymphoma 2 (Bcl2) family of proteins, which play key mediator roles of mitochondrial integrity and apoptosis (Czabotar et al. 2014). Significantly, embryonic sensory neurons lacking Bcl2, the prototypic member of this family, show reduced axon growth rates (Hilton et al. 1997), a phenotype we encountered in Tctp morphants. Particularly well defined is the association between Tctp and myeloid cell leukemia 1 (Mcl1) (Liu et al. 2005; Yang et al. 2005; Zhang et al. 2002), a neuroprotective Bcl2-related pro-survival factor (Mori et al. 2004), which prompted us to explore whether these two proteins shared a functional relationship in axons. We first showed that axonal Tctp interacts with Mcl1 using a proximity ligation assay, complementing previous biochemical data with an approach that allows the examination of protein–protein interactions with subcellular precision. Second, we looked for signs of unbalanced pro-survival signaling in Tctp-depleted axons[5] (Liu et al. 2005; Amson et al. 2012; Rho et al. 2011; Vaseva and Moll 2009; Leu et al. 2004). Both cleaved Caspase-3 and P53 levels were found to be elevated in axons in the absence of Tctp. Third, consistent with the idea that Tctp works via Mcl1 and the survival machinery to regulate axon development, Mcl1 morphants show similar, albeit milder, axon misprojection phenotypes.[6] Finally, since the N-terminal region of Tctp is required for its pro-survival properties

Fig. 10.2 (continued) blastomeres of four-cell stage *Xenopus laevis*, which give rise to the entire central nervous system. The retinotectal projection was labeled by intra-ocular delivery of a fluorescent lipophilic dye (DiI) at stage 40, when pioneer axons have completed their stereotyped growth through the optic tract (OT) and reached their target area. Whereas control embryos consistently developed compact axon profiles and had innervated the optic tectum, Tctp deficiency resulted in stunted and splayed projections that lagged in their development. The retinotectal projection is depicted in orange. Dashed contour represents the contralateral, dye-filled eye. (**b**) Tctp displays IgE-dependent histamine-releasing activity and other cytokine-like extracellular roles. Consequently, it could regulate axon development through its effects in the embryonic brain environment. To test this possibility, we devised an approach that generates embryos deficient in Tctp only in one-half of the nervous system. Because the retinotectal projection projects contralaterally (i.e., axons from the left eye extend towards the right side of the brain), this methodology allowed us to probe the effects of a Tctp-deficient optic tract pathway. Overall, normal axons developed unaffected through the Tctp morphant environment, suggesting that the observed axon phenotypes are independent of Tctp acting extracellularly

[5]Tctp stabilizes Mcl1 biological activity and promotes the degradation of P53, which itself counteracts the pro-survival actions of Mcl1 at the mitochondria. Hence, we speculated that Tctp deficiency resulted in compromised pro-survival signaling.

[6]This milder phenotype may be due to compensation by other members of the Bcl-2 family.

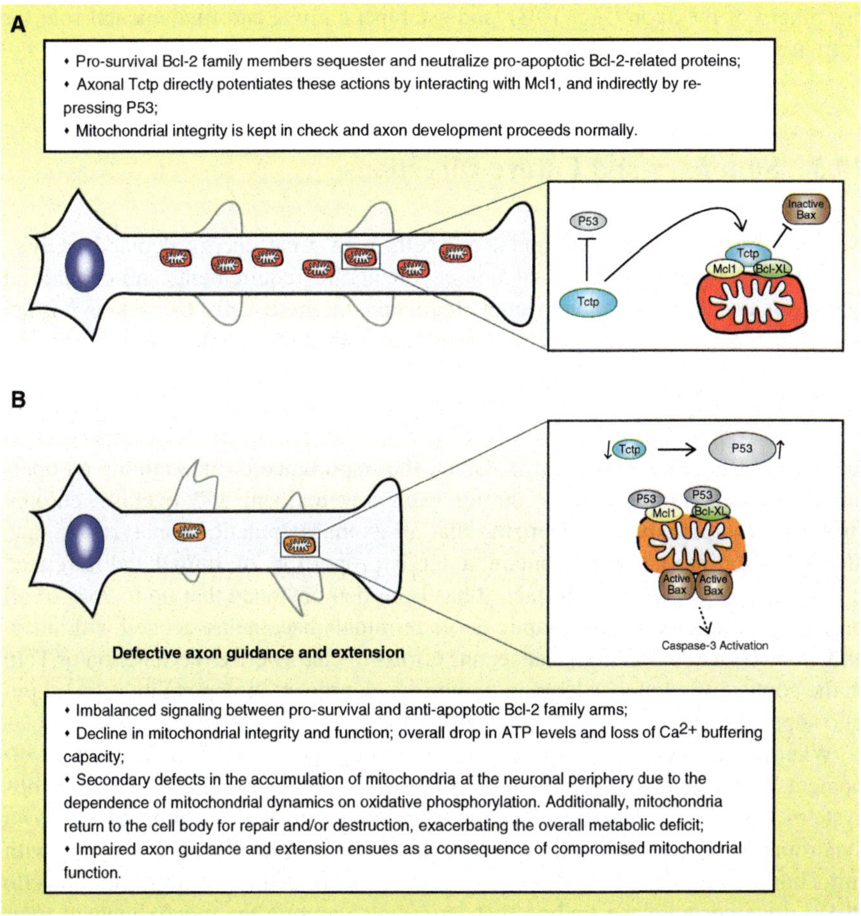

Fig. 10.3 Mechanistic insights into the role of Tctp in neuronal circuitry assembly. The normal physiologic scenario is illustrated in (**a**), whereas the consequences of Tctp deficiency on axon development programs uncovered by our study are shown in (**b**)

(Yang et al. 2005; Zhang et al. 2002), we were able to test whether Tctp pro-survival interactions are a requirement for normal axon development. To this end, we designed a transgene encoding a truncated Tctp protein that was devoid of pro-survival activity ($Tctp_{40\text{-}172aa}$). $Tctp_{40\text{-}172aa}$ retains Tctp's signature motifs, as well as the interactions domains required for the association with several known Tctp-interacting proteins, but lacks those necessary for the association with Mcl1 (Yang et al. 2005). Unlike full-length *tctp*, co-delivery of $tctp_{40\text{-}172aa}$ with a *tctp*-targeting morpholino failed to prevent the abnormal development of the retinotectal projection resulting from Tctp deficiency. Collectively, our various findings suggest that Tctp regulates axon development through its association with the survival machinery of the axon (Fig. 10.3), and establish a novel and fundamental role for Tctp in vertebrate neural circuitry assembly (Roque et al. 2016).

10.3 Summary and Future Directions

Neurons are highly compartmentalized cells with great energy demands. Given their elongated morphology and unique metabolic requirements, mitochondrial operation needs to be appropriately regulated in these cells to sustain normal neuronal functioning. This assumes particular relevance at distal axon terminals, which require the localized presence of mitochondria to support growth, maintenance, and synaptic transmission (Sheng and Cai 2012). Significantly, our study identified Tctp as a key checkpoint for normal axon development by impacting on axonal mitochondrial homeostasis. Given the importance of maintaining an operational mitochondrial network during axon development and overall neuronal function, it is perhaps not surprising that all axonal populations analyzed to date at the transcriptome level contain a large proportion of mitochondria-related mRNAs (Jung et al. 2012). In fact, it has been demonstrated that up to 25% of all proteins synthesized in presynaptic nerve terminals become associated with mitochondria (Gioio et al. 2001). Hence, our efforts to characterize the function of Tctp in the context of axon development typify the significant biological investment put into supporting these organelles subcellularly.

Whereas we focused exclusively on examining the role of Tctp in axon development, future work should aim at elucidating its implications in the adult nervous system. Indeed, the decreased Tctp protein expression levels observed in Down syndrome and Alzheimer's disease (Kim et al. 2001), pathologies associated with mitochondrial dysfunction (Friedman and Nunnari 2014; Pagano and Castello 2012), together with the finding that *tctp* is also among the most abundant transcripts in adult axons, prompt speculation that Tctp holds an important lifelong axonal function. However, given that Tctp is required for the assembly of neural circuitry, temporal control over its expression will be a key aspect of any future examination. This could be achieved with minimal effort by crossing the existing *tctp*-floxed heterozygous mice with an inducible, neuron-specific Cre recombinase strain (Chen et al. 2007; Susini et al. 2008). Considering that proper mitochondrial operation is an imperative of synaptic homeostasis (Sheng and Cai 2012), such strategy would, for example, allow one to study Tctp in the context of synaptic function independently of preceding defects in neural circuitry formation. Very much in line with these considerations, recent evidence collected from a neuron-like cell line suggests that Tctp can function as a promoter of dopamine release, a neurotransmitter involved in motor control and reward-motivated behaviors (Seo et al 2016).

Acknowledgements Our work was supported by Fundação para a Ciência e a Tecnologia [fellowship SFRH/BD/33891/2009 to Cláudio Gouveia Roque] and a Wellcome Trust Programme

Grant [085314/Z/08/Z to Christine Holt]. Cláudio Gouveia Roque is extremely grateful to the Doctoral Programme in Experimental Biology and Biomedicine hosted by the Center for Neuroscience and Cell Biology at University of Coimbra.

References

Alexander C et al (2000) OPA1, encoding a dynamin-related GTPase, is mutated in autosomal dominant optic atrophy linked to chromosome 3q28. Nat Genet 26(2):211–215
Amson R et al (2012) Reciprocal repression between P53 and TCTP. Nat Med 18(1):91–99
Andreassi C, Riccio A (2009) To localize or not to localize: mRNA fate is in 3'UTR ends. Trends Cell Biol 19(9):465–474
Andreassi C et al (2010) An NGF-responsive element targets myo-inositol monophosphatase-1 mRNA to sympathetic neuron axons. Nat Neurosci 13(3):291–301
Baleriola J et al (2014) Axonally synthesized ATF4 transmits a neurodegenerative signal across brain regions. Cell 158(5):1159–1172
Barnes AP, Polleux F (2009) Establishment of axon-dendrite polarity in developing neurons. Annu Rev Neurosci 32:347–381
Ben-Yaakov K et al (2012) Axonal transcription factors signal retrogradely in lesioned peripheral nerve. EMBO J 31(6):1350–1363
Biankin AV et al (2012) Pancreatic cancer genomes reveal aberrations in axon guidance pathway genes. Nature 491(7424):399–405
Brittis PA, Lu Q, Flanagan JG (2002) Axonal protein synthesis provides a mechanism for localized regulation at an intermediate target. Cell 110(2):223–235
Campbell DS, Holt CE (2001) Chemotropic responses of retinal growth cones mediated by rapid local protein synthesis and degradation. Neuron 32(6):1013–1026
Chen SH et al (2007) A knockout mouse approach reveals that TCTP functions as an essential factor for cell proliferation and survival in a tissue- or cell type-specific manner. Mol Biol Cell 18(7):2525–2532
Cox LJ et al (2008) Intra-axonal translation and retrograde trafficking of CREB promotes neuronal survival. Nat Cell Biol 10(2):149–159
Crossland WJ et al (1974) The specification of the retino-tectal projection in the chick. J Comp Neurol 155(2):127–164
Czabotar PE et al (2014) Control of apoptosis by the BCL-2 protein family: implications for physiology and therapy. Nat Rev Mol Cell Biol 15(1):49–63
Dekkers MP, Nikoletopoulou V, Barde YA (2013) Cell biology in neuroscience: death of developing neurons: new insights and implications for connectivity. J Cell Biol 203 (3):385–393
Delettre C et al (2000) Nuclear gene OPA1, encoding a mitochondrial dynamin-related protein, is mutated in dominant optic atrophy. Nat Genet 26(2):207–210
Fidler IJ (2003) The pathogenesis of cancer metastasis: the 'seed and soil' hypothesis revisited. Nat Rev Cancer 3(6):453–458. https://doi.org/10.1038/nrc1098
Fountoulakis M et al (2002) The rat liver mitochondrial proteins. Electrophoresis 23(2):311–328
Friedman JR, Nunnari J (2014) Mitochondrial form and function. Nature 505(7483):335–343
Gioio AE et al (2001) Local synthesis of nuclear-encoded mitochondrial proteins in the presynaptic nerve terminal. J Neurosci Res 64(5):447–453
Gumy LF et al (2011) Transcriptome analysis of embryonic and adult sensory axons reveals changes in mRNA repertoire localization. RNA 17(1):85–98
Hengst U et al (2009) Axonal elongation triggered by stimulus-induced local translation of a polarity complex protein. Nat Cell Biol 11(8):1024–1030

Hilton M, Middleton G, Davies AM (1997) Bcl-2 influences axonal growth rate in embryonic sensory neurons. Curr Biol 7(10):798–800

Hirokawa N, Takemura R (2005) Molecular motors and mechanisms of directional transport in neurons. Nat Rev Neurosci 6(3):201–214

Holt CE, Harris WA (1983) Order in the initial retinotectal map in Xenopus: a new technique for labelling growing nerve fibres. Nature 301(5896):150–152

Jung H, Yoon BC, Holt CE (2012) Axonal mRNA localization and local protein synthesis in nervous system assembly, maintenance and repair. Nat Rev Neurosci 13(5):308–324

Jung H et al (2014) Remote control of gene function by local translation. Cell 157(1):26–40

Kim SH et al (2001) Decreased brain histamine-releasing factor protein in patients with Down syndrome and Alzheimer's disease. Neurosci Lett 300(1):41–44

Lance-Jones C, Landmesser L (1981) Pathway selection by chick lumbosacral motoneurons during normal development. Proc R Soc Lond B Biol Sci 214(1194):1–18

Langley JN (1895) Note on regeneration of prae-ganglionic fibres of the sympathetic. J Physiol 18 (3):280–284

Lasek RJ, Dabrowski C, Nordlander R (1973) Analysis of axoplasmic RNA from invertebrate giant axons. Nat New Biol 244(136):162–165

Leu JI et al (2004) Mitochondrial p53 activates Bak and causes disruption of a Bak-Mcl1 complex. Nat Cell Biol 6(5):443–450

Leung KM et al (2006) Asymmetrical beta-actin mRNA translation in growth cones mediates attractive turning to netrin-1. Nat Neurosci 9(10):1247–1256

Leung LC et al (2013) Coupling of NF-protocadherin signaling to axon guidance by cue-induced translation. Nat Neurosci 16(2):166–173

Liu H et al (2005) Stabilization and enhancement of the antiapoptotic activity of mcl-1 by TCTP. Mol Cell Biol 25(8):3117–3126

Lowery LA, Van Vactor D (2009) The trip of the tip: understanding the growth cone machinery. Nat Rev Mol Cell Biol 10(5):332–343

Mehlen P, Delloye-Bourgeois C, Chedotal A (2011) Novel roles for slits and netrins: axon guidance cues as anticancer targets? Nat Rev Cancer 11(3):188–197

Meyer RL (1998) Roger Sperry and his chemoaffinity hypothesis. Neuropsychologia 36(10):957–980

Miller KE, Sheetz MP (2004) Axonal mitochondrial transport and potential are correlated. J Cell Sci 117(Pt 13):2791–2804

Mori M et al (2004) Expression of apoptosis inhibitor protein Mcl1 linked to neuroprotection in CNS neurons. Cell Death Differ 11(11):1223–1233

Nunnari J, Suomalainen A (2012) Mitochondria: in sickness and in health. Cell 148(6):1145–1159

O'Donnell M, Chance RK, Bashaw GJ (2009) Axon growth and guidance: receptor regulation and signal transduction. Annu Rev Neurosci 32:383–412

Ong CK et al (2012) Exome sequencing of liver fluke-associated cholangiocarcinoma. Nat Genet 44(6):690–693

Pagano G, Castello G (2012) Oxidative stress and mitochondrial dysfunction in Down syndrome. Adv Exp Med Biol 724:291–299

Paget S (1889) The distribution of secondary growths in cancer of the breast. Lancet 133 (3421):571–573

Pasquale EB (2010) Eph receptors and ephrins in cancer: bidirectional signalling and beyond. Nat Rev Cancer 10(3):165–180

Pease SE, Segal RA (2014) Preserve and protect: maintaining axons within functional circuits. Trends Neurosci 37(10):572–582

Perry RB et al (2012) Subcellular knockout of importin beta1 perturbs axonal retrograde signaling. Neuron 75(2):294–305

Raper JA, Bastiani M, Goodman CS (1983) Pathfinding by neuronal growth cones in grasshopper embryos. I. Divergent choices made by the growth cones of sibling neurons. J Neurosci 3 (1):20–30

Rezaul K et al (2005) A systematic characterization of mitochondrial proteome from human T leukemia cells. Mol Cell Proteomics 4(2):169–181
Rho SB et al (2011) Anti-apoptotic protein TCTP controls the stability of the tumor suppressor p53. FEBS Lett 585(1):29–35
Roque CG et al (2016) Tumor protein Tctp regulates axon development in the embryonic visual system. Development 143(7):1134–1148
Seo J, Maeng J, Kim HJ (2016) Translationally controlled tumor protein stimulates dopamine release from PC12 cells via Ca2+-independent phospholipase A(2) pathways. Int J Mol Sci 17(10)
Serafini T et al (1996) Netrin-1 is required for commissural axon guidance in the developing vertebrate nervous system. Cell 87(6):1001–1014
Sheng ZH, Cai Q (2012) Mitochondrial transport in neurons: impact on synaptic homeostasis and neurodegeneration. Nat Rev Neurosci 13(2):77–93
Sperry RW (1943) Visuomotor coordination in the newt (triturus viridescens) after regeneration of the optic nerve. J Comp Neurol 79(1):33–55
Sperry RW (1944) Optic nerve regeneration with return of vision in anurans. J Neurophysiol 7 (1):57–69
Sperry RW (1963) Chemoaffinity in the orderly growth of nerve fiber patterns and connections. Proc Natl Acad Sci USA 50(4):703–710
Susini L et al (2008) TCTP protects from apoptotic cell death by antagonizing bax function. Cell Death Differ 15(8):1211–1220
Taylor AM et al (2009) Axonal mRNA in uninjured and regenerating cortical mammalian axons. J Neurosci 29(15):4697–4707
Tessier-Lavigne M, Goodman CS (1996) The molecular biology of axon guidance. Science 274 (5290):1123–1133
Thiele H et al (2000) Expression of the gene and processed pseudogenes encoding the human and rabbit translationally controlled tumour protein (TCTP). Eur J Biochem 267(17):5473–5481
Valastyan S, Weinberg RA (2011) Tumor metastasis: molecular insights and evolving paradigms. Cell 147(2):275–292
van Kesteren RE et al (2006) Local synthesis of actin-binding protein beta-thymosin regulates neurite outgrowth. J Neurosci 26(1):152–157
Vaseva AV, Moll UM (2009) The mitochondrial p53 pathway. Biochim Biophys Acta 1787 (5):414–420
Verma P et al (2005) Axonal protein synthesis and degradation are necessary for efficient growth cone regeneration. J Neurosci 25(2):331–342
Weiss P (1936) Selectivity controlling the central-peripheral relations in the nervous system. Biol Rev 11(4):494–531
Wu KY et al (2005) Local translation of RhoA regulates growth cone collapse. Nature 436 (7053):1020–1024
Yang Y et al (2005) An N-terminal region of translationally controlled tumor protein is required for its antiapoptotic activity. Oncogene 24(30):4778–4788
Yoon BC et al (2012) Local translation of extranuclear lamin B promotes axon maintenance. Cell 148(4):752–764
Zhang D et al (2002) Physical and functional interaction between myeloid cell leukemia 1 protein (MCL1) and Fortilin. The potential role of MCL1 as a fortilin chaperone. J Biol Chem 277 (40):37430–37438
Zipursky SL, Sanes JR (2010) Chemoaffinity revisited: dscams, protocadherins, and neural circuit assembly. Cell 143(3):343–353
Zivraj KH et al (2010) Subcellular profiling reveals distinct and developmentally regulated repertoire of growth cone mRNAs. J Neurosci 30(46):15464–15478

Chapter 11
Elusive Role of TCTP Protein and mRNA in Cell Cycle and Cytoskeleton Regulation

Jacek Z. Kubiak and Malgorzata Kloc

Abstract Translationally Controlled Tumor-associated Protein (TCTP) is a small, 23 kDa multifunctional and ubiquitous protein localized both in the cytoplasm and in the nucleus of eukaryotic cells. It is evolutionarily highly conserved. Certain aspects of its structure show remarkable similarities to guanine nucleotide-free chaperons Mss4 and Dss4 suggesting that at least some functions of TCTP may depend on its chaperon-like action on other proteins. Besides other functions, TCTP is clearly involved in cell cycle regulation. It is also regulated in a cell-cycle-dependent manner suggesting a reciprocal interaction between this protein and the cell cycle-regulating machinery. TCTP also interacts with the cytoskeleton, mostly with actin microfilaments (MFs) and microtubules (MTs). It regulates the cytoskeleton organization and through this action it also influences cell shape and motility. The exact role of TCTP in cell cycle and cytoskeleton regulation is certainly not fully understood. In this chapter, we summarize recent data on cell cycle and cytoskeletal aspects of TCTP regulatory role.

J.Z. Kubiak (✉)
Institute of Genetics and Development of Rennes (IGDR), Cell Cycle Group, CNRS, UMR 6290, 35043 Rennes, France

Faculty of Medicine, University Rennes 1, UEB, IFR 140, 35043 Rennes, France

Laboratory of Regenerative Medicine and Cell Biology, Military Institute of Hygiene and Epidemiology (WIHE), Warsaw, Poland
e-mail: jacek.kubiak@univ-rennes1.fr

M. Kloc
The Houston Methodist Research Institute, Houston, TX, USA

Department of Surgery, The Houston Methodist Hospital, Houston, TX, USA

University of Texas, MD Anderson Cancer Center, Houston, TX, USA

A. Telerman, R. Amson (eds.), *TCTP/tpt1 - Remodeling Signaling from Stem Cell to Disease*, Results and Problems in Cell Differentiation 64,
DOI 10.1007/978-3-319-67591-6_11

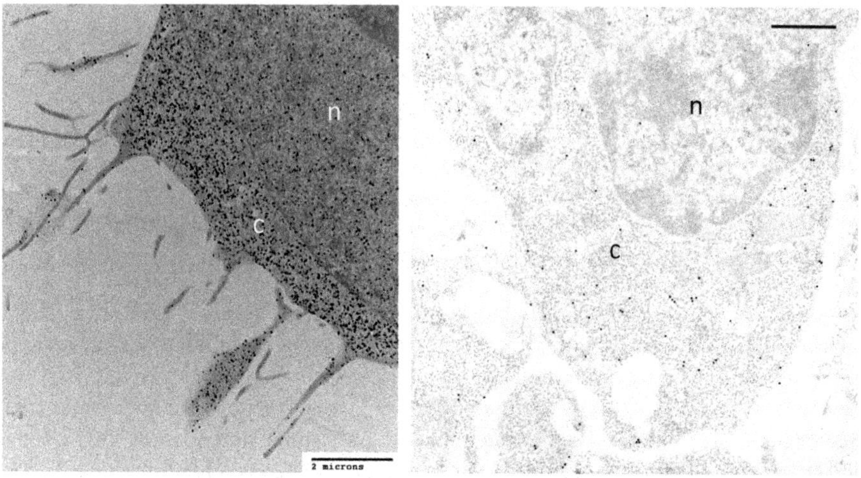

Fig. 11.1 Immunogold localization of TCTP in HIO180 cell from human ovary epithelium (*left*) and in *Xenopus laevis* ovary in oogonium (*right*). Note that TCTP is much more abundant in the cytoplasm (c) of the HIO180 cell than in the nucleus (n) and that TCTP-positive gold particles align along the actin filaments at the cell edge. In oogonium, in contrast to HIO180 cell, the nucleus (n) is almost completely devoid of TCTP

11.1 TCTP and Cell Cycle

11.1.1 The Role of TCTP Protein

TCTP is an abundant protein found in all eukaryotic cells studied so far (Bommer and Thiele 2004). It is predominantly cytoplasmic (Bazile et al. 2009), but may also localize to the nucleus (Kloc et al. 2012; Ma and Zhu 2012; Fig. 11.1). It is also a well known anti-apoptotic protein (Susini et al. 2008; Thébault et al. 2016). TCTP knockout, for instance by RNAi in human HeLa cells or *Xenopus laevis* XL2 cells, is accompanied by a rapid modification of cell shape followed by diminution of cell number in the culture because of the massive cell death (Bazile et al. 2009). The latter partially depends on TCTP capacity to downregulate p53 levels in the cells (Amson et al. 2011; Kloc et al. 2012). As p53 is also one of the major regulators of cell proliferation and cell cycle, the balance between TCTP and p53 influences these two aspects of cell physiology. However, there is an increasing body of evidences for more direct and more precise role of TCTP in cell cycle event regulation. One of them is its potential involvement in prophase checkpoint that controls the timing of M-phase entry (Burgess et al. 2008). The activation of this checkpoint via microtubule disassembly triggers transient return of early prophase cells to the G2 state (Matsusaka and Pines 2004; Mikhailov et al. 2004). One of the main actors of this checkpoint is the ubiquitin ligase Chfr (Checkpoint protein with FHA and RING finger domain). Chfr ubiquitinates Plk1 kinase and regulates the prophase checkpoint probably via Plk1 posttranslational modification since there is

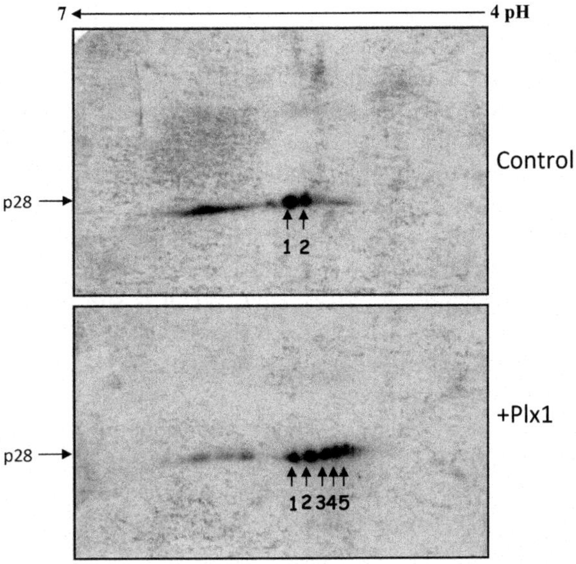

Fig. 11.2 Western blot of the 2 D gels of TCTP incubated in native conditions (control) and in the presence of active Plx1 (*Xenopus laevis* homologue of mammalian Plk1). Note the appearance of the series of new spots corresponding to novel forms of TCTP phosphorylated by Plx1

no massive Plk1 degradation upon prophase checkpoint activation (Matsusaka and Pines 2004). TCTP was shown to interact with Chfr in a microtubule-dependent manner (Burgess et al. 2008). The Chfr–TCTP interaction may be important for Chfr localization to the mitotic spindle and therefore for its proper action upon checkpoint activation. The MT disassembly provokes partial dissociation of TCTP and Chfr suggesting that the latter could be the checkpoint sensor detecting MT disassembly (Burgess et al. 2008). TCTP could play an important role by providing a link between Chfr and MTs. Despite these important data showing the involvement of TCTP in the prophase checkpoint, further analysis is necessary to fully understand its role in this process.

TCTP is phosphorylated by a cell cycle-regulating kinase Plk1 (Yarm 2002). Our unpublished results show that *Xenopus laevis* recombinant TCTP is phosphorylated by Xenopus homologue of Plk1, namely Plx1 (Bazile 2007; Fig. 11.2). This phosphorylation was proposed to decrease the microtubule-stabilizing activity of TCTP and to participate in regulation of MT dynamics during preparation for cell division and entry into the M-phase of cell cycle. As Plk1 is an important regulator of both M-phase entry (Mundt et al. 1997; Qian et al. 1999) and the mitotic spindle formation (Golsteyn et al. 1995), its kinase activity toward TCTP may allow better coordination of these processes.

The important role of TCTP in cell cycle progression and mitotic regulation was also found in human hepatocellular carcinoma (HCC) (Chan et al. 2012). The CHD1L (the chromodomain helicase/ATPase DNA-binding protein 1-like gene) is a specific oncogene of HCC. About a half of HCC cases carries CHD1L amplification. CHD1L protein directly binds to the promoter region of TCTP and activates its transcription. It has been shown that the overexpression of TCTP correlates with the increase in cell division failures and bad HCC patient prognosis.

At the molecular level, the TCTP overexpression promoted CDC25C phosphatase ubiquitin-mediated degradation by proteasomes. This in turn lowered the level of activation of the major M-phase kinase—CDK1, also known as a major enzymatic component of MPF, for M-phase Promoting Factor (Hunt 1989). CDC25C is a direct activator of CDK1 acting via dephosphorylation of two inhibitory sites on the kinase—Tyrosine 15 and Threonine 14 (Gautier et al. 1991). Therefore, proteolytic degradation of CDC25C diminishes its activation activity towards CDK1. This sequence of events results in shortening of mitosis and in consequence increases errors in chromosome separation leading to chromosome instability and an aneuploidy (Chan et al. 2012). However, it is still unclear how TCTP interacts with CDC25C phosphatase and by which mechanism it stimulates its degradation in the hepatocellular carcinoma cells. It is not clear whether TCTP acts in a similar way also in other cell types.

Knockout studies of TCTP-coding gene in the mouse has shown that in the absence of TCTP, the levels of cyclins D and E, necessary for the normal progression of the S-phase, drop dramatically resulting in lower cell proliferation (Chen et al. 2007). Thus, TCTP influences indirectly also the S-phase of the cell cycle.

11.1.2 The Role of TCTP mRNA

Several studies indicate that not only the TCTP protein, but also its transcript may play an important role in cell physiology and the cell cycle regulation. It has been shown that TCTP mRNA forms an elaborate secondary structure able to interact with protein targets. One of such proteins is protein kinase PKR, also called double-stranded RNA (dsRNA)-activated protein kinase (Bommer et al. 2002). PKR is mostly known for its involvement in viral infections. The binding of PKR to viral dsRNA triggers its autophosphorylation and activation. The phosphorylated form of PKR (pPKR) suppresses global translation via phosphorylation of the eukaryotic initiation factor 2α subunit (eIF2α). This regulation modifies multiple signaling pathways. Recent data show that among many pathways the pPKR directly regulates mitotic entry and hence the cell cycle progression (Kim et al. 2014). The pPKR is an upstream kinase for c-Jun N-terminal kinase (JNK), important regulator of the cell cycle and the cytoskeleton. It regulates the levels of mitotic cyclins and Plk1 as well as the level of histone H3 phosphorylation during mitosis. Thus, TCTP mRNA may indirectly modify cell cycle progression and cell proliferation (Fig. 11.3). Interestingly, PKR is not only activated by TCTP mRNA, but in turn it also regulates TCTP translation (Bommer et al. 2002). Thus, TCTP gene may act on the cell cycle progression using both of its gene products: the transcript and the protein. The nontranslational (structural) role of TCTP mRNA described above is a formidable example of recently described phenomenon of nontranslational roles of certain mRNAs in the organization of cytoskeleton, nucleation of subcellular organelles, and acting as a structural platform for the assembly of multiprotein complexes (Heasman et al. 2001; Jenny et al. 2006; Kanke et al. 2015; Ryu and

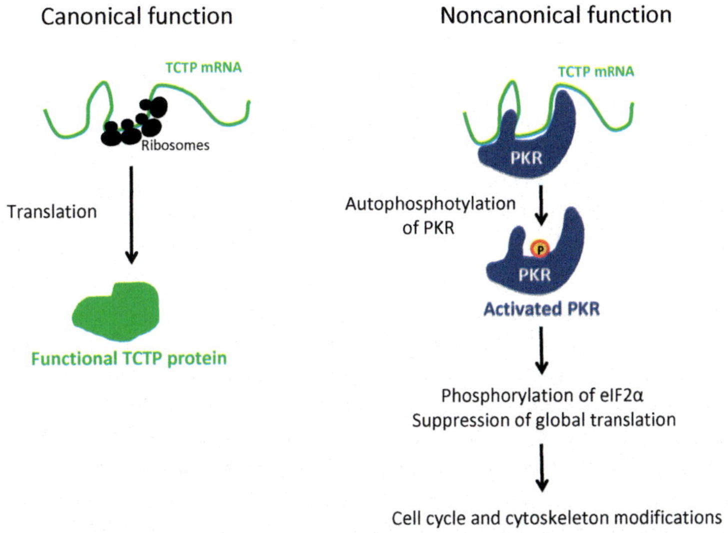

Fig. 11.3 Binary function of TCTP mRNA. *On the left*: canonical (translational) function of TCTP mRNA leading to the translation of functional TCTP protein. *On the right*: noncanonical (nontranslational) function, leading to PKR autophosphorylation and TCTP mRNA regulatory involvement in cell cycle and cytoskeleton modifications

Macdonald 2015; Blower et al. 2005, 2007; Kloc et al. 2007, 2011a, b; Kloc 2008, 2009; Kloc and Kubiak 2016, 2017).

So far, TCTP was not clearly linked to a precise cell cycle-regulating pathway. It does not seem to play a major role in a particular step of cell cycle regulation. However, on the basis of the above examples, it is very tempting to speculate that TCTP may act as a chaperon for various cell cycle regulators. Upon TCTP overexpression and in appropriate conditions, like those taking place, for example, in hepatocellular carcinoma cells, its modulatory role may be affected to such a degree that it profoundly modifies the cell cycle.

11.2 TCTP and Cytoskeleton

Although the association of TCTP with microtubules (MTs) and microfilaments (MFs) is still not very well characterized, it seems that TCTP has relatively low affinity for MTs, and much higher for MFs.

11.2.1 TCTP and Microtubules

TCTP clearly localizes to the mitotic spindle and to cytoplasmic MTs (Gachet et al. 1999; Yarm 2002; Burgess et al. 2008; Bazile et al. 2009; Jeon et al. 2016). Although TCTP is believed to interact with tubulin and MTs (Gachet et al. 1999) the TCTP pull-down experiments with purified MTs have shown very low affinity of TCTP for microtubules (Bazile et al. 2009). Similar experiments with actin MFs have shown that TCTP has high affinity for MFs when the experiment is done in the presence of cell-free extract from *Xenopus laevis* embryos (ibid.). This suggests that the cell-free extract provides either intermediate proteins that attach TCTP to F-actin or it modifies TCTP or actin in a posttranslational manner (e.g., via phosphorylation/dephosphorylation) allowing their mutual interaction. Precise localization of TCTP and tubulin in human HeLa cells and *Xenopus laevis* XL2 cells has shown that some fibers (thick filaments) positive for TCTP do not stain with anti-tubulin antibody and vice versa (Bazile et al. 2009). This suggests that TCTP may decorate other intracellular fibers than MTs. Also a co-injection of anti-TCTP antibody with rhodamine-tubulin into cells has shown that the tubulin-positive network does not entirely correspond to the TCTP-positive array (Bazile 2007; Fig. 11.4). It is still not clear what fibers are decorated by TCTP. Our electron microscopy studies using immune-gold method have shown at the ultrastructural level that TCTP localizes in the vicinity of MTs, but does not decorate them directly (Jaglarz et al. 2012). This again supports a hypothesis that the association of TCTP with MTs is indirect and more elusive than with classical MT-Associated Proteins. Moreover, we found TCTP localized in centrosomes (Bazile et al. 2009; Jaglarz et al. 2012) and more precisely, at their outer part, outside of the gamma-tubulin-containing part of the centrosome (Jaglarz et al. 2012). These data strongly suggest that TCTP is associated with MTs, but in a special way and in addition it may influence MT cytoskeleton organization via modification of certain functions of centrosomes.

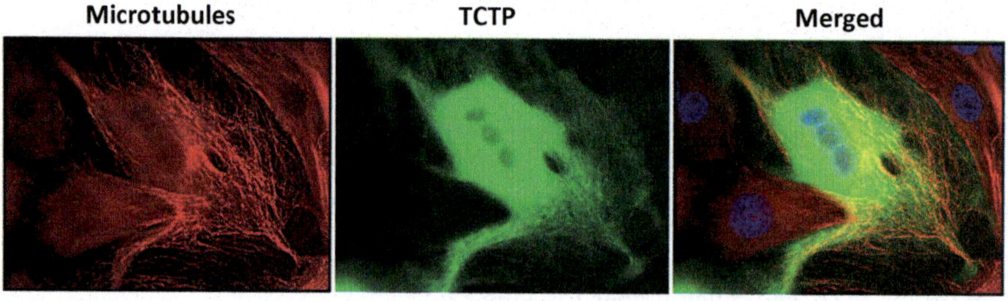

Fig. 11.4 TCTP-positive network does not exactly correspond to MT arrays. Non-overlapping of MTs (*left*; *red*) and TCTP-positive filaments (*middle*; *green*) in a cell co-injected with rhodamine-tubulin and anti-TCTP (subsequently stained green with FITC conjugated secondary antibody). The merged image of *red* and *green* is shown in the *right* panel. Nuclei (*blue*) are stained with Dapi

11.2.2 TCTP and Actin

TCTP co-localizes clearly with curly actin filaments at the cell border (Bazile et al. 2009). TCTP localization on actin filaments can be also seen in Fig. 11.1 where TCTP-positive gold particles align along the actin filaments at the border of HIO180 cell. TCTP has an actin-binding site with high homology to ADF/cofilin (Tsarova et al. 2010). This may suggest that TCTP and cofilin compete for actin binding and this competition influence actin stability. However, TCTP peptide, which is homologue to cofilin, has a higher affinity for G-actin than F-actin and does not block actin-filament depolymerization by cofilin (ibid.). Thus, TCTP seems to favor the sequestration of G-actin and in consequence targets more cofilin to F-actin. Inversely, the loss of TCTP could result in increased sequestration of cofilin by monomeric actin. Indeed, TCTP downregulation results in profound alteration of actin cytoskeleton (Liu et al. 2015). At least a part of this action may depend on the competition between TCTP and cofilin for G-actin-binding sites.

Finally, TCTP is a calcium-binding protein (Kim et al. 2000; Arcuri et al. 2004) and thus may participate in the cytoskeleton organization through local modifications of calcium homeostasis in the cytoplasm. Interestingly, it is known that actin-filament polymerization depends on calcium ions and that any disturbance in calcium homeostasis may influence actin-dependent cell shape (Ishida et al. 2017). Thus, it will be very interesting to study TCTP and actin interactions during cell shape changes related to cell polarization, cell movement, and interaction with extracellular matrix and neighboring cells.

11.3 Conclusions

TCTP plays multiple roles in cell physiology. It is a pro-proliferative and anti-apoptotic protein. Besides its direct role in the negative regulation of apoptosis, it participates in cell cycle regulation. It is involved in prophase checkpoint and in the control of M-phase duration. Its association with microtubules, actin, and centrosomes suggests that it participates in the coordination of the organization of the cytoskeleton. Its calcium-binding properties suggest that it may regulate calcium homeostasis and actin-related functions in the cell. In addition, potential structural (nontranslational functions) of TCTP mRNA should be an extremely interesting subject for future studies.

References

Amson R, Pece S, Lespagnol A, Vyas R, Mazzarol G, Tosoni D, Colaluca I, Viale G, Rodrigues-Ferreira S, Wynendaele J, Chaloin O, Hoebeke J, Marine JC, Di Fiore PP, Telerman A (2011) Reciprocal repression between P53 and TCTP. Nat Med 18:91–99

Arcuri F, Papa S, Carducci A, Romagnoli R, Liberatori S, Riparbelli MG, Sanchez JC, Tosi P, del Vecchio MT (2004) Translationally controlled tumor protein (TCTP) in the human prostate and prostate cancer cells: expression, distribution, and calcium binding activity. Prostate 60:130–140

Bazile F (2007) Recherche et étude de fonction de protéines ubiquitinées et/ou dégradées pendant la première mitose embryonnaire chez *Xenopus laevis*. Thesis of the University Rennes 1, Rennes, France, pp 1–148

Bazile F, Pascal A, Arnal I, Le Clainche C, Chesnel F, Kubiak JZ (2009) Complex relationship between TCTP, microtubules and actin microfilaments regulates cell shape in normal and cancer cells. Carcinogenesis 30:555–565

Blower MD, Nachury M, Heald R, Weis K (2005) A Rae1-containing ribonucleo- protein complex is required for mitotic spindle assembly. Cell 121:223–345

Blower MD, Feric E, Weis K, Heald R (2007) Genome-wide analysis demonstrates conserved localization of messenger RNAs to mitotic microtubules. J Cell Biol 179:1365–1373

Bommer UA, Thiele BJ (2004) The translationally controlled tumour protein (TCTP). Int J Biochem Cell Biol 36:379–385

Bommer UA, Borovjagin AV, Greagg MA, Jeffrey IW, Russell P, Laing KG, Lee M, Clemens MJ (2002) The mRNA of the translationally controlled tumor protein P23/TCTP is a highly structured RNA, which activates the dsRNA-dependent protein kinase PKR. RNA 8 (4):478–496

Burgess A, Labbé JC, Vigneron S, Bonneaud N, Strub JM, Van Dorsselaer A, Lorca T, Castro A (2008) Chfr interacts and colocalizes with TCTP to the mitotic spindle. Oncogene 27:5554–5566

Chan TH, Chen L, Liu M, Hu L, Zheng BJ, Poon VK, Huang P, Yuan YF, Huang JD, Yang J, Tsao GS, Guan XY (2012) Translationally controlled tumor protein induces mitotic defects and chromosome missegregation in hepatocellular carcinoma development. Hepatology 55 (2):491–505

Chen SH, Wu PS, Chou CH, Yan YT, Liu H, Weng SY, Yang-Yen HF (2007) A knockout mouse approach reveals that TCTP functions as an essential factor for cell proliferation and survival in a tissue- or cell type-specific manner. Mol Biol Cell 18(7):2525–2532

Gachet Y, Tournier S, Lee M, Lazaris-Karatzas A, Poulton T, Bommer UA (1999) The growth-related, translationally controlled protein P23 has properties of a tubulin binding protein and associates transiently with microtubules during the cell cycle. J Cell Sci 112:1257–1271

Gautier J, Solomon MJ, Booher RN, Bazan JF, Kirschner MW (1991) cdc25 is a specific tyrosine phosphatase that directly activates p34cdc2. Cell 67:197–211

Golsteyn RM, Mundt KE, Fry AM, Nigg EA (1995) Cell cycle regulation of the activity and subcellular localization of Plk1, a human protein kinase implicated in mitotic spindle function. J Cell Biol 129(6):1617–1628

Heasman J, Wessely O, Langland R, Craig EJ, Kessler DS (2001) Vegetal localization of maternal mRNAs is disrupted by VegT depletion. Dev Biol 240:377–386

Hunt T (1989) Maturation promoting factor, cyclin and the control of M-phase. Curr Opin Cell Biol 1:268–274

Ishida H, Jensen KV, Woodman AG, Hyndman ME, Vogel HJ (2017) The calcium-dependent switch helix of L-plastin regulates actin bundling. Sci Rep 7:40662

Jaglarz MK, Bazile F, Laskowska K, Polanski Z, Chesnel F, Borsuk E, Kloc M, Kubiak JZ (2012) Association of TCTP with centrosome and microtubules. Biochem Res Int 2012:541906

Jenny A, Hachet O, Závorszky P, Cyrklaff A, Weston MD, Johnston DS, Erdélyi M, Ephrussi A (2006) A translation independent role of oskar RNA in early drosophila oogenesis. Development 133:2827–2833

Jeon HJ, You SY, Park YS, Chang JW, Kim JS, Oh JS (2016) TCTP regulates spindle microtubule dynamics by stabilizing polar microtubules during mouse oocyte meiosis. Biochim Biophys Acta 1863(4):630–637

Kanke M, Jambor H, Reich J, Marches B, Gstir R, Ryu YH, Ephrussi A, Macdonald PM (2015) Oskar RNA plays multiple noncoding roles to support oogenesis and maintain integrity of the germline/soma distinction. RNA 21:1096–1109

Kim M, Jung Y, Lee K, Kim C (2000) Identification of the calcium binding sites in translationally controlled tumor protein. Arch Pharm Res 23:633–636

Kim Y, Lee JH, Park JE, Cho J, Yi H, Kim VN (2014) PKR is activated by cellular dsRNAs during mitosis and acts as a mitotic regulator. Genes Dev 28(12):1310–1322

Kloc M (2008) Emerging novel functions of RNAs, and binary phenotype? Dev Biol 317:401–404

Kloc M (2009) Teachings from the egg: new and unexpected functions of RNAs. Mol Rep Dev 76:922–932

Kloc M, Kubiak JZ (2016) Are morpholino technology dilemmas an affidavit of structural function of mRNA? Trends Dev Biol 9:11–169

Kloc M, Kubiak JZ (2017) Structural mRNA. In: Choi S (ed) Encyclopedia of signaling molecules, 2nd edn. Springer, Heidelberg

Kloc M, Bilinski S, Dougherty MT (2007) Organization of cytokeratin cytoskeleton and germ plasm in the vegetal cortex of Xenopus laevis oocytes depends on coding and non-coding RNAs: three-dimensional and ultrastructural analysis. Exp Cell Res 313:1639–1651

Kloc M, Dallaire P, Reunov A, Major F (2011a) Structural messenger RNA contains cytokeratin polymerization and depolymerization signals. Cell Tissue Res 346:209–222

Kloc M, Foreman V, Reddy SA (2011b) Binary function of mRNA. Biochimie 93:1955–1961

Kloc M, Tejpal N, Sidhu J, Ganachari M, Flores-Villanueva P, Jennings NB, Sood AK, Kubiak JZ, Ghobrial RM (2012) Inverse relationship between TCTP/RhoA and p53/cyclin a/actin expression in ovarian cancer cells. Folia Histochem Cytobiol 50(3):358–367

Liu Y, Zhang L, Tejpal N, Kubiak JZ, Ghobrial RM, Li XC, Kloc M (2015) TCTP silencing in ovarian cancer cells results in actin cytoskeleton remodeling and motility increase. J Anal Oncol 4:122–131

Ma YP, Zhu WL (2012) Cytoplasmic and nuclear localization of TCTP in normal and cancer cells. Biochem Res Int 2012:871728

Matsusaka T, Pines J (2004) Chfr acts with the p38 stress kinases to block entry to mitosis in mammalian cells. J Cell Biol 166:507–516

Mikhailov A, Shinohara M, Rieder CL (2004) Topoisomerase II and histone deacetylase inhibitors delay the G2/M transition by triggering the p38 MAPK checkpoint pathway. J Cell Biol 166:517–526

Mundt KE, Golsteyn RM, Lane HA, Nigg EA (1997) On the regulation and function of human polo-like kinase 1 (PLK1): effects of overexpression on cell cycle progression. Biochem Biophys Res Commun 239(2):377–385

Qian YW, Erikson E, Maller JL (1999) Mitotic effects of a constitutively active mutant of the xenopus polo-like kinase Plx1. Mol Cell Biol 19(12):8625–8632

Ryu YH, Macdonald PM (2015) RNA sequences required for the noncoding function of Oskar RNA also mediate regulation of oskar protein expression by bicoid stability factor. Dev Biol 407:211–223

Susini L, Besse S, Duflaut D, Lespagnol A, Beekman C, Fiucci G, Atkinson AR, Busso D, Poussin P, Marine JC, Martinou JC, Cavarelli J, Moras D, Amson R, Telerman A (2008) TCTP protects from apoptotic cell death by antagonizing bax function. Cell Death Differ 15:1211–1220

Thébault S, Agez M, Chi X, Stojko J, Cura V, Telerman SB, Maillet L, Gautier F, Billas-Massobrio I, Birck C, Troffer-Charlier N, Karafin T, Honoré J, Senff-Ribeiro A, Montessuit S, Johnson CM, Juin P, Cianférani S, Martinou JC, Andrews DW, Amson R, Telerman A, Cavarelli J (2016) TCTP contains a BH3-like domain, which instead of inhibiting, activates bcl-xL. Sci Rep 6:19725

Tsarova K, Yarmola EG, Bubb MR (2010) Identification of a cofilin-like actin-binding site on translationally controlled tumor protein (TCTP). FEBS Lett 584(23):4756–4760

Yarm FR (2002) Plk phosphorylation regulates the microtubule-stabilizing protein TCTP. Mol Cell Biol 22:6209–6221

Chapter 12
The Translationally Controlled Tumor Protein and the Cellular Response to Ionizing Radiation-Induced DNA Damage

Jie Zhang, Grace Shim, Sonia M. de Toledo, and Edouard I. Azzam

Abstract The absorption of ionizing radiation by living cells can directly disrupt atomic structures, producing chemical and biological changes. It can also act indirectly through radiolysis of water, thereby generating reactive chemical species that may damage nucleic acids, proteins, and lipids. Together, the direct and indirect effects of radiation initiate a series of biochemical and molecular signaling events that may repair the damage or culminate in permanent physiological changes or cell death. In efforts to gain insight into the mechanisms underlying these effects, we observed a prominent upregulation of the Translationally Controlled Tumor Protein (TCTP) in low dose/low dose rate ^{137}Cs γ-irradiated cells that was associated with adaptive responses that reduced chromosomal damage to a level lower than what occurs spontaneously. Therefore, TCTP may support the survival and genomic integrity of irradiated cells through a role in the DNA damage response. Consistent with this postulate, TCTP was shown to physically interact with ATM, an early sensor of DNA damage, and to exist in a complex with γH2A.X, in agreement with its distinct localization with the foci of the DNA damage marker proteins γH2A.X, 53BP1, and P-ATM. Cells lacking TCTP failed to repair chromosomal damage induced by γ-rays. Further, TCTP was shown to interact with the DNA-binding subunits, Ku70 and Ku80, of DNA-PK, a major participant in nonhomologous end joining of DNA double strand breaks. Moreover, TCTP physically interacted with p53, and its knockdown attenuated the radiation-induced G_1 delay, but prolonged the G_2 delay. Here, we briefly review the biochemical events leading to DNA damage by ionizing radiation and to its sensing and repair,

J. Zhang (✉)
Institute of Preventive Medicine, Fourth Military Medical University, Xi'an Shannxi, People's Republic of China
e-mail: zhangjie78@fmmu.edu.cn

G. Shim • S.M. de Toledo • E.I. Azzam (✉)
Department of Radiology, New Jersey Medical School, Cancer Center, Rutgers University, Newark, NJ, USA
e-mail: graceshim1@gmail.com; detolesm@njms.rutgers.edu; edouard.azzam@rutgers.edu

A. Telerman, R. Amson (eds.), *TCTP/tpt1 - Remodeling Signaling from Stem Cell to Disease*, Results and Problems in Cell Differentiation 64,
DOI 10.1007/978-3-319-67591-6_12

and highlight TCTP's critical role in maintaining genomic integrity in response to DNA-damaging agents.

12.1 Introduction

Cells in the human body are challenged by a variety of harmful compounds generated by normal physiological processes and through exposure to environmental agents. For example, we are constantly exposed to low levels of ionizing radiation from natural sources, and we may also be exposed to radiation released to the environment from nuclear fallouts and man-made sources, including consumer products, and discharges of radioactive waste. In addition, humans may become exposed to radiation during occupational activities related to nuclear technology, mining, high altitude airline travel, and deep space exploration. In particular, with the explosive growth in the use of diagnostic radiology, an increasing numbers of individuals are being repeatedly exposed to low and moderate doses of radiation during conventional as well as computed tomography, and nuclear medicine imaging procedures (NCRP 2009). In addition to low dose exposures, the use of different irradiation modalities (e.g., external beam therapies, brachytherapy with various sources of radiation) remains an effective and widely used means to treat cancer and other pathological conditions such as arteriovenous malformations (Coutard 1937; Hacein-Bey et al. 2014). Currently, 20–60% of all new cancer cases worldwide are treated with external photon beam radiotherapy as a standard option (Tyldesley et al. 2001; Delaney et al. 2005).

Exposures to ionizing radiation are, therefore, an inevitable part of the environment and increasingly of modern life. Hence, elucidation of the mechanisms underlying the cellular responses to low dose radiation is essential for estimating long-term health risks of low level exposures, for understanding the basis of normal tissue toxicities that arise following therapeutic exposures, and for enhancing the efficacy of radiotherapy through combined treatment modalities. Importantly, the study of the biochemical and molecular events underlying cellular and tissue responses to radiation have been major contributors to our knowledge of the stress response to radiation exposure, in particular the DNA damage response.

When cells are exposed to ionizing radiation, nucleic acids, proteins, lipids and other cellular constituents undergo chemical modifications. Of all the cellular constituents, stability of DNA is undoubtedly critical. Damage to the nuclear genome by direct interaction of DNA with radiation, or indirectly by its interaction with reactive compounds (e.g., reactive oxygen and nitrogen species) generated following cellular exposure to radiation, can cause a variety of adverse chemical/structural changes. Depurinations, depyrimidinations, base oxidation and strand breaks are some of the DNA damages that may occur (Dizdaroglu 2012). If unrepaired or misrepaired and propagated to progeny cells, these changes can lead to mutations that promote the development of cancer and degenerative diseases (Georgakilas 2011; Hall and Giaccia 2006; Little 2000).

The early radiation effects on the nuclear genome may be enhanced by excess oxidative stress resulting from persistent perturbations in oxidative metabolism, which can result in excess generation of reactive chemical species that exacerbate harmful conditions (Azzam et al. 2012; Spitz et al. 2004). In response to these stresses, cells have evolved multiple mechanisms to prevent the chemical/structural changes to DNA from occurring and/or repair the damage once it has occurred, thus maintaining genomic integrity. Cells react to genotoxic stress by activating a large network of signaling pathways that sense specific types of damage and trigger a coordinated and complex response to the chromosomal insults by activation of DNA damage sensing and repair pathways that restitute the damage or attenuate its level (Harper and Elledge 2007; Jeggo et al. 2016). In addition, cells activate antioxidant defenses to scavenge reactive chemical species (Petkau 1987) and cell cycle checkpoints to provide additional time for the defense mechanisms to operate (Little 1968; Zhou and Elledge 2000). In exposed tissues, immune responses may be triggered to promote repair and/or eliminate damaged cells (Jackson and Bartek 2009). In addition, apoptosis, necrosis, or mitotic death pathways may also be induced, particularly when the damage is severe (Okada and Mak 2004). These protective pathways have been under intense investigation and have been a fertile field for discovery of new players in the cellular defense against exogenous and endogenous stresses.

The Translationally Controlled Tumor Protein (TCTP) was identified over 25 years ago in Ehrlich's ascites tumor and in mouse erythroleukemia (Yenofsky et al. 1983; Chitpatima et al. 1988; Thomas et al. 1981). The protein was also named histamine releasing factor (HRF), fortilin, tpt1, Q23, P21, and P23 by independent groups, based on its multiple characteristics and activities (Thomas et al. 1981; Yenofsky et al. 1982; Bohm et al. 1989; MacDonald et al. 1995; Li et al. 2001). TCTP is an evolutionarily conserved molecule expressed in many eukaryotic cells (Bommer and Thiele 2004), which highlights its central biological importance. The complex nature of TCTP regulation during normal physiological functions and in response to stress continues to evolve. The various chapters in this book comprehensively describe the role of TCTP in various physiological functions and in cancer. TCTP has been implicated in transcription (Koziol et al. 2007), protein synthesis (Cans et al. 2003), cell cycle control (Brioudes et al. 2010), cytoskeleton regulation (Gachet et al. 1999; Burgess et al. 2008; Tsarova et al. 2010), immune responses (Macdonald 2012; Kashiwakura et al. 2012), development (Chen et al. 2007a, b; Hsu et al. 2007), viability (Susini et al. 2008), and cancer induction and reversion (Chan et al. 2012; Jung et al. 2011; Tuynder et al. 2002, 2004). Our emerging work has identified a novel role for TCTP in the repair of radiation-induced DNA damage, which is critical to health and disease (Zhang et al. 2012): Unrepaired DNA damage leads to genomic instability, which is a hallmark of cancer (Huang et al. 2003). As a background to this role, we briefly review the cellular effects of ionizing radiation. In particular, we discuss the induction of radiation-induced DNA damage and its repair, and the role of TCTP in the repair of DNA damage. Throughout, we identify areas where research may further illuminate the role of TCTP in this critical field.

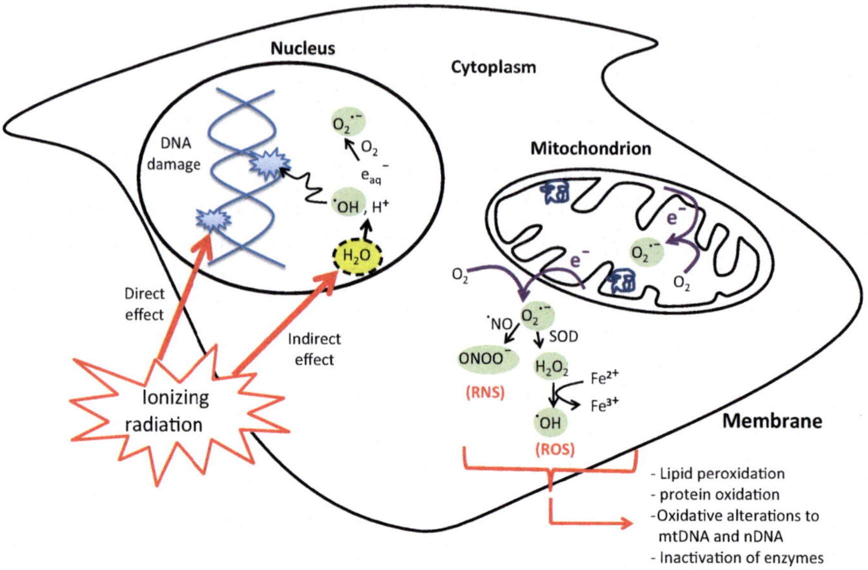

Fig. 12.1 The direct and indirect cellular effects of ionizing radiation on macromolecules. Absorption of ionizing radiation by living cells directly disrupts atomic structures, producing chemical and biological changes and indirectly through radiolysis of cellular water and generation of reactive chemical species by stimulation of oxidases and nitric oxide synthases. Ionizing radiation may also disrupt mitochondrial functions significantly contributing to persistent alterations in lipids, proteins, nuclear DNA (nDNA), and mitochondrial DNA (mtDNA)

12.2 Primary Effects of Ionizing Radiation

12.2.1 Direct and Indirect Effects of Ionizing Radiation

In mammalian cells, significant chemical modifications take place during or shortly after (within 10^{-15} to 10^{-6} s) the radiation exposure (Barendsen 1964; O'Neill and Wardman 2009; Meesungnoen et al. 2001). These modifications occur through direct interaction of the radiation with components of the exposed cells (e.g., DNA) or indirectly through generation of oxidizing species from water radiolysis (Fig. 12.1). Water is the major (~80%) constituent of cells. A thorough knowledge of water radiolysis is therefore critical for understanding radiobiological effects (Zimbrick 2002; LaVerne and Pimblott 1993).

The absorption of energetic radiations by water results in both excitations and ionizations leading to production of free radicals that in turn can attack other critical molecules (indirect effect). The schematic in Fig. 12.1 describes the complex events that accompany the absorption of high-energy photons or the passage of fast charged particles through water. These events can be divided into four, more or less clearly demarcated, consecutive, temporal stages (Platzman 1958). During the first or "physical" stage, the energy deposition is caused by the incident radiation and secondary electrons are generated. The resulting species are extremely unstable and undergo fast reorganization in the second or "physicochemical" stage. These processes produce radical and molecular products of radiolysis that are distributed in a nonhomogeneous track structure. The initial ($\sim 10^{-12}$ s) spatial distribution of reactants is then directly used as the starting point for the so-called stage of "nonhomogeneous chemistry." During this third stage, the various chemically reactive species diffuse and react with one another or with the environment ($\sim 10^{-6}$ s). Finally, in a physiologic system, the "biological" stage follows, in which the cells respond to the damage resulting from the products formed in the preceding stages. During this stage ($\sim 10^{-3}$ s or longer, depending very much upon the medium), the biological responses affecting the long-term consequences of radiation exposure are induced. In an aerobic cellular environment at physiological pH, the major reactive species at homogeneity ($\sim 10^{-6}$ s) include $O_2^{\bullet -}$, $^{\bullet}OH$, and H_2O_2 [reviewed in Azzam et al. (2012)]. Whereas ~1/3 of DNA damage from sparsely ionizing radiation (low linear energy transfer type such as energetic X- or γ-rays) emanates from direct interaction of the DNA with radiation, ~2/3 results from indirect effects involving water radiolysis products (Fig. 12.1) (Hall and Giaccia 2006). In the case of densely ionizing radiations such as alpha particles emanating from environmental radon gas, or high atomic number (Z) and high energy (E) (HZE) particles used in modern radiotherapy (Newhauser and Durante 2011) or encountered by astronauts during deep space travel (Li et al. 2014), it is commonly accepted that the cellular effects of HZE particles on macromolecules are mainly due to direct rather than indirect effects involving water radiolysis products (Hall and Giaccia 2006). Regardless, in cells exposed to such particulate radiations, the concentration of radiolytic species is very dense in and around the particle track (Goodhead 1988; Muroya et al. 2006; Chatterjee and Schaefer 1976), causing extensive covalent modifications in affected macromolecules (Li et al. 2014). Therefore, it would be of great interest to investigate the protective role of TCTP in the cellular defense against the damages induced by either sparsely or densely ionizing radiations.

In addition to a role of protecting against DNA damage induced through the direct effect of radiation, TCTP may have a role in promoting the scavenging of the reactive/DNA-damaging species generated during the biological stage of water radiolysis, which is consistent with TCTP's antioxidant role and the general cellular stress response (Gnanasekar and Ramaswamy 2007). While TCTP may act at early stages following irradiation, it could also exert an antioxidant role for longer times to help alleviate oxidative stress induced as a result of activation of oxidases/perturbations in oxidative metabolism (Spitz et al. 2004). Interestingly, although the family of TCTP proteins showed no primary sequence homology to any other protein family, the core domain of TCTP displays remarkable structural similarity with three families of proteins: Mss4/Dss4 proteins, which bind to the GDP/GTP-free form of Rabs proteins (Thaw et al. 2001), methionine sulfoxide reductases, and RNA helicases (Amson et al. 2013). Notably, methionine sulfoxide reductases play an important role in antioxidant defense, protein regulation, and survival (Moskovitz 2005). It is, therefore, attractive to speculate that TCTP may protect cells against oxidative stress by a mechanism that is yet to be discovered due to its structural similarity with methionine-R-sulfoxide reductase B1. Furthermore, it is of interest that one of the TCTP genes deposited in GenBank is designated as a PO_2 related protein (accession no. AAM51565) (Oikawa et al. 2002).

12.3 Endogenous and Radiation-Induced DNA Alterations

A strong emphasis thus far has been on the effect of exogenous agents such as ionizing radiation on DNA damage. However, improvements in the sensitivity of analytic methods to measure oxidative damage (Cadet et al. 2011) have revealed altered bases and nucleotides in the DNA of normal cells that have not been exposed to ionizing radiation or other mutagens (Weinberg 2007). The analyses have shown that endogenous biochemical processes greatly contribute to genome mutations. The reactive oxygen species (ROS) produced during normal cellular metabolic processes (mainly $O_2^{\bullet -}$ and H_2O_2) cause extensive depurinations and, to a lesser extent, depyrimidinations. In addition, ROS can oxidize bases in DNA, such as the oxidation of deoxyguanosine (dG) to 8-hydroxyguanine (8-oxodG), with ~100–500 of such lesions being formed per day in a human cell (Lindahl 1993). The rate of occurrence of these alterations has been closely linked to the rate of oxidative metabolism: higher oxygen consumption in different species were correlated with an increased rate of base oxidation in DNA (Ames 1989). A failure to repair oxidized bases creates a risk of mutation during DNA replication. For example, 8-oxodG mispairs with deoxyadenosine (dA) rather than deoxycytosine (dC), resulting in a C–A point mutation. Notably, oxidatively induced DNA lesions and DNA repair proteins have been suggested as potential biomarkers for early detection, cancer risk assessment, prognosis, and for monitoring therapy (Dizdaroglu 2012).

Several cellular defenses act to restore DNA integrity. Interestingly, the knock-down of TCTP by RNA interference (RNAi) in *normal* unirradiated human diploid fibroblasts led to an increase in the spontaneous rate of DNA damage in the form of micronuclei, which was validated by an increase in the average number of γH2A.X foci per cell [(Zhang et al. 2012); see results described in Fig. 12.5]. Micronuclei arise mainly from DNA double strand breaks (DSBs), a serious DNA lesion that leads to cell death (Fenech and Morley 1985; Baumstark-Khan 1993). Therefore, it is attractive to speculate that the increase in micronuclei upon knockdown of TCTP is likely due to abrogation of its antioxidant function, leading to the accumulation of oxidized bases that eventually result in DNA breaks.

12.3.1 DNA Damage Response Pathways and DNA Damage Repair Mechanisms

As highlighted above, DNA is continuously exposed to damaging agents from endogenous and external environmental stresses, along with lifestyle factors. This constant assault on DNA yields tens of thousands of DNA lesions per day in every human cell (Weinberg 2007). These DNA lesions must be repaired to prevent loss or incorrect transmission of genetic material, which can lead to tumorigenesis and other pathologies (Jackson and Bartek 2009; Lin et al. 2012).

As illustrated in Fig. 12.2, the direct interaction of DNA with ionizing radiation and the radiation-induced ROS induces a wide range of DNA damage of varying levels of complexity, such as base damage, single strand breaks (SSBs), abasic sites, DNA–protein cross-links, and DSBs (Nikjoo et al. 2001). Figure 12.3 shows that the choice of the repair system depends on the type of DNA lesion. Single strand breaks or single-base damage (i.e., DNA lesions on a single strand that do not significantly disrupt the helical structure) are generally repaired by base excision repair (BER) (Chou et al. 2015), whereas DNA damage that significantly distorts the DNA helix (e.g., bulky lesions and crosslinks) is repaired by nucleotide excision repair (NER) (Petruseva et al. 2014). Small chemical changes affecting a single base are repaired via direct repair (DR) (Yi and He 2013), and mismatches in base pairing caused by DNA replication errors are repaired by mismatch repair (MMR) (Larrea et al. 2010). Finally, DSBs are repaired via homologous recombination (HR) and/or nonhomologous end joining (NHEJ). The choice of repair system for DSB repair depends on the phase of the cell cycle and the expression, availability, and activation of DNA repair proteins (Lieber 2008; Shah and Mahmoudi 2015). Regardless of the type of lesion and the mechanisms required for its repair, cells initiate a complex signaling cascade that includes activation of DNA repair pathways, cell cycle arrest to allow time for repair of DNA damage, and in certain cases, initiation of senescence or apoptosis. These series of coordinated events are known as the DNA damage response (DDR) pathways (Jackson and Bartek 2009). Upon induction of DSBs, the central components of DDR activation are ATM, ATR, and DNA-dependent protein kinase (DNA-PK), members of the phosphatidyl inositol 3-kinase-like kinase (PIKK) family. ATM and DNA-PK are predominantly activated by DNA DSBs, whereas other types of DNA damage (e.g., replication-induced DSBs, base adducts, and cross-links) activate ATR (Branzei and Foiani 2008; Nam and Cortez 2011; Shiloh and Ziv 2013). Our studies have shown that TCTP interacts with components of both HR and NHEJ to promote repair of DNA damage (Zhang et al. 2012).

As shown in Fig. 12.3, DNA-PK and ATM are activated by the recruitment of Ku70/Ku80 and the MRN complex, respectively, to DSBs. Ku70/Ku80 and DNA-PK promote NHEJ repair of DSBs. The DNA-PK catalytic subunit (DNA-PKcs) keeps the broken DNA ends in close proximity during NHEJ repair and recruits end-processing factors (e.g., Artemis, PNKP, APE1, and TDP1), which prepare the DNA ends for re-ligation by the XRCC4–XLF–LIG4 complex

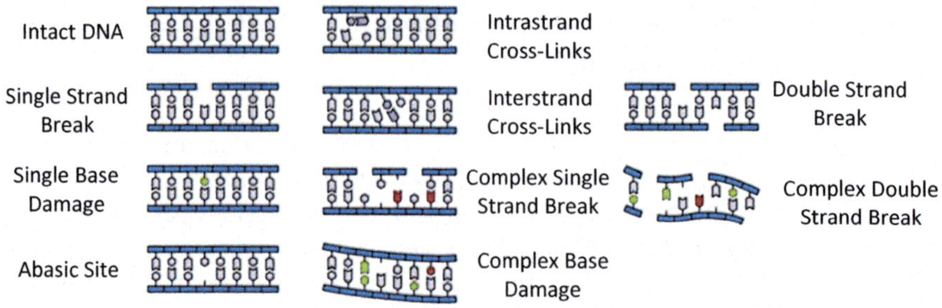

Fig. 12.2 DNA lesions induced by ionizing radiation. Cellular exposure to ionizing radiation induces a wide range of damage in DNA including single strand breaks (SSB), base damage, abasic sites, DNA-protein cross-links, and double strand breaks (DSB)

(Fig. 12.4) (Postel-Vinay et al. 2012; Panier and Boulton 2014). In recent years, alternative end-joining pathways that repair DSBs independently of one or more core components of this classical-NHEJ machinery have been described (Decottignies 2013; Badie et al. 2015). Our work has indicated that cells lacking TCTP failed to repair chromosomal damage induced by γ-rays (Fig. 12.5), which as will be shown below, perhaps as a result of decreased binding of the Ku proteins to damaged DNA. Significantly, defects in DNA repair genes have long been associated with human disease and in cellular sensitivity to DNA-damaging agents (Jackson and Bartek 2009; McKinnon 2009; Hoeijmakers 2009; Jasin 2015; Weichselbaum et al. 1980).

As a consequence of DSB induction, ATM is activated (Shiloh 2003) and phosphorylates the histone H2A.X (to form γH2A.X), which leads to both structural alterations to the chromatin around the damaged site to allow repair proteins access to the damaged regions of the DNA and the recruitment and retention of key DDR factors (Stucki and Jackson 2006). In addition, accumulating evidence indicates that γH2A.X may also be involved in functions that are not directly related to its function as a DNA DSB marker [reviewed in detail in Turinetto and Giachino (2015)]. γH2A.X foci are formed within minutes after exposure to ionizing radiation in a dose-dependent manner, peak at <1 h post-irradiation, and then rapidly decay to baseline levels within one to several days, depending on the dose received (Rogakou et al. 1998). H2A.X phosphorylation leads to recruitment of many checkpoint and repair factors, such as MDC1, MRN, and the ubiquitin ligases RNF8 and UBC13 (Postel-Vinay et al. 2012; Panier and Boulton 2014). These factors promote the recruitment of 53BP1, BRCA1, and more ATM to facilitate the spreading of the DDR signal through the nucleus. These proteins go on to initiate the phosphorylation and dimerization of checkpoint kinases CHK2/CHK1, which targets effectors including p53, CDC25A, and CDC25C that in turn activates cell cycle checkpoints or induce apoptosis (Raynaud et al. 2008; Thompson 2012). While NHEJ is active in all phases of the cell cycle, HR is restricted to the S and G2 phases when sister chromatids are available in close proximity as repair templates (Branzei and Foiani 2008; Symington and Gautier 2011). Significantly, we have

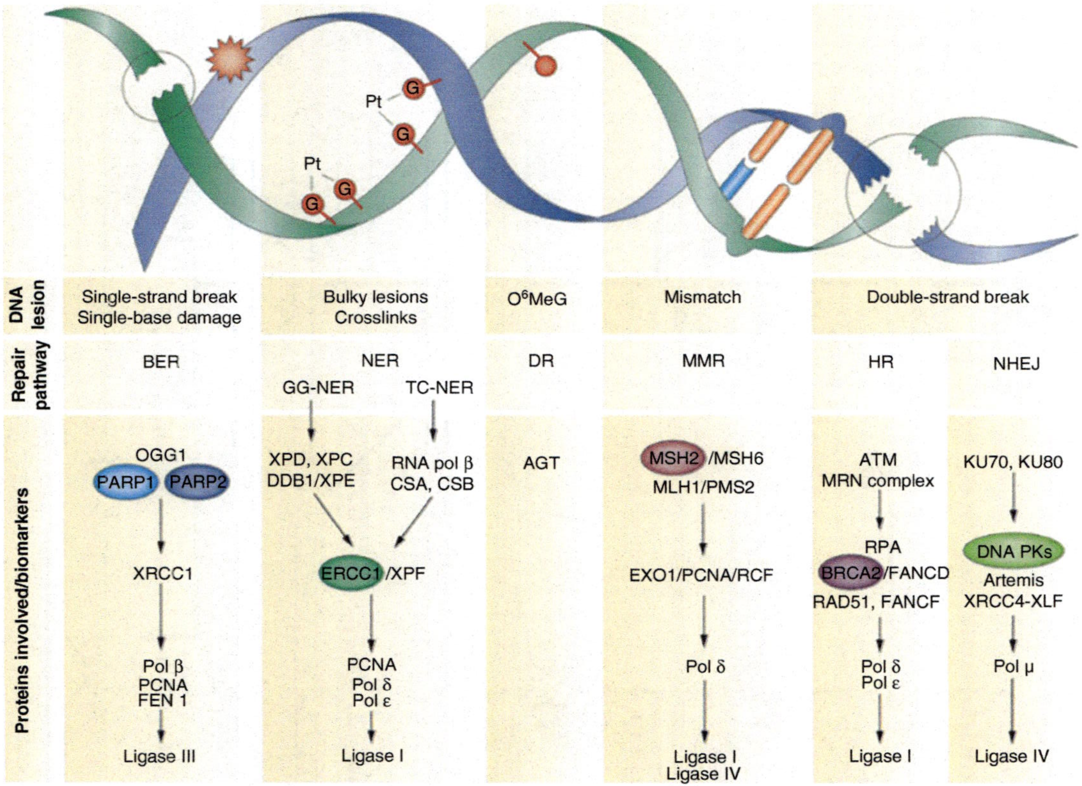

Fig. 12.3 Types of DNA lesions and their corresponding DDR pathways. SSBs or single-base damage (i.e., DNA lesions on a single strand that do not significantly disrupt the helical structure) are generally repaired by base excision repair (BER), whereas DNA damage that significantly distorts the DNA helix (e.g., bulky lesions and crosslinks) is repaired by nucleotide excision repair (NER). Small chemical changes affecting a single base are repaired via direct repair (DR), and mismatches in base pairing caused by DNA replication errors are repaired by mismatch repair (MMR). Finally, DSBs are repaired via homologous recombination (HR) and nonhomologous end joining (NHEJ). AGT = O6-alkylguanine-DNA alkyltransferase; GG-NER = global genome NER; O6MeG = O6-methylguanine; TC-NER = transcription-coupled NER [adapted from Postel-Vinay et al. (2012)]

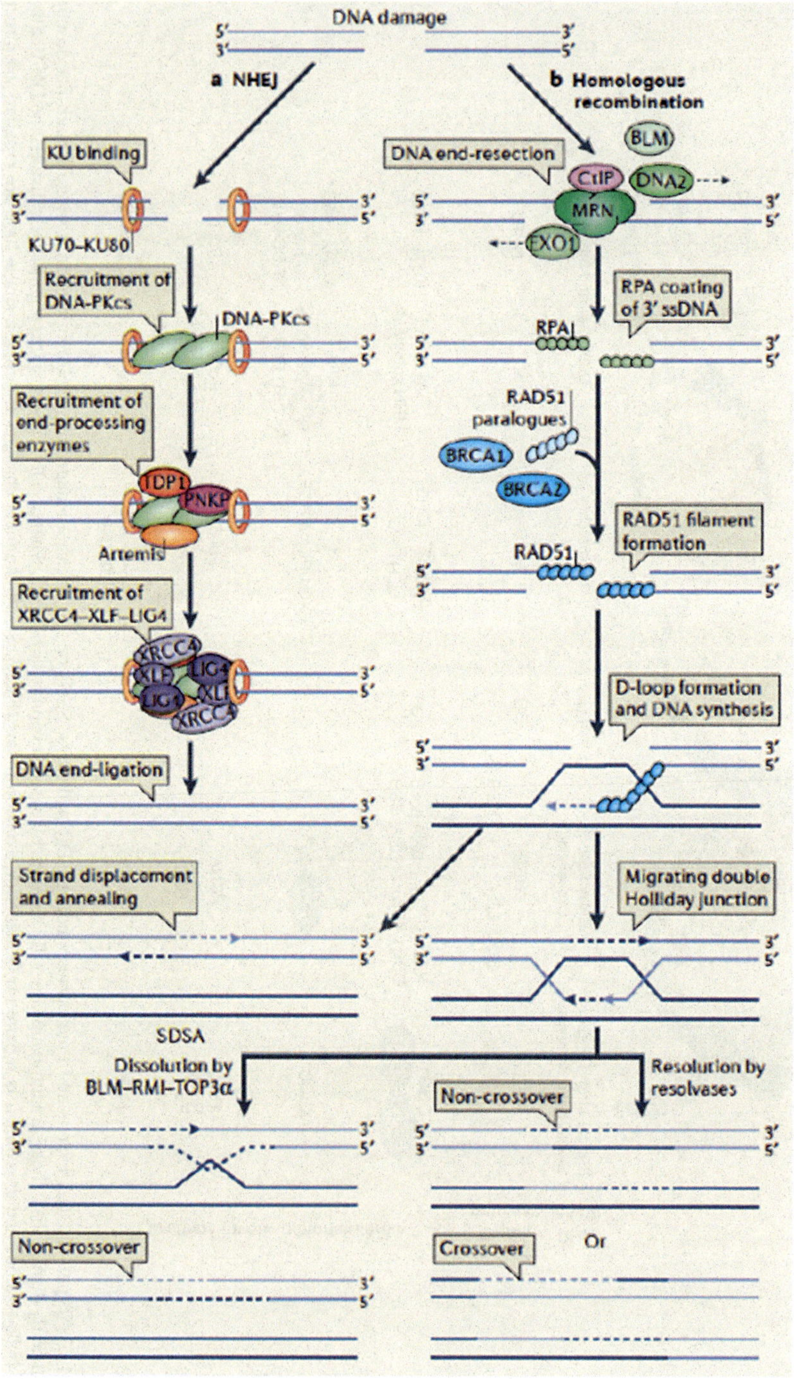

Fig. 12.4 DNA DSB repair pathways. The two main DNA DSB repair pathways in eukaryotic cells: nonhomologous end joining (NHEJ; *part a*) and homologous recombination (*part b*).

shown that upon exposure of normal human cells that are in different phases of the cell cycle to low dose γ-rays, the TCTP protein level was greatly increased, with a significant enrichment in nuclei. TCTP upregulation occurred in a manner dependent on ATM and DNA-PK (cells deficient in ATM or DNA-PKcs function failed to upregulate TCTP) and was associated with protective effects against DNA damage and cell killing (Fig. 12.5). In chromatin of irradiated cells, TCTP was found to physically interact with ATM and to exist in a complex with γH2A.X, in agreement with its distinct localization with the foci of the DNA damage marker proteins γH2A.X, 53BP1, and p-ATM (Fig. 12.6). Importantly, compared to cells transfected with Scr siRNA, depletion of TCTP by siRNA resulted in opposite abundance patterns of Ku proteins in cytoplasm and nucleus of γ-irradiated cells. The dramatic reduction in TCTP level was associated with radiation dose-dependent decrease in Ku70 and Ku80 in the nucleus. Furthermore, relative to Scr siRNA-treated cells, the decreases in Ku70 and Ku80 abundance in nuclei of irradiated siTCTP-transfected cells were associated with significant attenuation (>50%, $p < 0.001$) in the DNA-binding activity of Ku70 and Ku80 from extracts of irradiated cells (Fig. 12.7) (Zhang et al. 2012). These findings constitute previously unrecognized roles for TCTP in maintaining genome integrity under stressful conditions (Zhang et al. 2012; Bommer et al. 2010; Nagano-Ito et al. 2009; Gnanasekar et al. 2009). However, the exact function of TCTP in DNA damage sensing and the different modes of DNA repair still remains to be clearly elucidated.

12.4 TCTP and the Sensing of Genotoxic Stress

There is extensive evidence indicating that TCTP is abundantly expressed in eukaryotes and interacts with several proteins to exert various physiological functions (Amson et al. 2012). Following exposure to environmental insults, the cells in most tissues dramatically increase the production of a small group of proteins that are collectively known as "heat-shock" or stress proteins. Their increased expression in tissues that are subjected to various proteotoxic stressors is an adaptive response that enhances cell survival (Whitesell and Lindquist 2005). In numerous

Fig. 12.4 (continued) APE1 = AP endonuclease 1; BLM = Bloom's syndrome helicase; BRCA1/2 = breast cancer 1/2; CtIP = CtBP-interacting protein; DNA2 = DNA replication ATP-dependent helicase; DNAPKcs = DNA-PK catalytic subunit; EXO1 = exonuclease 1; LIG4 = DNA ligase 4; MRN (MRE11–RAD50–NBS1) complex; PNKP = polynucleotide kinase/phosphatase; RMI = RecQ-mediated genome instability protein 1; RPA = replication protein A; SDSA = synthesis-dependent strand annealing; ssDNA = single-stranded DNA; TDP1 = tyrosyl–DNA phosphodiesterase 1; TOP3α = topoisomerase 3α; XLF = XRCC4-like factor; XRCC4 = X-ray repair cross complementing protein 4 [adapted from Panier and Boulton (2014)]

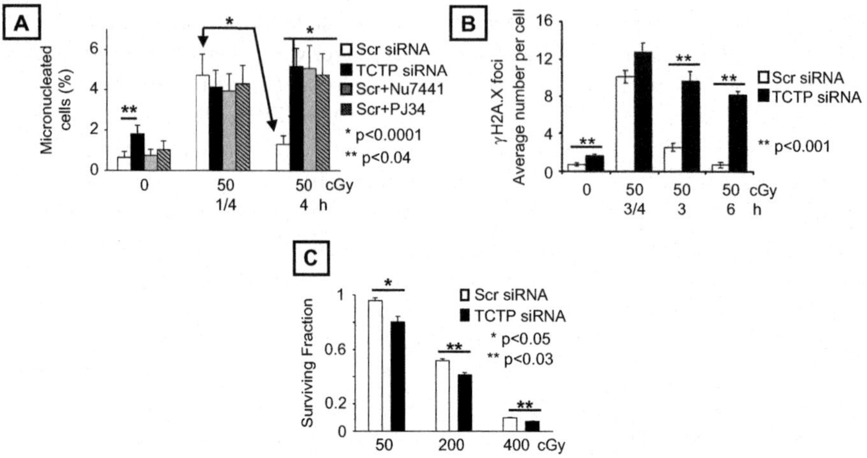

Fig. 12.5 Involvement of TCTP in protection against DNA damage in γ-irradiated AG1522 cells. (**a**) Micronucleus formation in confluent cells that were subcultured at 0.25 or 4 h after acute 50 cGy exposure. The cells were transfected with scrambled (Scr) or TCTP siRNA. Scrambled siRNA-transfected cells were also treated with PJ34 or Nu7441 that inhibit DNA repair. (**b**) Analyses of γH2A.X foci in scrambled or TCTP siRNA transfected cells at different time point after irradiation with 50 cGy. 75–100 cells were counted for each time point. Data represent mean ± SD of three independent experiments. (**c**) Clonogenic survival of γ-irradiated cells transfected with Scr or TCTP siRNA [adapted from Zhang et al. (2012)]

experimental settings and biological systems, it was established that TCTP levels are highly regulated in response to a wide range of extracellular signals and cellular conditions, including heat, heavy metals, hypoxia, and oxidative stress (Bommer and Thiele 2004). Since both human TCTP (HuTCTP) and a TCTP homolog from *Schistosoma mansoni* (SmTCTP) can bind to a variety of denatured proteins and protect them from the harmful effects of thermal shock, it has been suggested that TCTP may belong to a novel heat shock protein with chaperone-like activity (Gnanasekar et al. 2009). Our immuno-precipitation (IP) and mass spectrometry (MS) analyses indicated that TCTP interacts with heat shock 90 kDa, 70 kDa, and 60 kDa proteins in irradiated cells (Zhang et al. 2012), which may also imply a chaperone role of TCTP.

Low doses of toxic agents often induce protective mechanisms that enhance the ability of the organism to cope with stress from normal metabolism or from exogenous agents (Azzam 2011; Azzam et al. 2016). A study by Lucibello et al. proposed TCTP as a "stress hallmark" in cancer cells (Lucibello et al. 2011): TCTP levels were upregulated in cells surviving mild oxidative stress and were downregulated when cells were treated with severe oxidative stress, which was followed by cell death. In our studies, we showed that doses of γ-rays as low as even 1 cGy, particularly when delivered at very low dose-rate (0.2 cGy/h), upregulate TCTP expression and decrease the frequency of micronuclei formation to below the spontaneous rate in normal human fibroblasts (Zhang et al. 2012). These significant changes in TCTP expression were also detected when normal cells were treated

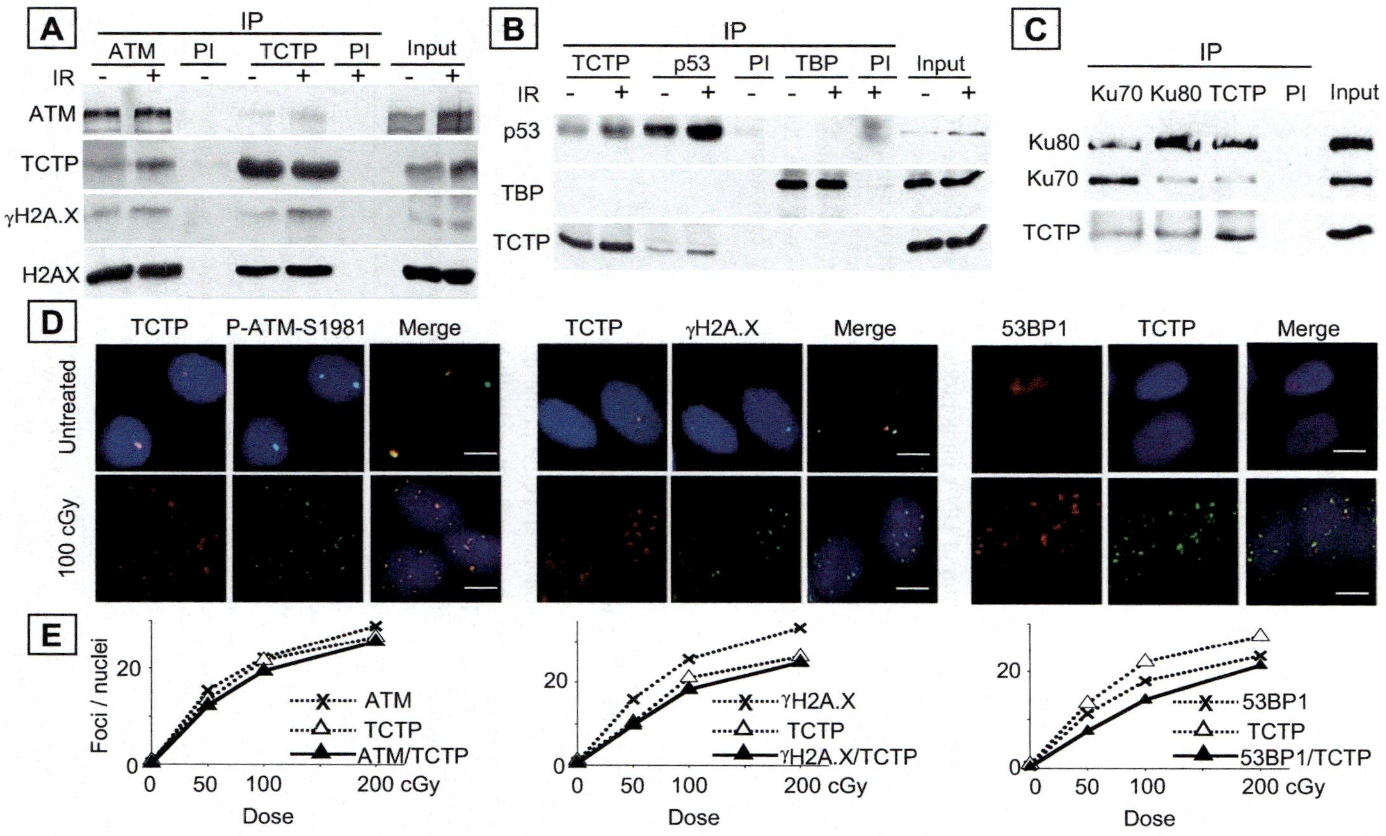

Fig. 12.6 TCTP interacts with components of DNA damage sensing and repair. (**a**) Benzonase-treated chromatin-enriched fractions that were isolated 30 min after irradiation (IR) with 0 (−) or acute 50 cGy (+) from confluent U2OS cells were immunoprecipitated with anti-ATM or anti-TCTP antibodies, or normal

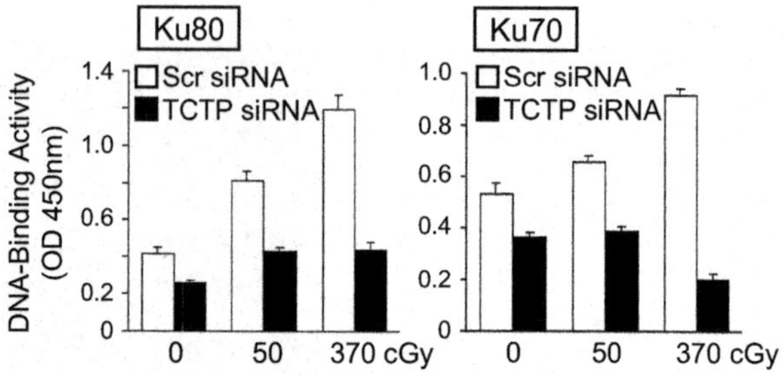

Fig. 12.7 TCTP is involved in regulation of DNA-binding activity of Ku70 and Ku80 after γ-irradiation. Ku70 or Ku80 DNA-binding activity in irradiated U2OS confluent cells expressing scrambled or TCTP siRNA at 30 min after irradiation

with low levels of t-butyl hydroperoxide and hyperthermia (data unpublished). Like ionizing radiation and t-butyl hydroperoxide, some effects of hyperthermia are mediated by ROS (Katschinski et al. 2000). In contrast, cells exposed to UVC, which induces the formation of DNA photoproducts, did not change TCTP levels (data unpublished). Thus, these results support the association of TCTP upregulation by low levels of environmental stresses with oxidative stress. Cells, especially embryonic mouse stem cells exposed to dioxin, a potent toxic synthetic environmental pollutant that induces production of ROS, experience significant upregulation of the expression and secretion of TCTP (Oikawa et al. 2002), which also support the role of TCTP in sensing and defending cells against oxidative stress.

12.5 TCTP and the Repair of DNA Damage

Presence of abundant amounts of TCTP in the nucleus strongly suggests that this ubiquitous molecule may have a role in protecting the DNA from insults. Our data showed the induction of TCTP occurred in both the soluble nuclear fraction and the chromatin-enriched fraction of irradiated cells. When nucleoplasmic proteins were

Fig. 12.6 (continued) mouse or rabbit serum (PI). Immunoblots were then reacted with antibodies against ATM, TCTP, γH2A.X, or H2AX. (**b**) Benzonase-treated nuclear extracts isolated 30 min after exposure of U2OS confluent cells to 0 or acute 50 cGy were immunoprecipitated with anti-TCTP, anti-p53 or control anti-TBP antibodies. Mouse or rabbit preimmune serum (PI) was used as a control. Immunoblotting was performed using antibodies against p53, TCTP, or TBP. (**c**) Immunoblotting of TCTP, Ku70, and Ku80 in benzonase-treated nuclear extracts of control unirradiated U2OS confluent cells after immunoprecipitation with either normal serum (PI) or antibodies against TCTP, Ku70, or Ku80. (**d**) Untreated or γ-irradiated (acute 100 cGy) AG1522 asynchronous cells were pre-extracted, fixed 1 h later, and immunostained in situ with anti-TCTP, anti-P-ATM (S1981), anti-γH2A.X, or anti 53BP1 antibodies. Bars, 10 μm. (**e**) Quantitative assessment of co-localization of TCTP foci with those of P-ATM (S1981) (*left panel*), γH2A (*middle panel*), and 53BP1 (*right panel*) in AG1522 asynchronous cells at 1 h after exposure to 50, 100, or 200 cGy [adapted from Zhang et al. (2012)]

removed by detergent treatment, nuclear TCTP foci were clearly visible, and their number increased as a function of radiation dose, strongly supporting a role for TCTP in repair and/or sensing of DNA damage induced by radiation (Zhang et al. 2012). Studies by Ramaswamy demonstrated that the entry of TCTP into the nucleus is important for its antioxidant function, and TCTP transport into the nucleus is mediated by sumoylation (Munirathinam and Ramaswamy 2012). Also, Rid et al. showed that H_2O_2-dependent translocation of TCTP into the nucleus enables its interaction with the vitamin D receptor (VDR) in human keratinocytes (Rid et al. 2010). However, how TCTP translocates into the nucleus soon after cellular exposure of normal or tumor cells to ionizing radiation remains unknown. It is attractive to speculate that this may occur either through its chaperone function or by sumoylation.

As discussed above, eukaryotes have evolved complex mechanisms to repair DSBs through coordinated actions of protein sensors, transducers, and effectors. Our data established a functional relation between TCTP and several key proteins such as ATM, Ku70, Ku80, and p53 that participate in DSB sensing and repair. Exposure to ionizing radiation induced a significant TCTP protein enrichment in nuclei, which was dependent on early sensors of DNA damage, specifically ATM and DNA-PK (Bakkenist and Kastan 2004). Importantly, this induction was associated with protective effects against DNA damage. Like in the case of cells treated with DNA repair inhibitors, repair of γ-ray-induced chromosomal damage was compromised in TCTP-deficient cells (Fig. 12.5). In chromatin of irradiated cells, TCTP was found to exist in complex with ATM and γH2AX, in agreement with its distinct localization with the foci of the DNA damage marker proteins γH2A.X, 53BP1, and p-ATM (Fig. 12.6) (Zhang et al. 2012). However, the exact nature of the interaction (direct or indirect) between TCTP and ATM kinase remains unknown. Although TCTP contains putative PI3K phosphorylation sites (e.g., T39, S46 and S53), in experiments involving IP of TCTP from AG1522 normal human fibroblasts (wildtype *ATM*), U2OS human bone osteosarcoma cells (wildtype *ATM*) or AG4405 human fibroblasts (mutated *ATM*) exposed to γ-rays, followed by in-gel trypsin digestion, and analysis by liquid chromatography/mass spectrometry (LC-MS/MS) on "Orbitrap velos instrument," we did not detect any phosphorylation site within the protein. Therefore, the issue of how TCTP might be recruited to the ends of DBSs remains open.

Ku proteins are of central importance to DNA repair in eukaryotes. Ku70/86 heterodimer is the first component of NHEJ as it directly binds DNA and recruits other NHEJ factors to promote the repair of the broken ends (Downs and Jackson 2004). TCTP is required for the DNA-binding activity of Ku70 and Ku80 in response to irradiation. Such important effects of TCTP on Ku proteins also highlight the role of TCTP in NHEJ; furthermore, inactivation of Ku proteins leads to defects in telomere maintenance and chromosomal end fusion (Williams et al. 2009). Further, Ku has a key role in a number of other fundamental cellular processes such as transcription and apoptosis (Downs and Jackson 2004). Therefore, a possible chaperone role of TCTP in Ku translocation may imply additional functions of TCTP (Gnanasekar et al. 2009).

Interestingly, IP/MS experiments also indicate that TCTP interacts with filamin-A (Zhang et al. 2012). Filamin-A interacts with the DNA damage response proteins BRCA1 and BRCA2 and therefore may be required for efficient HRR. Defects in filamin-A impair the repair of DSBs resulting in sensitization of cells to ionizing radiation (Yue et al. 2009). The role of TCTP in such critical mechanisms that maintain genomic integrity may explain in part why homozygous mutation in TCTP is embryonically lethal (Chen et al. 2007a). However, more work will be required to establish how the TCTP/filamin-A complex functions in DNA repair.

12.6 TCTP and Control of Cell Cycle Progression Under Normal and Stress Conditions

TCTP is a conserved mitotic growth integrator in animals and plants (Brioudes et al. 2010). Overexpression of TCTP resulted in growth retardation of cells and affected microtubule stabilization and cell morphology (Gachet et al. 1999). Several nuclear proteins involved in mitotic progression have been proposed to interact with TCTP, either regulating or being regulated by TCTP (Bommer 2012). Association of TCTP with microtubules, the important apparatus of the mitotic spindle, has been demonstrated through binding tubulin (Gachet et al. 1999) and actin (Bazile et al. 2009) in a cell cycle-dependent manner. While TCTP is bound to the mitotic spindle, predominantly to the poles, to stabilize spindle microtubules, it is detached from the spindle during metaphase–anaphase transition (Gachet et al. 1999; Burgess et al. 2008; Yarm 2002). Two phosphorylation sites, for mitotic polo-like kinase (Plk-1) in the flexible loop of the TCTP structure, have been identified (Yarm 2002). Phosphorylation decreases the microtubule-stabilizing activity of TCTP and promotes the increase in microtubule dynamics that occurs after metaphase. Expression of a TCTP protein mutated in these sites led to severe disturbance of mitotic progression and to the formation of multinucleated cells (Yarm 2002; Johnson et al. 2008). Phospho-TCTP-ser46 was even confirmed as a marker for Plk-1 activity in vivo (Cucchi et al. 2010). Notably, Johansson et al. described the interaction of TCTP with the two nuclear proteins, nucleophosmin and nucleolin, in embryonic stem cells (Johansson et al. 2010a). In the case of nucleophosmin, the interaction was shown to be independent of phosphorylation by Plk-1 (Johansson et al. 2010a, b), suggesting the involvement of additional unrecognized mechanisms.

Deregulated microtubule dynamics and chromosome segregation enhances genomic instability (Rao et al. 2009). *CHFR* (checkpoint protein with FHA and RING domains) is a modulator of the mitotic stress checkpoint that delays entry into metaphase (Scolnick and Halazonetis 2000). Chfr is a tumor suppressor that ensures chromosomal stability by controlling the expression levels of key mitotic proteins such as Aurora A (Yu et al. 2005). Interestingly, the interaction of Chfr with TCTP occurs throughout the cell cycle, but it could be diminished by depolymerization of the microtubules (Burgess et al. 2008). Although Chfr could be the sensor that detects microtubule disruption and then activates the prophase checkpoint, it remains to be examined whether TCTP binding to Chfr protects the latter from being degraded. Recently, a novel pathway CHD1L/TCTP/Cdc25C/Cdk1 involved in hepatocellular carcinoma development has been identified. Overexpression of TCTP transcriptionally induced by CHD1L promoted the ubiquitin–proteasome degradation of Cdc25C during mitotic progression, which caused a failure in dephosphorylation of Cdk1 and decreased Cdk1 activity. Consequently, a faster mitotic exit and chromosome mis-segregation led to chromosomal instability (Chan et al. 2012).

To maintain genome integrity, cells need to adequately respond to various modes of genotoxic stress. DNA damage is known to trigger cell cycle arrest in the G_1, S, or G2 phases of the cell cycle through activation of DNA-damage checkpoints (Iliakis et al. 2003). This arrest can be reversed once the damage has been repaired, but irreparable damage can promote apoptosis or senescence. Alternatively, cells can reenter the cell cycle before repair has been completed, giving rise to mutations (Medema and Macurek 2012). Our study (Zhang et al. 2012) showed that TCTP physically associates with p53, a protein with essential function in radiation-induced G_1 checkpoint (Sengupta and Harris 2005). Depletion of TCTP greatly attenuated the magnitude of radiation-induced G_1 delay in normal human fibroblasts, as well as the induction of $p21^{Waf1}$, a cyclin-dependent kinase inhibitor (Abbas and Dutta 2009). The loss of normal G_1 checkpoint control could disrupt DNA repair and is an early step in carcinogenesis (Syljuasen et al. 1999), which highlights the role for TCTP in p53-dependent mechanisms that maintain genome integrity under stressful conditions.

Further, our studies have shown that knockdown of TCTP modulates the γ-ray-induced G_2 checkpoint. While an earlier entry into G_2 phase may be a consequence of faster progression through G_1 to S phase in irradiated cells with downregulated levels of TCTP, the longer delay in G_2 following exposure to ionizing radiation is likely due to a greater level of DNA damage (Zhang et al. 2012). Clearly, additional studies are needed to clarify the role of TCTP in checkpoint control in irradiated cells. As mentioned previously, Phospho-TCTP-ser46 is a marker for Plk-1 activity in vivo (Cucchi et al. 2010). Plk-1 activation can promote mitotic entry in an unperturbed cell cycle, but following a DNA-damaging insult, cells come to completely rely on Plk1 to reenter the mitotic cycle following G_2 arrest (van Vugt et al. 2010). Indeed, it has been shown that Plk-1 phosphorylates G_2- and S-phase-expressed protein-1, which acts as a negative regulator of p53, thus suggesting that Plk1 activity contributes to suppression of p53 during checkpoint recovery (Liu et al. 2010). Therefore, it will be of interest to see if phosphorylated TCTP would also be a new negative regulator of p53. Interestingly, overexpression of TCTP was shown in lung carcinoma cells to destabilize p53 (Rho et al. 2011). Our results in normal human fibroblasts therefore open an exciting possibility that TCTP effects on p53 may differ in different cell lines/strains subjected to genotoxic stress.

12.7 TCTP and Cell Death

Several models have been proposed to explain how TCTP functions as a mediator of programmed cell death. During early mammalian development, it plays a role in anti-apoptotic activity through functional antagonism of the BMP4 pathway (Koide et al. 2009). As a Ca^{++}-scavenger and molecular chaperone, TCTP protects cells under Ca^{++}-stress (Graidist et al. 2007) or heat shock conditions (Gnanasekar et al. 2009). In addition, it has been shown that TCTP blocks the cleavage of poly (ADP-ribose)-polymerase (PARP) (Tuynder et al. 2002), a key event in apoptosis (Lazebnik et al. 1994). Also, TCTP protects ovarian carcinoma cells against TSC-22-mediated apoptosis (Lee et al. 2008).

TCTP proteins contain a H2–H3 helices structural similarity to channel-forming helices (Petros et al. 1644) of the pro-apoptotic protein Bax (Suzuki et al. 2000). Investigations showed that TCTP exerts its anti-apoptotic function by insertion into the mitochondrial membrane and inhibiting the dimerization of Bax (Susini et al. 2008). Also, it has been shown that TCTP interacts with Bcl-xL (Yang et al. 2005) and Mcl-1 (Liu et al. 2005; Zhang et al. 2002), two other anti-apoptotic proteins of the Bcl-2 family. The N-terminal region of TCTP and the BH3 domain of Bcl-xL are thought to mediate the interaction between these two proteins (Yang et al. 2005), which may inhibit T-cell apoptosis by preventing the phosphorylation/inactivation of Bcl-xL. However, the interaction between TCTP and Mcl-1 is debatable; while some suggested that the two proteins stabilize each other (Liu et al. 2005; Zhang et al. 2002), others showed that they exert their anti-apoptotic function independently of each other (Graidist et al. 2004).

Recently, a novel function of TCTP in intercellular signaling leading to anti-apoptotic effects was proposed by Sirois et al. (2011), which sheds light on a new direction in bystander effects research, which is under intense investigation in radiation studies (Azzam et al. 2003; Mothersill and Seymour 2004; Prise et al. 2005). Interestingly, TCTP was identified on the surface of extracellular vesicles purified from medium conditioned by apoptotic endothelial cells, and caspase-3 activation plays a key role for the release of TCTP when these cells are dying by apoptosis (Sirois et al. 2011). Further, the nanovesicles, which are different from apoptotic blebs, induced an extracellular signal-regulated kinase 1/2 (ERK 1/2)-dependent anti-apoptotic phenotype in vascular smooth muscle cells (VSMC) (Sirois et al. 2011). During cancer radiotherapy, activated caspase-3 in dying tumor cells has been shown to regulate the release of prostaglandin E2 (PGE2), which can potently stimulate growth of surviving tumor cells (Huang et al. 2011). It would be interesting to determine if TCTP also plays an anti-apoptotic function by intercellular signaling in radiation-induced bystander effects.

12.8 Perspective

In humans, DNA can be damaged by various endogenous and environmental agents, leading to various disorders. Mechanisms must, therefore, exist to protect or repair DNA. At high doses, ionizing radiation is known to cause excessive DNA damage, often followed by cancer or degenerative diseases. We studied cellular responses to low doses of ionizing radiation that are typical of certain occupational activities or diagnostic radiography. Surprisingly, we observed significant adaptive responses when normal human cells were exposed to low doses of cesium-137 γ-rays (Azzam et al. 1996; de Toledo et al. 2006) and identified TCTP as a specific protein involved in this response (Zhang et al. 2012).

In our initial irradiation tests, we found that irradiated cells harbored lower levels of chromosomal damage than what occurred spontaneously at the basal level (de Toledo et al. 2006). This unexpected finding prompted us to use a proteomic approach to identify proteins that are differentially expressed in cells after exposure to 10 cGy of cesium-137 γ-rays delivered at a low dose over 50 h. TCTP was found to be upregulated and appeared most sensitive in this context. The precise pro-survival mechanism mediated by TCTP remains poorly understood.

To this end, we tested the hypothesis that TCTP plays a critical role in response to DNA damage and that this function is essential particularly for the survival and genomic integrity of irradiated cells. We found that upon exposure to doses as low as 1 cGy of cesium-137 γ-rays (a dose received in many diagnostic procedures), the TCTP level was greatly increased in normal human cells, with a significant enrichment in the nuclei. The protein level was similarly upregulated in tissues of low-dose-irradiated mice. Moreover, this upregulation was induced by moderate and high doses of different types of ionizing radiation.

Interestingly, TCTP upregulation was dependent on the early sensors of DNA damage, specifically the protein ATM and the enzyme DNA-PK. Importantly, this upregulation was associated with protective effects against DNA damage. As shown in the case of cells treated with DNA repair inhibitors in previous experiments, repair of γ-ray-induced chromosomal damage was compromised in TCTP-deficient cells. In the chromatin of irradiated cells, TCTP was found to exist in complex with ATM and γH2A.X, a protein that marks the sites of DNA damage. This finding is in agreement with TCTP's distinct localization with the foci of the DNA damage marker proteins γH2A.X, 53BP1, and p-ATM. Furthermore, TCTP was shown to interact with the DNA-binding subunits Ku70 and Ku80 of DNA-PK, a protein with a major role in repair of DNA DSBs, a particularly harmful form of DNA damage (Zhang et al. 2012).

Our findings are consistent with the observation that TCTP knockdown led to decreased levels of Ku70 and Ku80 in the nuclei of irradiated cells and attenuated the DNA-binding activity of DNA-PK. Interestingly, the protective effects of TCTP were not confined to low-dose-irradiated cells, but were observed even against the lethal effects of therapeutic doses of γ-rays. This may explain why knockdown of TCTP increased the failure of normal human cells to divide, or reproduce, when

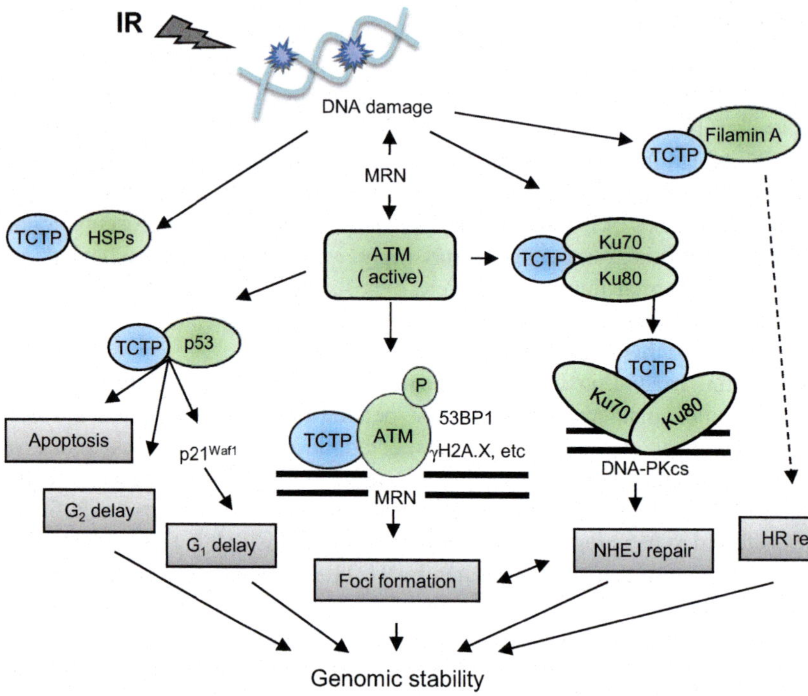

Fig. 12.8 The role of TCTP in DNA damage sensing and repair. TCTP is upregulated by ionizing radiation; it interacts with elements of DNA damage sensing and repair and modulates radiation-induced cell cycle checkpoints (*IR* ionizing radiation, *P* phosphorylation, *MRN* MRE11–RAD50–NBS1, *NHEJ* nonhomologous end joining, *HR* homologous recombination, *HSPs* heat shock proteins)

exposed to 200 or 400 cGy (doses received during cancer radiotherapy) (Zhang et al. 2012).

In normal cells, TCTP did not affect such cell cycle progression towards division under normal, homeostatic conditions. However, TCTP had a prominent effect on stress-induced cell cycle checkpoints, which ensure that the cell cycle progresses without any DNA damage. We found that TCTP interacted with p53, a critical protein component of such checkpoints that maintains genomic integrity. Furthermore, TCTP knockdown shortened the radiation-induced delay in the G_1 phase of the cell cycle, which is the pre-DNA synthesis phase. The latter effect was associated with attenuated induction of $p21^{Waf1}$, an inhibitor of master regulators of the cell cycle. The loss of the normal G_1 checkpoint control disrupts DNA repair and is an early step in carcinogenesis, thus highlighting the role of TCTP in maintaining healthy survival. In addition, TCTP has a role in the post-DNA synthesis (G_2) phase, where it modulates the duration of the radiation-induced G_2 checkpoint. Cells with downregulated TCTP entered G_2 phase faster than control cells and were arrested longer in G_2 phase (Zhang et al. 2012).

Together, our results identify TCTP as a new member of a group of proteins involved in DNA damage response (Fig. 12.8). Our results also point to a chaperone-like role of TCTP, where it interacts with several stress-induced molecular chaperones/heat-shock proteins in irradiated cells. The new role of TCTP in sensing and repairing radiation-induced DNA damage will aid in understanding the system responses to low-dose radiation exposures and in turn help in estimating health risks of such exposures. It may also aid in understanding the molecular events induced by therapeutic doses of radiation. Clearly, future studies need to address the exact role of TCTP in HR, NHEJ, and other modes of DNA repair. To this end, the use of cells that are proficient or deficient in either of the latter DNA repair mechanisms and where TCTP levels are altered should be informative.

Acknowledgements This work was supported by grant CA049062 from the National Institutes of Health, grant NNX15AD62G from the National Aeronautics and Space Administration, and the Program for Changjiang Scholars and Innovative Research Team, University of Ministry of Education, China.

References

Abbas T, Dutta A (2009) p21 in cancer: intricate networks and multiple activities. Nat Rev Cancer 9:400–414

Ames BN (1989) Endogenous oxidative DNA damage, aging, and cancer. Free Radic Res Commun 7:121–128

Amson R, Pece S, Lespagnol A et al (2012) Reciprocal repression between P53 and TCTP. Nat Med 18:91–99

Amson R, Pece S, Marine JC et al (2013) TPT1/TCTP-regulated pathways in phenotypic reprogramming. Trends Cell Biol 23:37–46

Azzam EI (2011) Exposure to low level environmental agents: the induction of hormesis. Mutat Res 726:89–90

Azzam EI, de Toledo SM, Raaphorst GP et al (1996) Low-dose ionizing radiation decreases the frequency of neoplastic transformation to a level below the spontaneous rate in C3H 10T1/2 cells. Radiat Res 146:369–373

Azzam EI, de Toledo SM, Little JB (2003) Oxidative metabolism, gap junctions and the ionizing radiation-induced bystander effect. Oncogene 22:7050–7057

Azzam EI, Jay-Gerin JP, Pain D (2012) Ionizing radiation-induced metabolic oxidative stress and prolonged cell injury. Cancer Lett 327:48–60

Azzam EI, Colangelo NW, Domogauer JD et al (2016) Is ionizing radiation harmful at any exposure? An echo that continues to vibrate. Health Phys 110:249–251

Badie S, Carlos AR, Folio C et al (2015) BRCA1 and CtIP promote alternative non-homologous end-joining at uncapped telomeres. EMBO J 34:410–424

Bakkenist CJ, Kastan MB (2004) Initiating cellular stress responses. Cell 118:9–17

Barendsen GW (1964) Impairment of the proliferative capacity of human cells in cultures by a particles of differing linear energy transfer. Int J Radiat Biol 8:453–466

Baumstark-Khan C (1993) X-ray-induced DNA double-strand breaks as lethal lesions in diploid human fibroblasts compared to Chinese hamster ovary cells. Int J Radiat Biol 63:305–311

Bazile F, Pascal A, Arnal I et al (2009) Complex relationship between TCTP, microtubules and actin microfilaments regulates cell shape in normal and cancer cells. Carcinogenesis 30:555–565

Bohm H, Benndorf R, Gaestel M et al (1989) The growth-related protein P23 of the Ehrlich ascites tumor: translational control, cloning and primary structure. Biochem Int 19:277–286
Bommer UA (2012) Several nuclear proteins involved in mitotic progression have been proposed to interact with TCTP, either regulating or being regulated by TCTP. Open Allergy J 5:19–32
Bommer UA, Thiele BJ (2004) The translationally controlled tumour protein (TCTP). Int J Biochem Cell Biol 36:379–385
Bommer UA, Heng C, Perrin A et al (2010) Roles of the translationally controlled tumour protein (TCTP) and the double-stranded RNA-dependent protein kinase, PKR, in cellular stress responses. Oncogene 29:763–773
Branzei D, Foiani M (2008) Regulation of DNA repair throughout the cell cycle. Nat Rev 9:297–308
Brioudes F, Thierry AM, Chambrier P et al (2010) Translationally controlled tumor protein is a conserved mitotic growth integrator in animals and plants. Proc Natl Acad Sci USA 107:16384–16389
Burgess A, Labbe JC, Vigneron S et al (2008) Chfr interacts and colocalizes with TCTP to the mitotic spindle. Oncogene 27:5554–5566
Cadet J, Douki T, Ravanat JL (2011) Measurement of oxidatively generated base damage in cellular DNA. Mutat Res 711:3–12
Cans C, Passer BJ, Shalak V et al (2003) Translationally controlled tumor protein acts as a guanine nucleotide dissociation inhibitor on the translation elongation factor eEF1A. Proc Natl Acad Sci USA 100:13892–13897
Chan TH, Chen L, Liu M et al (2012) Translationally controlled tumor protein induces mitotic defects and chromosome missegregation in hepatocellular carcinoma development. Hepatology 55:491–505
Chatterjee A, Schaefer HJ (1976) Microdosimetric structure of heavy ion tracks in tissue. Radiat Environ Biophys 13:215–227
Chen SH, Wu PS, Chou CH et al (2007a) A knockout mouse approach reveals that TCTP functions as an essential factor for cell proliferation and survival in a tissue- or cell type-specific manner. Mol Biol Cell 18:2525–2532
Chen Z, Zhang H, Yang H et al (2007b) The expression of AmphiTCTP, a TCTP orthologous gene in amphioxus related to the development of notochord and somites. Comp Biochem Physiol B Biochem Mol Biol 147:460–465
Chitpatima ST, Makrides S, Bandyopadhyay R et al (1988) Nucleotide sequence of a major messenger RNA for a 21 kilodalton polypeptide that is under translational control in mouse tumor cells. Nucleic Acids Res 16:2350
Chou WC, Hu LY, Hsiung CN et al (2015) Initiation of the ATM-Chk2 DNA damage response through the base excision repair pathway. Carcinogenesis 36:832–840
Coutard H (1937) The results and methods of treatment of cancer by radiation. Ann Surg 106:584–598
Cucchi U, Gianellini LM, De Ponti A et al (2010) Phosphorylation of TCTP as a marker for polo-like kinase-1 activity in vivo. Anticancer Res 30:4973–4985
de Toledo SM, Asaad N, Venkatachalam P et al (2006) Adaptive responses to low-dose/low-dose-rate gamma rays in normal human fibroblasts: the role of growth architecture and oxidative metabolism. Radiat Res 166:849–857
Decottignies A (2013) Alternative end-joining mechanisms: a historical perspective. Front Genet 4:48
Delaney G, Jacob S, Featherstone C et al (2005) The role of radiotherapy in cancer treatment: estimating optimal utilization from a review of evidence-based clinical guidelines. Cancer 104:1129–1137
Dizdaroglu M (2012) Oxidatively induced DNA damage: mechanisms, repair and disease. Cancer Lett 327:26–47
Downs JA, Jackson SP (2004) A means to a DNA end: the many roles of Ku. Nat Rev 5:367–378
Fenech M, Morley AA (1985) Measurement of micronuclei in lymphocytes. Mutat Res 147:29–36

Gachet Y, Tournier S, Lee M et al (1999) The growth-related, translationally controlled protein P23 has properties of a tubulin binding protein and associates transiently with microtubules during the cell cycle. J Cell Sci 112(Pt 8):1257–1271
Georgakilas AG (2011) From chemistry of DNA damage to repair and biological significance. Comprehending the future. Mutat Res 711:1–2
Gnanasekar M, Ramaswamy K (2007) Translationally controlled tumor protein of Brugia malayi functions as an antioxidant protein. Parasitol Res 101:1533–1540
Gnanasekar M, Dakshinamoorthy G, Ramaswamy K (2009) Translationally controlled tumor protein is a novel heat shock protein with chaperone-like activity. Biochem Biophys Res Commun 386:333–337
Goodhead DT (1988) Spatial and temporal distribution of energy. Health Phys 55:231–240
Graidist P, Phongdara A, Fujise K (2004) Antiapoptotic protein partners fortilin and MCL1 independently protect cells from 5-fluorouracil-induced cytotoxicity. J Biol Chem 279:40868–40875
Graidist P, Yazawa M, Tonganunt M et al (2007) Fortilin binds Ca2+ and blocks Ca2+-dependent apoptosis in vivo. Biochem J 408:181–191
Hacein-Bey L, Konstas AA, Pile-Spellman J (2014) Natural history, current concepts, classification, factors impacting endovascular therapy, and pathophysiology of cerebral and spinal dural arteriovenous fistulas. Clin Neurol Neurosurg 121:64–75
Hall EJ, Giaccia AJ (2006) Radiobiology for the radiologist, 6th edn. Lippincott Williams & Wilkins, Philadelphia, PA
Harper JW, Elledge SJ (2007) The DNA damage response: ten years after. Mol Cell 28:739–745
Hoeijmakers JH (2009) DNA damage, aging, and cancer. N Engl J Med 361:1475–1485
Hsu YC, Chern JJ, Cai Y et al (2007) Drosophila TCTP is essential for growth and proliferation through regulation of dRheb GTPase. Nature 445:785–788
Huang L, Snyder AR, Morgan WF (2003) Radiation-induced genomic instability and its implications for radiation carcinogenesis. Oncogene 22:5848–5854
Huang Q, Li F, Liu X et al (2011) Caspase 3-mediated stimulation of tumor cell repopulation during cancer radiotherapy. Nat Med 17:860–866
Iliakis G, Wang Y, Guan J et al (2003) DNA damage checkpoint control in cells exposed to ionizing radiation. Oncogene 22:5834–5847
Jackson SP, Bartek J (2009) The DNA-damage response in human biology and disease. Nature 461:1071–1078
Jasin M (2015) Accolades for the DNA damage response. N Engl J Med 373:1492–1495
Jeggo PA, Pearl LH, Carr AM (2016) DNA repair, genome stability and cancer: a historical perspective. Nat Rev Cancer 16:35–42
Johansson H, Svensson F, Runnberg R et al (2010a) Phosphorylated nucleolin interacts with translationally controlled tumor protein during mitosis and with Oct4 during interphase in ES cells. PloS One 5:e13678
Johansson H, Vizlin-Hodzic D, Simonsson T et al (2010b) Translationally controlled tumor protein interacts with nucleophosmin during mitosis in ES cells. Cell Cycle 9:2160–2169
Johnson TM, Antrobus R, Johnson LN (2008) Plk1 activation by Ste20-like kinase (Slk) phosphorylation and polo-box phosphopeptide binding assayed with the substrate translationally controlled tumor protein (TCTP). Biochemistry 47:3688–3696
Jung J, Kim HY, Kim M et al (2011) Translationally controlled tumor protein induces human breast epithelial cell transformation through the activation of Src. Oncogene 30:2264–2274
Kashiwakura JC, Ando T, Matsumoto K et al (2012) Histamine-releasing factor has a proinflammatory role in mouse models of asthma and allergy. J Clin Invest 122:218–228
Katschinski DM, Boos K, Schindler SG et al (2000) Pivotal role of reactive oxygen species as intracellular mediators of hyperthermia-induced apoptosis. J Biol Chem 275:21094–21098
Koide Y, Kiyota T, Tonganunt M et al (2009) Embryonic lethality of fortilin-null mutant mice by BMP-pathway overactivation. Biochim Biophys Acta 2009:326–338

Koziol MJ, Garrett N, Gurdon JB (2007) Tpt1 activates transcription of oct4 and nanog in transplanted somatic nuclei. Curr Biol 17:801–807

Larrea AA, Lujan SA, Kunkel TA (2010) SnapShot: DNA mismatch repair. Cell 141:730–e731

LaVerne JA, Pimblott SM (1993) Yields of hydroxyl radical and hydrated electron scavenging reactions in aqueous solutions of biological interest. Radiat Res 135:16–23

Lazebnik YA, Kaufmann SH, Desnoyers S et al (1994) Cleavage of poly(ADP-ribose) polymerase by a proteinase with properties like ICE. Nature 371:346–347

Lee JH, Rho SB, Park SY et al (2008) Interaction between fortilin and transforming growth factor-beta stimulated clone-22 (TSC-22) prevents apoptosis via the destabilization of TSC-22. FEBS Lett 582:1210–1218

Li F, Zhang D, Fujise K (2001) Characterization of fortilin, a novel antiapoptotic protein. J Biol Chem 276:47542–47549

Li M, Gonon G, Buonanno M et al (2014) Health risks of space exploration: targeted and non-targeted oxidative injury by high charge and high energy particles. Antioxid Redox Signal 20:1501–1523

Lieber MR (2008) The mechanism of human nonhomologous DNA end joining. J Biol Chem 283:1–5

Lin J, Epel E, Blackburn E (2012) Telomeres and lifestyle factors: roles in cellular aging. Mutat Res 730:85–89

Lindahl T (1993) Instability and decay of the primary structure of DNA. Nature 362:709–715

Little JB (1968) Delayed initiation of DNA synthesis in irradiated human diploid cells. Nature 218:1064–1065

Little JB (2000) Radiation carcinogenesis. Carcinogenesis 21:394–404

Liu H, Peng HW, Cheng YS et al (2005) Stabilization and enhancement of the antiapoptotic activity of mcl-1 by TCTP. Mol Cell Biol 25:3117–3126

Liu XS, Li H, Song B et al (2010) Polo-like kinase 1 phosphorylation of G2 and S-phase-expressed 1 protein is essential for p53 inactivation during G2 checkpoint recovery. EMBO Rep 11:626–632

Lucibello M, Gambacurta A, Zonfrillo M et al (2011) TCTP is a critical survival factor that protects cancer cells from oxidative stress-induced cell-death. Exp Cell Res 317:2479–2489

Macdonald SM (2012) Potential role of histamine releasing factor (HRF) as a therapeutic target for treating asthma and allergy. J Asthma Allergy 5:51–59

MacDonald SM, Rafnar T, Langdon J et al (1995) Molecular identification of an IgE-dependent histamine-releasing factor. Science 269:688–690

McKinnon PJ (2009) DNA repair deficiency and neurological disease. Nat Rev Neurosci 10:100–112

Medema RH, Macurek L (2012) Checkpoint control and cancer. Oncogene 31:2601–2613

Meesungnoen J, Benrahmoune M, Filali-Mouhim A et al (2001) Monte Carlo calculation of the primary radical and molecular yields of liquid water radiolysis in the linear energy transfer range 0.3–6.5 keV/um: application to ^{137}Cs gamma rays. Radiat Res 155:269–278. Erratum Published in Radiat Res 155 (2001) 2873

Moskovitz J (2005) Roles of methionine suldfoxide reductases in antioxidant defense, protein regulation and survival. Curr Pharm Des 11:1451–1457

Mothersill C, Seymour CB (2004) Radiation-induced bystander effects--implications for cancer. Nat Rev Cancer 4:158–164

Munirathinam G, Ramaswamy K (2012) Sumoylation of human translationally controlled tumor protein is important for its nuclear transport. Biochem Res Int 2012:831940

Muroya Y, Plante I, Azzam EI et al (2006) High-LET ion radiolysis of water: visualization of the formation and evolution of ion tracks and relevance to the radiation-induced bystander effect. Radiat Res 165:485–491

Nagano-Ito M, Banba A, Ichikawa S (2009) Functional cloning of genes that suppress oxidative stress-induced cell death: TCTP prevents hydrogen peroxide-induced cell death. FEBS Lett 583:1363–1367

Nam EA, Cortez D (2011) ATR signalling: more than meeting at the fork. Biochem J 436:527–536
NCRP (2009) Ionizing radiation exposure of the population of the United States. National Council on Radiation Protection and Measurements, Bethesda
Newhauser WD, Durante M (2011) Assessing the risk of second malignancies after modern radiotherapy. Nat Rev Cancer 11:438–448
Nikjoo H, O'Neill P, Wilson WE et al (2001) Computational approach for determining the spectrum of DNA damage induced by ionizing radiation. Radiat Res 156:577–583
Oikawa K, Ohbayashi T, Mimura J et al (2002) Dioxin stimulates synthesis and secretion of IgE-dependent histamine-releasing factor. Biochem Biophys Res Commun 290:984–987
Okada H, Mak TW (2004) Pathways of apoptotic and non-apoptotic death in tumour cells. Nat Rev Cancer 4:592–603
O'Neill P, Wardman P (2009) Radiation chemistry comes before radiation biology. Int J Radiat Biol 85:9–25
Panier S, Boulton SJ (2014) Double-strand break repair: 53BP1 comes into focus. Nat Rev 15:7–18
Petkau A (1987) Role of superoxide dismutase in modification of radiation injury. Br J Cancer Suppl 8:87–95
Petros AM, Olejniczak ET, Fesik SW (1644) Structural biology of the Bcl-2 family of proteins. Biochim Biophys Acta 2004:83–94
Petruseva IO, Evdokimov AN, Lavrik OI (2014) Molecular mechanism of global genome nucleotide excision repair. Acta Naturae 6:23–34
Platzman R (1958) The physical and chemical basis of mechanisms in radiation biology. In: Claus W (ed) Radiation biology and medicine. Selected reviews in the life sciences. Addison-Wesley, Reading, MA, pp 15–72
Postel-Vinay S, Vanhecke E, Olaussen KA et al (2012) The potential of exploiting DNA-repair defects for optimizing lung cancer treatment. Nature reviews. Clin Oncol 9:144–155
Prise KM, Schettino G, Folkard M et al (2005) New insights on cell death from radiation exposure. Lancet Oncol 6:520–528
Rao CV, Yamada HY, Yao Y et al (2009) Enhanced genomic instabilities caused by deregulated microtubule dynamics and chromosome segregation: a perspective from genetic studies in mice. Carcinogenesis 30:1469–1474
Raynaud CM, Sabatier L, Philipot O et al (2008) Telomere length, telomeric proteins and genomic instability during the multistep carcinogenic process. Crit Rev Oncol Hematol 66:99–117
Rho SB, Lee JH, Park MS et al (2011) Anti-apoptotic protein TCTP controls the stability of the tumor suppressor p53. FEBS Lett 585:29–35
Rid R, Onder K, Trost A et al (2010) H2O2-dependent translocation of TCTP into the nucleus enables its interaction with VDR in human keratinocytes: TCTP as a further module in calcitriol signalling. J Steroid Biochem Mol Biol 118:29–40
Rogakou EP, Pilch DR, Orr AH et al (1998) DNA double-stranded breaks induce histone H2AX phosphorylation on serine 139. J Biol Chem 273:5858–5868
Scolnick DM, Halazonetis TD (2000) Chfr defines a mitotic stress checkpoint that delays entry into metaphase. Nature 406:430–435
Sengupta S, Harris CC (2005) p53: traffic cop at the crossroads of DNA repair and recombination. Nat Rev 6:44–55
Shah NR, Mahmoudi M (2015) The role of DNA damage and repair in atherosclerosis: a review. J Mol Cell Cardiol 86:147–157
Shiloh Y (2003) ATM and related protein kinases: safeguarding genome integrity. Nat Rev Cancer 3:155–168
Shiloh Y, Ziv Y (2013) The ATM protein kinase: regulating the cellular response to genotoxic stress, and more. Nat Rev 14:197–210
Sirois I, Raymond MA, Brassard N et al (2011) Caspase-3-dependent export of TCTP: a novel pathway for antiapoptotic intercellular communication. Cell Death Differ 18:549–562

Spitz DR, Azzam EI, Li JJ et al (2004) Metabolic oxidation/reduction reactions and cellular responses to ionizing radiation: a unifying concept in stress response biology. Cancer Metastasis Rev 23:311–322

Stucki M, Jackson SP (2006) GammaH2AX and MDC1: anchoring the DNA-damage-response machinery to broken chromosomes. DNA Repair (Amst) 5:534–543

Susini L, Besse S, Duflaut D et al (2008) TCTP protects from apoptotic cell death by antagonizing bax function. Cell Death Differ 15:1211–1220

Suzuki M, Youle RJ, Tjandra N (2000) Structure of Bax: coregulation of dimer formation and intracellular localization. Cell 103:645–654

Syljuasen RG, Krolewski B, Little JB (1999) Loss of normal G1 checkpoint control is an early step in carcinogenesis, independent of p53 status. Cancer Res 59:1008–1014

Symington LS, Gautier J (2011) Double-strand break end resection and repair pathway choice. Annu Rev Genet 45:247–271

Thaw P, Baxter NJ, Hounslow AM et al (2001) Structure of TCTP reveals unexpected relationship with guanine nucleotide-free chaperones. Nat Struct Biol 8:701–704

Thomas G, Thomas G, Luther H (1981) Transcriptional and translational control of cytoplasmic proteins after serum stimulation of quiescent Swiss 3T3 cells. Proc Natl Acad Sci USA 78:5712–5716

Thompson LH (2012) Recognition, signaling, and repair of DNA double-strand breaks produced by ionizing radiation in mammalian cells: the molecular choreography. Mutat Res 751:158–246

Tsarova K, Yarmola EG, Bubb MR (2010) Identification of a cofilin-like actin-binding site on translationally controlled tumor protein (TCTP). FEBS Lett 584:4756–4760

Turinetto V, Giachino C (2015) Multiple facets of histone variant H2AX: a DNA double-strand-break marker with several biological functions. Nucleic Acids Res 43:2489–2498

Tuynder M, Susini L, Prieur S et al (2002) Biological models and genes of tumor reversion: cellular reprogramming through tpt1/TCTP and SIAH-1. Proc Natl Acad Sci USA 99:14976–14981

Tuynder M, Fiucci G, Prieur S et al (2004) Translationally controlled tumor protein is a target of tumor reversion. Proc Natl Acad Sci USA 101:15364–15369

Tyldesley S, Boyd C, Schulze K et al (2001) Estimating the need for radiotherapy for lung cancer: an evidence-based, epidemiologic approach. Int J Radiat Oncol Biol Phys 49:973–985

van Vugt MA, Gardino AK, Linding R et al (2010) A mitotic phosphorylation feedback network connects Cdk1, Plk1, 53BP1, and Chk2 to inactivate the G(2)/M DNA damage checkpoint. PLoS Biol 8:e1000287

Weichselbaum RR, Nove J, Little JB (1980) X-ray sensitivity of fifty-three human diploid fibroblast cell strains from patients with characterized genetic disorders. Cancer Res 40:920–925

Weinberg RA (2007) The biology of cancer. Garland Science, Taylor & Francis Group, LLC, New York

Whitesell L, Lindquist SL (2005) HSP90 and the chaperoning of cancer. Nat Rev Cancer 5:761–772

Williams ES, Klingler R, Ponnaiya B et al (2009) Telomere dysfunction and DNA-PKcs deficiency: characterization and consequence. Cancer Res 69:2100–2107

Yang Y, Yang F, Xiong Z et al (2005) An N-terminal region of translationally controlled tumor protein is required for its antiapoptotic activity. Oncogene 24:4778–4788

Yarm FR (2002) Plk phosphorylation regulates the microtubule-stabilizing protein TCTP. Mol Cell Biol 22:6209–6221

Yenofsky R, Bergmann I, Brawerman G (1982) Messenger RNA species partially in a repressed state in mouse sarcoma ascites cells. Proc Natl Acad Sci USA 79:5876–5880

Yenofsky R, Cereghini S, Krowczynska A et al (1983) Regulation of mRNA utilization in mouse erythroleukemia cells induced to differentiate by exposure to dimethyl sulfoxide. Mol Cell Biol 3:1197–1203

Yi C, He C (2013) DNA repair by reversal of DNA damage. Cold Spring Harb Perspect Biol 5: a012575

Yu X, Minter-Dykhouse K, Malureanu L et al (2005) Chfr is required for tumor suppression and Aurora A regulation. Nat Genet 37:401–406

Yue J, Wang Q, Lu H et al (2009) The cytoskeleton protein filamin-A is required for an efficient recombinational DNA double strand break repair. Cancer Res 69:7978–7985

Zhang D, Li F, Weidner D et al (2002) Physical and functional interaction between myeloid cell leukemia 1 protein (MCL1) and Fortilin. The potential role of MCL1 as a fortilin chaperone. J Biol Chem 277:37430–37438

Zhang J, de Toledo SM, Pandey BN et al (2012) Role of the translationally controlled tumor protein in DNA damage sensing and repair. Proc Natl Acad Sci USA 109:E926–E933

Zhou BB, Elledge SJ (2000) The DNA damage response: putting checkpoints in perspective. Nature 408:433–439

Zimbrick JD (2002) Radiation chemistry and the Radiation Research Society: a history from the beginning. Radiat Res 158:127–140

Chapter 13
TCTP Has a Crucial Role in the Different Stages of Prostate Cancer Malignant Progression

Virginie Baylot, Sara Karaki, and Palma Rocchi

Abstract Prostate cancer (PC) is the second most common cause of cancer-related mortality in men in the western world after lung cancer. Many patients are not candidates for resection given the advanced stage of their cancer. The primary treatment for advanced PC is the castration therapy which supresses the production of androgens, hormone that promotes PC growth. Despite the efficiency of the castration therapy, most patients develop castration resistant disease which remains uncurable. Clearly, novel approaches are required to effectively treat castration resistant PC (CRPC). New strategies that identify the molecular mechanisms by which PC becomes resistant to conventional therapies may enable the identification of novel therapeutic targets that could improve clinical outcome. Recent studies have demonstrated the implication of TCTP's over-expression in PC and CRPC, and its role in resistance to treatment. TCTP's interaction with p53 and their negative feedback loop regulation have also been described to be causal for PC progression and invasion. A novel nanotherapy that inhibits TCTP has been developed as a new therapeutical strategy in CRPC. This chapter will highlight the role of TCTP as new therapeutic target in PC, in particular, therapy-resistant advanced PC and report the development of novel nanotherapy against TCTP that restore treatment-sensitivity in CRPC that deserve to be tested in clinical trial.

V. Baylot
Division of Oncology, Departments of Medicine and Pathology, Stanford University School of Medicine, Stanford, CA 94305, USA

S. Karaki • P. Rocchi (✉)
UMR 1068, Inserm, CRCM, BP30059, Cedex9, 27 Boulevard Leï Roure, Marseille 13273, France

Institut Paoli-Calmettes, Marseille 13009, France

Aix-Marseille Université, 13284 Marseille, France

CNRS, UMR7258, CRCM, Marseille 13009, France
e-mail: palma.rocchi@inserm.fr

A. Telerman, R. Amson (eds.), *TCTP/tpt1 - Remodeling Signaling from Stem Cell to Disease*, Results and Problems in Cell Differentiation 64,
DOI 10.1007/978-3-319-67591-6_13

13.1 TCTP Is Upregulated in Prostate Cancer

Although Prostate cancer (PC) is a deadly cancer with a rapidly increasing frequency in the western countries localized PC is usually treated with surgery and radiation (American cancer society: prostate cancer statistics http://www.cancer.org) (Jemal et al. 2010; Siegel et al. 2012). The importance of androgens for the initiation and progression of PC has been shown early in the twentieth century. Androgens bind to their specific receptors (AR) and are well known to supply the PC cell growth.

In 2004, a first study has shown that, while TCTP appears to be essential for prostate gland normal physiologic functions, its expression is increased in PC cells (Arcuri et al. 2004). The authors described TCTP as one of the main calcium binding protein in PC cells in which it regulates key processes like apoptosis and cellular differentiation. More recently, Gnanasekar et al. have examined if TCTP was a good potential new therapeutic target in androgen-dependent PC cells. Specifically, it has been demonstrated that silencing TCTP gene with a siRNA dramatically increases the androgen-sensitive PC cell death due to apoptosis (Gnanasekar et al. 2009). TCTP has further been identified as a novel androgen regulated gene whose expression is induced at both mRNA and protein level by androgens (Kaarbo et al. 2013).

13.2 TCTP: A Promising Target in Castration-Resistant Prostate Cancer

The first-line treatment for advanced or metastatic PC is the castration therapy consisting in androgen deprivation. Castration therapy cut off the supply of androgens that encourage PC growth (McLeod 2003; Theodore 2004). Despite the dramatic tumor regression that follows the castration therapy, the patients will ultimately become unresponsive and the prostate tumors will relapse within 1–3 years in a more aggressive castration-resistant mode (Fusi et al. 2004).

13.2.1 TCTP Mediates Heat Shock Protein 27 Cytoprotective Function in CRPC

In 2005, Rocchi et al. found that one of the most common genetic events in castration-resistant prostate cancer (CRPC) is the activation of heat shock protein 27 (Hsp27) expression (Rocchi et al. 2005). But the mechanism by which Hsp27 induces a multi-drugs resistance to CRPC tumors was unknown. Thus, in order to elucidate the pathways by which Hsp27 imposes its cytoprotective effect and find new therapeutic targets specific of CRPC tumors, Dr. Rocchi laboratory has screened all Hsp27 interactors using a two-hybrid system (Katsogiannou et al. 2014) and identified TCTP as a new Hsp27 protein partner (Baylot et al. 2012). This work pinpoints for the first time TCTP as a potential therapeutic target in CRPC (Baylot et al. 2012; Acunzo et al. 2014).

This study has demonstrated that TCTP protein level correlates with PC cells aggressiveness. In castration-resistant (CR) prostate tumor cells, TCTP protein expression is strongly increased compared to its expression in castration naïve PC cells. Furthermore, a tissue microarray experiment performed on 211 clinical specimens showed that TCTP is highly uniformly overexpressed in 75% of the CRPC samples. These observations highlight its association with the aggressiveness of the human disease. Additionally, no or weak TCTP expression has been detected in normal or benign tissues, suggesting that targeting TCTP in human CRPC may cause only weak undesirable toxicity in normal tissues.

Further mechanistic investigations showed that in castration-sensitive cells, overexpressing Hsp27 is sufficient to increase TCTP protein level but not TCTP mRNA level. Additionally, it has been demonstrated that Hsp27 is a direct upstream regulator of TCTP and that this chaperone protects TCTP from its ubiquitination and proteasomal degradation.

In CRPC cells, TCTP inhibition leads to cell viability reduction, cell cycle arrest, and caspase-*3*-dependent apoptosis activation. Moreover, in castration naïve PC cells stably overexpressing Hsp27, TCTP downregulation increases apoptosis *via* caspase-*3* activation and enhances chemotherapy. These data show that TCTP silencing suppresses the chemo-resistance of CRPC cells due to high Hsp27 levels and suggest that TCTP is a mediator of Hsp27 cytoprotective function in CR prostate tumors. Furthermore, targeting TCTP in vivo with an antisense oligonucleotide, developed by Dr. Rocchi laboratory (Baylot et al. Patent PCT10306447.3 2010), suppresses the growth of PC cell xenografts and significantly enhances chemotherapy activity upon systemic delivery. These findings open up the possibility for using TCTP knockdown in combination with other established therapeutic approaches to increase treatment efficacy in CRPC.

13.2.2 TCTP and P53 in CRPC: "Neither Can Live While the Other Survives"

Very interestingly, Baylot et al. also found that CR progression correlates with the loss of the tumor suppressor P53. In a prior study, Amson et al. has shown that TCTP and P53 are involved in a reciprocal negative-feedback loop in breast cancer (Amson et al. 2012). But the role of P53 in the PC progression was still elusive. Baylot et al. has demonstrated for the first time a link between TCTP, P53, and the CR progression of PC (Fig. 13.1).

On one hand, after castration, the prostate tumors that have progressed from a castration-sensitive state to a CR state overexpress TCTP and loose P53 expression.

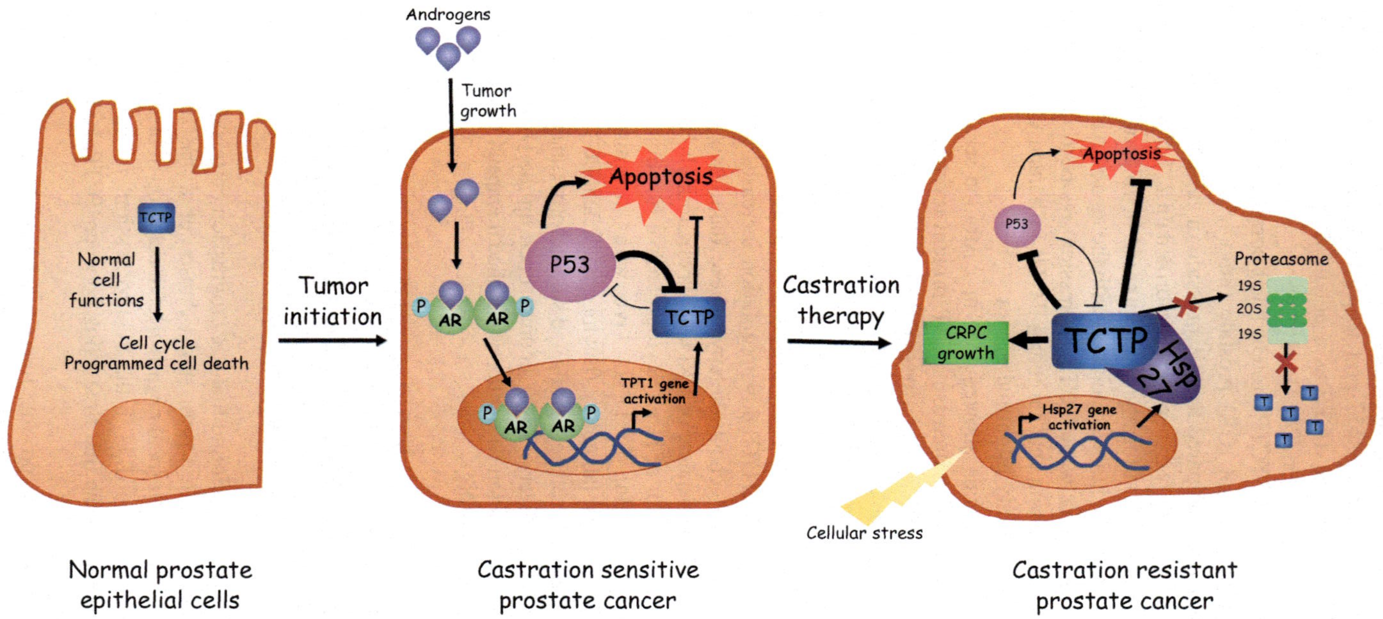

Fig. 13.1 The central role of TCTP in Prostate Cancer (PC) progression. Despite the essential role of TCTP in normal prostate epithelial cells (*first panel*), TCTP expression is highly increased in prostate tumor cells (*second and third panels*). In castration-sensitive PC (CSPC), the androgen receptor (AR) directly activates the transcription of TCTP by binding to the promoter of this gene (TPT-1). In early stages of PC, the tumor suppressor P53 is highly expressed and represses TCTP transcription. These opposite functions of the AR and P53 maintain a low level of TCTP in CSPC leading to a high sensitivity of the CS tumors to therapies-induced apoptosis (*second panel*). Following the castration therapy, the tumors will initially regress but within 1 to 3 years they will relapse in a castration-resistant mode. In castration-resistant PC (CRPC), heat shock protein 27 (Hsp27) is highly expressed and its cytoprotective function is crucial for CRPC growth, invasion, and chemo-resistance. By directly binding to TCTP protein in the cytoplasm of CRPC cells, Hsp27 protects TCTP from its degradation by the ubiquitin–proteasome pathway, leading to a massive increase of TCTP protein level in CRPC tumors. In late stages of PC, the high level of TCTP protein level promotes P53 degradation and therapies-induced apoptosis resistance (*third panel*)

TCTP silencing using TCTP antisense oligonucleotide is able to restore P53 expression and function in CRPC tumors, suggesting that castration sensitivity is directly linked to P53. On the other hand, P53 downregulation in castration-sensitive PC cells significantly inhibits chemotherapy-induced apoptosis compared to the control cells, suggesting an important link between P53 status and PC tumors chemotherapy resistance.

These data show that TCTP is upregulated in CRPC tumors leading to the loss of P53 expression and function together with castration- and chemo-therapies resistance. This work importantly highlights the crucial role of TCTP/P53 axis in CR progression of PC.

13.2.3 Development of a TCTP Antisense Oligonucleotide for Clinical Applications

As mentioned above, Dr. Rocchi laboratory has developed a TCTP inhibitor that can be used for human therapy and has screened by gene walk all antisense oligonucleotide (ASO) sequences targeting TCTP full-length mRNA (Karaki et al. 2017). Initially, 28 ASOs have been designed. Finally, three ASO lead sequences, that potently inhibited TCTP expression, have been furthered examined for their ability to affect CRPC cells and tumor growth. Thus, it has been reported that TCTP-ASOs enter to the cells *via* macropinocytosis, increased caspase-*3*-dependent apoptosis, blocked cell cycle, and enhanced chemotherapy in CRPC cells in vitro. And, consistent with these in vitro data, systemic administration of TCTP-ASOs in immunocompromised mice suppressed CRPC tumor growth and also significantly enhanced castration and chemotherapy activities in vivo. Additionally, TCTP-ASO treated mice showed a significant decrease of Ki-67 levels, a proliferation marker, compared to the control group. Moreover, possible toxic effects resulting from oligonucleotide administration have been checked. The animals did not show any change neither in their behavior, nor in their body weight during the experiment. Furthermore, treated mice did not show any signs for hepatic damage, since their aspartate aminotransferase (AST) and alanine aminotransferase (ALT) levels were within the range of normal values, and they didn't show significant difference compared to the control group. Finally, no sign of renal damage was observed, since the creatinine levels were normal and the biochemical analyses of urine were normal for the tested parameters.

13.3 Discussion

Taken together, these studies demonstrate that TCTP has a crucial role in the different stages of PC malignant progression from the tumor initiation to the multi-drugs resistant stage of the disease.

Arcuri et al. were the first to investigate TCTP expression and function in human prostate normal and cancer tissues. The authors have notably found that TCTP is the most highly expressed calcium binding protein in the human prostate cancer cells. Subsequent studies from different laboratories confirmed that TCTP is upregulated in early stages of the disease, in which the prostate tumor growth is fueled by androgens (Gnanasekar et al. 2009; Kaarbo et al. 2013). TCTP has since been identified as an androgen-regulated gene (Kaarbo et al. 2013). Furthermore, TCTP has been reported to be causal for the resistance to androgen withdrawal and chemotherapy in PC (Baylot et al. 2012; Acunzo et al. 2014). This work has validated, using 211 clinical specimens, that TCTP is slightly over-expressed in the castration naïve specimens compared to normal specimen, confirming his implication in PC initiation, and that its expression is abolished upon castration therapy. Most importantly this study has shown for the first time, that TCTP is highly over-expressed in multi-drugs resistant prostate tumors and metastases, pinpointing TCTP as a key protein in the late stages of PC in which the tumors grow in a castration-resistant mode (CRPC). Currently, there is no effective therapy for patients with CRPC and existing novel therapies only have a modest impact on the overall survival of these patients (McKeage 2012; de Bono et al. 2010). Clearly, novel approaches were required to effectively treat CRPC, in particular new strategies that identify the molecular mechanisms by which CRPC becomes chemoresistant, as well as the identification of novel therapeutic targets that could improve clinical outcome. Thus, identifying TCTP as a new therapeutic target for the treatment of CRPC represents a major advance in the field. Additionally, altogether these findings also strongly suggest that TCTP is highly prognostic in human PC.

A TCTP inhibitor has been developed by Dr. Rocchi laboratory (Baylot et al., Patent PCT10306447.3; 2010) for clinical applications. Recent results showed that the TCTP inhibitor can suppress CRPC tumor growth and enhance castration- and chemo-therapies in vitro and in vivo. The stability, biodisponibility, and delivery improvement of this TCTP inhibitor for human treatment is currently under investigation in Dr. Rocchi laboratory and represents today a great hope for the patients with CRPC.

References

Acunzo J, Baylot V, So A, Rocchi P (2014) TCTP as therapeutic target in cancers. Cancer Treat Rev 40(6):760–769

Amson R, Pece S, Lespagnol A, Vyas R, Mazzarol G, Tosoni D et al (2012) Reciprocal repression between P53 and TCTP. Nat Med 18:91–99

Arcuri F, Papa S, Carducci A, Romagnoli R, Liberatori S, Riparbelli MG et al (2004) Translationally controlled tumor protein (TCTP) in the human prostate and prostate cancer cells: expression, distribution, and calcium binding activity. Prostate 60:130–140

Baylot V, Andrieu C, Acunzo J, Katsogiannou M, Rocchi P (2010) Nucleic acid targeting TCTP for use in the treatment of cancer. Patent PCT10306447.3

Baylot V, Katsogiannou M, Andrieu C, Taieb D, Acunzo J, Giusiano S et al (2012) Targeting TCTP as a new therapeutic strategy in castration-resistant prostate cancer. Mol Ther 20(12): 2244–2256

de Bono JS, Oudard S, Ozguroglu M, Hansen S, Machiels JP, Kocak I et al (2010) Prednisone plus cabazitaxel or mitoxantrone for metastatic castration-resistant prostate cancer progressing after docetaxel treatment: a randomised open-label trial. Lancet 376:1147–1154

Fusi A, Procopio G, Della Torre S, Ricotta R, Bianchini G, Salvioni R et al (2004) Treatment options in hormone-refractory metastatic prostate carcinoma. Tumori 90:535–546

Gnanasekar M, Thirugnanam S, Zheng GX, Chen AS, Ramaswamy K (2009) Gene silencing of translationally controlled tumor protein (TCTP) by siRNA inhibits cell growth and induces apoptosis of human prostate cancer cells. Int J Oncol 34:1241–1246

Jemal A, Siegel R, Xu J, Ward E (2010) Cancer statistics, 2010. CA Cancer J Clin 60:277–300

Kaarbo M, Storm M, Qu S, Wæhre H, Risberg B, Danielsen HE et al (2013) TCTP is an androgen regulated gene implicated in prostate cancer. PLoS One 8(7):e69398

Karaki S et al (2017) Lipid-oligonucleotide conjugates improve cellular uptake and efficiency of TCTP-antisense in castration-resistant prostate cancer. J Control Release 258:1–9

Katsogiannou M, Andrieu C, Baylot V, Baudot A, Dusetti NJ, Gayet O et al (2014) The functional landscape of Hsp27 reveals new cellular processes such as DNA repair and alternative splicing and proposes novel anticancer targets. Mol Cell Proteomics 13(12):3585–3601

McKeage K (2012) Docetaxel: a review of its use for the first-line treatment of advanced castration-resistant prostate cancer. Drugs 72:1559–1577

McLeod DG (2003) Hormonal therapy: historical perspective to future directions. Urology 61:3–7

Rocchi P, Beraldi E, Ettinger S, Fazli L, Vessella RL, Nelson C et al (2005) Increased Hsp27 after androgen ablation facilitates androgen-independent progression in prostate cancer via signal transducers and activators of transcription 3-mediated suppression of apoptosis. Cancer Res 65:11083–11093

Siegel R, Naishadham D, Jemal A (2012) Cancer statistics, 2012. CA Cancer J Clin 62:10–29

Theodore C (2004) Hormone therapy for prostate cancer. Bull Cancer 91:69–74

Chapter 14
Role of TCTP for Cellular Differentiation and Cancer Therapy

Ean-Jeong Seo, Nicolas Fischer, and Thomas Efferth

Abstract The translationally controlled tumor protein (TCTP) is a highly conserved protein that is regulated due to a high number of extracellular stimuli. TCTP has an important role for cell cycle and normal development. On the other side, tumor reversion and malignant transformation have been associated with TCTP. TCTP has been found among the 12 genes that are differentially expressed during mouse oocyte maturation, and an overexpression of this gene was reported in a wide variety of different cancer types. Its antiapoptotic effect is indicated by the interaction with several proapoptotic proteins of the Bcl-2 family and the p53 tumor suppressor protein. In this article, we draw attention to the role of TCTP in cancer, especially, focusing on cell differentiation and tumor reversion, a biological process by which highly tumorigenic cells lose their malignant phenotype. This protein has been shown to be the most strongly downregulated protein in revertant cells compared to the parental cancer cells. Decreased expression of TCTP results either in the reprogramming of cancer cells into reversion or apoptosis. As conventional chemotherapy is frequently associated with the development of drug resistance and high toxicity, the urge for the development of new or additional scientific approaches falls into place. Differentiation therapy aims at reinducing differentiation backward to the nonmalignant cellular state. Here, different approaches have been reported such as the induction of retinoid pathways and the use of histone deacetylase inhibitors. Also, PPARγ agonists and the activation of the vitamin D receptor have been reported as potential targets in differentiation therapy. As TCTP is known as the histamine-releasing factor, antihistaminic drugs have been shown to target this protein. Antihistaminic compounds, hydroxyzine and promethazine, inhibited cell growth of cancer cells and decreased TCTP expression of breast cancer and leukemia cells. Recently, we found that two antihistaminics, levomepromazine and buclizine, inhibited cancer cell growth by direct binding to TCTP and induction of cell differentiation. These data confirmed that TCTP is an exquisite target for anticancer differentiation therapy and antihistaminics have

E.-J. Seo • N. Fischer • T. Efferth (✉)
Department of Pharmaceutical Biology, Institute of Pharmacy and Biochemistry, Johannes Gutenberg University, Staudinger Weg 5, 55128 Mainz, Germany
e-mail: efferth@uni-mainz.de

A. Telerman, R. Amson (eds.), *TCTP/tpt1 - Remodeling Signaling from Stem Cell to Disease*, Results and Problems in Cell Differentiation 64,
DOI 10.1007/978-3-319-67591-6_14

potential to be lead compounds for the direct interaction with TCTP as new inhibitors of human TCTP and tumor growth.

Abbreviations

ADDS	Adenylosuccinate synthase
AIF	Apoptosis-inducing factor
APL	Acute promyelocytic leukemia
ATRA	All-trans retinoic acid
Bcl-2	B-cell lymphoma 2
BMP	Bone morphogenic proteins
CH60	Mitochondrial 60 kDa heat shock protein
CHFR	Checkpoint with forkhead and ring finger domains
COF1	Cofilin-1
ENOA	α-Enolase
ER60	Probable protein disulfide isomerase
ES	Embryonic stem
FABP	Liver fatty acid-binding protein
GTA1	Glutathione S-transferase alpha
HDAC	Histone deacetylase
HSP105	Heat shock protein 105
KCRB	Creatine kinase B
Mcl-1	Myeloid cell leukemia 1
MDM2	Murine double minute 2
MPSS	Megasort and massively parallel signature sequencing
NDKA	Nucleoside diphosphate kinase A
NPM2	Nucleoplasmin 2
Oct4	Octamer-binding transcription factor 4
PDCD6IP	Programmed cell death six-interacting protein
PPARγ	Peroxisome proliferator-activated receptor-γ
PS1	Presenilin 1
RARs	Retinoic acid receptors
RMS	Rhabdomyosarcoma
SAHA	Suberoylanilide hydroxamic acid
SIAH1	Seven in absentia homologue 1
Sox2	Sex-determining region Y-box 2
STAT3	Signal transducer and activator of transcription 3
STI1	Stress-inducible phosphoprotein 1
TACC3	Transforming acidic coiled-coil protein 3
TCTP	Translationally controlled tumor protein
TSAP	Tumor suppressor-activated pathway
VDR	Vitamin D receptor

14.1 Introduction

The translationally controlled tumor protein (TCTP) encoded by the *TPT1* gene is a highly conserved protein that can be found within all eukaryotic organisms, all tissues and cell types (Oh et al. 2013; Acunzo et al. 2014). It was initially identified in Ehrlich ascites tumor cells, and no homologies to other proteins have been found at that time (Oh et al. 2013). Due to its numerous functions, it is also known as histamine-releasing factor, TPT1, p23, or fortilin (Acunzo et al. 2014; Nagano-Ito and Ichikawa 2012). It can be found in both, in the nucleus and the cytoplasm of cells, and it is regulated by a high number of extracellular stimuli. The protein interacts with itself and with a wide variety of other proteins [such as myeloid cell leukemia 1 (Mcl-1) and p53] (Acunzo et al. 2014). Furthermore, TCTP is expressed extra- and intracellularly, implicating that it is involved in many biological processes such as development (Kubiak et al. 2008; Hsu et al. 2007; Chen et al. 2007a), cell cycle (Brioudes et al. 2010; Gachet et al. 1999), cellular growth (Koziol and Gurdon 2012), protein synthesis (Cans et al. 2003), cytoskeleton (Burgess et al. 2008; Tsarova et al. 2010), immune response (MacDonald et al. 1995), cell death (Li et al. 2001; Liu et al. 2005; Yang et al. 2005), and induction of pluripotent stem cells (Acunzo et al. 2014) (Fig. 14.1). On the other hand, it has also been associated with tumor reversion and malignant transformation (Tuynder et al. 2002; Rho et al. 2011).

Its crucial biological functions have been demonstrated in several studies. Knockout of TCTP was lethal in utero in mice (Chen et al. 2007a; Susini et al. 2008), pointing to its role in normal development of organisms. Within the cell cycle, TCTP binds to tubulin and associates with the microtubules during different phases of the cell cycle (Gachet et al. 1999). In metaphase, TCTP is bound to the mitotic spindle. During the transition from metaphase to anaphase, it is dissociated from the spindle (Acunzo et al. 2014).

Phosphorylation of TCTP by PLK1 is necessary for cell cycle. Additionally, TCTP interacts with CHFR and thus prevents the entry into mitosis (Acunzo et al. 2014).

TCTP is able to interact with Mcl-1 (Li et al. 2001; Liu et al. 2005) and other members of the B-cell lymphoma 2 (Bcl-2) family (Yang et al. 2005; Susini et al. 2008) and thus plays an important role in the regulation of apoptosis. Destabilizing the tumor suppressor p53 leads to an additional antiapoptotic effect (Rho et al. 2011; Amson et al. 2011), indicating an involvement in tumorigenesis.

Besides the direct effect on the cell cycle, TCTP reduces cellular stress and protects the cell from thermal shock by binding to denatured proteins and refolding them by acting as a molecular chaperone (Gnanasekar et al. 2009). Additionally, it acts as an antioxidant by neutralizing radicals directly (Bini et al. 1997).

Due to the involvement in numerous biological cell processes, its relevance as target for cancer therapy has been discussed. Overexpression of TCTP in a large variety of cancers was reported (Acunzo et al. 2014; Nagano-Ito and Ichikawa 2012; Amson et al. 2013b; Telerman and Amson 2009; Miao et al. 2013).

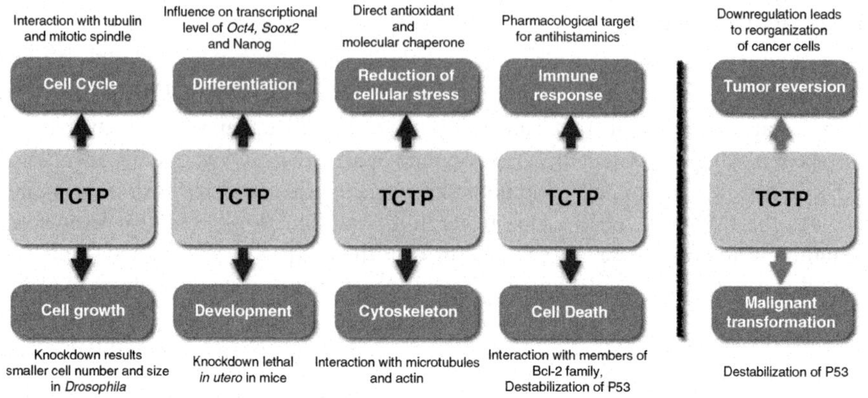

Fig. 14.1 Multiple functions of TCTP

Telerman's group demonstrated that downregulation of TCTP in MCF7 and T47D cells caused a reorganization of cells (Acunzo et al. 2014; Tuynder et al. 2002). Additionally, they showed that the protein was drastically downregulated within the process of tumor reversion and correlated with clinical and pathological parameters of the aggressiveness of breast cancer (Acunzo et al. 2014; Amson et al. 2011). This indicates a role of TCTP as prognostic factor for breast cancer (Amson et al. 2013b). A change in the expression of TCTP was associated with in vivo colon carcinogenesis. A decrease of TCTP by 2.7-fold during cell differentiation in Caco2 cells has been observed (Stierum et al. 2003), indicating the importance of the protein for the differentiation process. The limited number of tumor markers especially in colon cancer makes the retrieval for new targets more urgent (Stierum et al. 2003; Williams et al. 1996).

In other cancer types, an involvement of TCTP in tumor reversion was also found, e.g., prostate cancer, lung cancer, leukemia, erythroleukemia, glioma, lymphoma, squamous cell carcinoma, colon cancer, hepatocellular cancer, liver cancer, larynx cancer, and melanoma (Acunzo et al. 2014; Amson et al. 2013b; Miao et al. 2013). In this wide variety of cancers, higher TCTP expression levels were found in tumors compared to the corresponding normal tissues (Acunzo et al. 2014; Tuynder et al. 2002; Amson et al. 2013b; Sinha et al. 2000).

In conclusion, the involvement of TCTP in cell differentiation and tumor reversion makes it an interesting target for anticancer therapy (Acunzo et al. 2014).

14.1.1 TCTP in Differentiation Processes

Cellular differentiation is a process in the development of immature cells to more complex states. TCTP plays an important role in cell differentiation, not only in humans but also in microorganisms, plants, and animals.

TcpA, a protein with high similarity to TCTP, can be found in *Aspergillus nidulans*. TcpA plays an important role in cell cycle progression and development of this model organism. Furthermore, TcpA expression influences the balance between asexual and sexual differentiation in *A. nidulans*, further stressing its role in cell differentiation (Oh et al. 2013).

Also in plants, TCTP or homologues can be identified. In *Robinia pseudoacacia*, the TCTP homologue Rpf41 regulates symbiotic nodulation in legume, indicating an involvement of TCTP in symbiotic cell differentiation. Functional parallels in the regulation of cell division by TCTP between *Arabidopsis* and *Drosophila* were found. Here, the TCTP expression levels in the plant varied due to stress conditions such as darkness, cold, salt, drought, or heavy metals. Especially high *TCTP* gene expression was observed in physically active tissues (Chou et al. 2016). Low *TCTP* expression was found during the reversion of cells from the transformed to the normal phenotype (Chou et al. 2016). In *Arabidopsis* two isoforms AtTCTP1 and AtTCTP2 have been detected. The two TCTP homologues reveal different functions. AtTCTP1 is an important regulator of mitosis, while AtTCTP2 plays a role in vegetative reproduction (Toscano-Morales et al. 2015).

AmphiTCTP is a TCTP orthologous gene in *Amphioxus*. The expression pattern of AmphiTCTP correlated with differentiation of notochord and somite, implying a role in embryonic development (Chen et al. 2007b). In *Hydra vulgaris*, another TCTP homologue showed an expression pattern coinciding with the proliferation status (Yan et al. 2000).

TCTP is among the 12 genes that are differentially expressed during mouse oocyte maturation. The other genes are the transforming acidic coiled-coil protein 3 (TACC3), heat shock protein 105 (HSP105), programmed cell death six-interacting protein (PDCD6IP), stress-inducible phosphoprotein 1 (STI1), importin α2, adenylsuccinate synthase (ADDS), nudix, spindlin, lipocalin, lysozyme, and nucleoplasmin 2 (NPM2) (Vitale et al. 2007). Mouse embryonic stem cells represent a good model system for studying stem cell biology since murine and human embryonic stem cells share many conserved pathways in self-renewal and differentiation (Sato et al. 2003; Ginis et al. 2004). They can differentiate into two different types of neurons. Proteomic analysis of E14 cells and neurons showed that TCTP was significantly downregulated. In motor neurons, a stronger downregulation was found than in dopaminergic neurons. The TCTP expression levels were independent of the extracellular Ca^{2+}-concentrations during neuronal differentiation. This indicates an involvement of TCTP in neurogenesis through modulating tubulin expression and Ca^{2+} binding (Wang and Gao 2005). In mouse cells, TCTP expression is highly regulated at both transcriptional and translational levels by a broad range of extracellular signals (Bommer and Thiele 2004). Human embryonic stem cells represent pluripotent cells. They are able to self-renew and proliferate without limitations. Originating from the inner cell mass of the human blastocysts of the embryo, they differentiate into any fetal or adult cell type (Donovan and Gearhart 2001). A small number of key transcription factors were described for self-renewal and suppression of differentiation (Chambers et al. 2003). In embryotic stem cells, several genes are upregulated, among them octamer-binding transcription factor 4 (*Oct4*) acting as marker of pluripotency (Pesce and Schöler 2001) and differentiation, as it is able to repress and activate the expression of different genes by directly binding to their promotor regions or indirectly by neutralizing their transcription activators (Pan et al. 2002). *Oct4* becomes silent after gastrulation in mouse and human mammalian somatic cells (Kirchhof et al. 2000). *Oct4* is highly conserved (Koziol and Gurdon 2012). For *Oct4* activation, nuclear actin polymerization is necessary in *Xenopus laevis* oocytes (Miyamoto et al. 2011). As TCTP has an actin binding site, interaction can be expected here (Koziol and Gurdon 2012). Another central player in pluripotency is Nanog, which is required for maintaining the undifferentiated state of early postimplantation embryos and ES cells (Chambers et al. 2003; Mitsui et al. 2003). It leads embryonic stem (ES) cells into self-renewal by acting in parallel with cytokine stimulation of signal transducer and activator of transcription 3 (STAT3). Nanog has exclusively been identified in ES cells (Chambers et al. 2003; Mitsui et al. 2003). In human nuclei, a change in the transcriptional level of *Oct4* and Nanog has been demonstrated under the influence of TCTP (Koziol et al. 2007). Due to its highly conserved function, an effect of TCTP on the activation of pluripotency can be predicted (Tani et al. 2007). Also, bone morphogenic proteins (BMP), which inhibit differentiation, can be seen as pluripotency markers (Masui et al. 2007). They suppress differentiation and thus lead to a higher pluripotency (Ying et al. 2003). One more important marker of pluripotency is sex-determining region Y-box 2 (Sox2), which binds Oct4 and activates genes promoting pluripotency (Nishimoto et al. 1999) and controls its inhibitors (Niwa et al. 2000). Sox2 regulates several transcription factors affecting the expression of Oct3/4 making Sox2 an essential factor for maintaining ES cells in a pluripotent state (Masui et al. 2007).

TCTP is expected to promote pluripotency in two different ways, one by directly activating pluripotency genes such as *Sox2*, *Nanog*, *Oct4*, and *Klf4* and one indirect way by inhibiting the expression of somatic genes (Koziol and Gurdon 2012).

14.2 TCTP in Cancer

14.2.1 TCTP and Tumor Reversion

Tumor reversion is a biological process, by which highly tumorigenic cells lose their malignant phenotype (Telerman and Amson 2009; Amson et al. 2013a). For instance, teratoma cells differentiated into normal somatic tissues, and tumor cells acquired the molecular circuitry that resulted in the negation of chromosomal instability, translocations, oncogene activation, and loss of tumor suppressor genes (Telerman and Amson 2009; Askanazy 1907). Telemann's group developed a series of revertants using the H1 parvovirus, which is a small DNA virus that preferentially kills tumor cells, but keeps their normal counterparts alive. They analyzed changes in gene expression (Tuynder et al. 2002, 2004; Telerman et al. 1993; Toolan 1967; Mousset and Rommelaere 1982; Nemani et al. 1996; Amson et al. 1996; Roperch et al. 1999) and identified about 300 genes as gene tumor reversion by mRNA differential display, Megasort and massively parallel signature sequencing (MPSS) (Tuynder et al. 2002; Amson et al. 1996; Roperch et al. 1999; Liang and Pardee 1992; Brenner et al. 2000a, b; Israeli et al. 1997). P53-regulated proteins were seven in absentia homologue 1 (SIAH1), an E3 ligase and a transcriptional target of p53 (Nemani et al. 1996; Amson et al. 1996; Roperch et al. 1999; Fiucci et al. 2004), presenilin 1 (PS1), a predisposition gene for familial Alzheimer's disease (Roperch et al. 1998), tumor suppressor-activated pathway (TSAP), a transcriptional target of p53 controlling the secretion of proteins (Amson et al. 1996; Passer et al. 2003; Amzallag et al. 2004; Lespagnol et al. 2008), and translationally controlled tumor protein (TCTP), the inhibitor of p53 activity (Cans et al. 2003; Tuynder et al. 2002, 2004; Susini et al. 2008; Amson et al. 2011). Inhibition of TCTP expression increased the number of revertant cells, which regained sensitivity to contact inhibition and decreased tumor-forming capability (Telerman and Amson 2009; Tuynder et al. 2004). As TCTP was the most strongly downregulated protein in the revertant cells compared to the parental cancer cells, the effects of its inhibition in several biological and genetic models have been studied (Tuynder et al. 2002, 2004; Amson et al. 2011; Telerman and Amson 2009). Decreasing TCTP expression resulted in either reprogramming of cancer cells into revertants or apoptosis.

14.2.2 TCTP as Antiapoptotic Protein

TCTP is known to play a key role in the regulation of apoptosis. TCTP regulated antiapoptotic activity by suppressing Mcl-1 degradation through blocking its ubiquitination (Li et al. 2001; Liu et al. 2005; Yang et al. 1995). However, TCTP and Mcl-1 could independently protect cells from apoptosis (Graidist et al. 2004). TCTP interacted with other antiapoptotic proteins from the Bcl-2 family such as Bcl-xL (Yang et al. 2005) or Bax (Susini et al. 2008) (Fig. 14.2). Yang et al. identified the interaction site to the N-terminal region of TCTP and the Bcl-2 homology domain 3 of Bcl-xL and demonstrated that the TCTP N-terminal region mediates inhibition of apoptosis (Yang et al. 2005). This result corresponds to data from Zhang et al., who showed that Arg21 in the N-terminal region of TCTP was critical for TCTP binding to Mcl-1 (Zhang et al. 2002). The homodimerization of proapoptotic Bax is required for its apoptotic activity. TCTP prevented the apoptotic effect of Bax by inserting into the mitochondrial membrane and inhibiting Bax dimerization. Unlike Mcl-1 and Bcl-xL, TCTP did not directly bind Bax (Susini et al. 2008).

P53 protein is well-known as tumor suppressor. It is a transcription factor and regulates the transcription of numerous genes. It activates the transcription of DNA repair genes upon DNA damage by regulating genes involved in cell cycle and apoptosis such as Bax and Bcl-2 (Riley et al. 2008). P53 promotes apoptosis in cancer cells, whereas TCTP prevents apoptosis by repressing the transcription of p53 (Rho et al. 2011; Amson et al. 2011) (Fig. 14.2). TCTP bound p53 and prevented

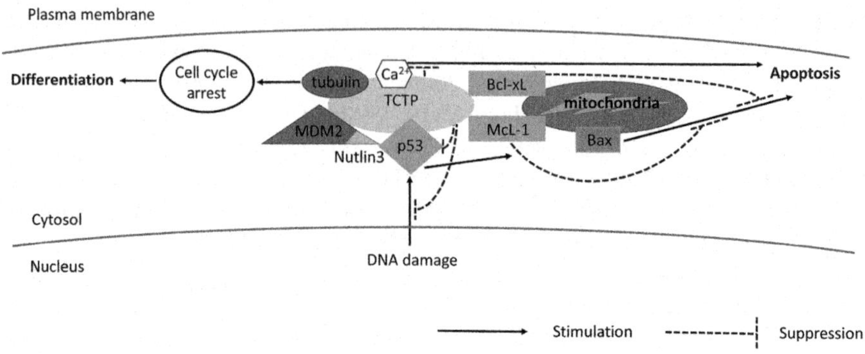

Fig. 14.2 Interaction partners of TCTP to differentiation and apoptosis regulation

apoptosis by destabilizing the protein (Rho et al. 2011). The murine double minute 2 (MDM2) is a transcriptional target of p53. If it is overexpressed, MDM2 ubiquitinates and degrades p53. TCTP directly associates with the E3 ubiquitin ligase MDM2, increasing MDM2-mediated ubiquitination of p53 and promoting its degradation (Amson et al. 2011; Amit et al. 2009) (Fig. 14.2). Nutlin-3, a protein that promotes apoptosis, blocked the interaction between MDM2 and TCTP (Funston et al. 2012) (Fig. 14.2).

14.2.3 Cell Cycle Regulation of TCTP

TCTP is involved in the cell cycle (Gachet et al. 1999). It has a tubulin-binding site that allows binding to microtubules in a cell-cycle-dependent way (Gachet et al. 1999) (Fig. 14.2). TCTP is recruited to the mitotic spindle during metaphase, but is released at the M/G1 transition (Gachet et al. 1999). TCTP interacts with the protein checkpoint by the forkhead and ring finger domains (CHFR) that binds to microtubules (Burgess et al. 2008). If microtubules are depolymerized, CHFR and TCTP interaction is reduced. This interaction senses microtubule abnormalities by CHFR that results in CHFR activation, polo-like kinase 1 (PLK1) degradation, and finally cell cycle arrest (Burgess et al. 2008). If PLK1 phosphorylation sites on TCTP were blocked, increased numbers of multinucleated cells were observed, indicating that the completion of mitosis was inhibited (Yarm 2002). This result demonstrates that TCTP is crucial for cell cycle regulation and that its phosphorylation by PLK1 is required for the precise exit from mitosis (Yarm 2002).

14.2.4 TCTP Reduces Cellular Stress

Cell death can be induced by Ca^{2+} influx. The level of TCTP is controlled by the intracellular Ca^{2+} concentration, and elevation of Ca^{2+} increased *TCTP* mRNA in cells (Xu et al. 1999). Binding of TCTP to Ca^{2+} was first reported using *Trypanosoma brucei* protein and later on using the human protein (Haghighat and Ruben 1992; Sanchez et al. 1997). Thapsigargin increased cytosolic levels by blocking the ability of the cells to pump calcium into the ER, which depletes its Ca^{2+} stores. This activated plasma membrane calcium channels allowing Ca^{2+} influx into the cytosol, thereby initiating apoptosis. The lack of TCTP resulted in exaggerated increases of Ca^{2+} in thapsigargin-challenged cells (Graidist et al. 2007). Increasing the intracellular Ca^{2+} levels beyond the normal range could damage the mitochondrial membranes and leads to the release of cytochrome C and apoptosis-inducing factor (AIF), resulting in apoptosis. Ca^{2+} binding of TCTP was required for cellular protection against thapsigargin-induced apoptosis (Graidist et al. 2007) (Fig. 14.2). TCTP binds to and scavenges Ca^{2+}, thus preventing the ion from activating downstream apoptotic execution pathways (Graidist et al. 2007) (Fig. 14.2). Thapsigargin also induced ER stress, in which unfolded proteins were accumulated in the organelle (Nagano-Ito and Ichikawa 2012). Thapsigargin decreased Ca^{2+} concentration in the ER and suppressed small molecule Ca^{2+}-dependent chaperones in the organelle, allowing accumulation of abnormal proteins, which eventually drove cells to undergo apoptosis (Nagano-Ito and Ichikawa 2012).

Therefore, it can be concluded that TCTP protects cells from ER stress-induced apoptosis by inhibiting the corresponding signal pathways.

14.3 TCTP for Differentiation Therapy

14.3.1 Approaches of Differentiation Therapy in General

Cancer cells fail to differentiate into functional mature cells, and differentiation therapy aims to reinducing differentiation backward to nonmalignant cellular states. This process is termed tumor reversion (Spira and Carducci 2003; Pierce and Wallace 1971). Differentiation therapy is based on the assumption that specific neoplastic cells exhibit aberrant patterns of differentiation and that treatment with appropriate agents results in tumor reprogramming, ultimately leading to a loss in proliferative capacity and induction of differentiation (Leszczyniecka et al. 2001). Conventional chemotherapy is frequently associated with the development of drug resistance and high toxicity, both of which limit its therapeutic efficacy (Lal et al. 1993). Stierum et al. studied protein expression changes in differentiating Caco-2 cells by proteomics approach (Stierum et al. 2003). Eleven proteins were identified including TCTP, liver fatty acid-binding protein (FABL), three forms of α-enolase (ENOA), nucleoside diphosphate kinase A (NDKA), cofilin-1 (COF1), mitochondrial 60 kDa heat shock protein (CH60), probable protein disulfide isomerase (ER60), creatine kinase B (KCRB), and glutathione S-transferase alpha (GTA1) (Stierum et al. 2003). This differentiation-related change in phenotype of Caco-2 cells involved changes in a variety of distinct biochemical pathways (Stierum et al. 2003). The processes were related to protein folding and disulfide bridge formation, cytoskeleton formation and maintenance, nucleotide metabolism, glycolysis, as well as tumorigenesis-associated proteins (Stierum et al. 2003).

14.3.2 Retinoids

Retinoids are a class of compounds derived from vitamin A possessing the ability to regulate cell proliferation, differentiation, and apoptosis in normal and cancer cells (Garattini et al. 2007). Retinoids play a fundamental role in chemoprevention of carcinogenesis and in differentiation therapy (Hansen et al. 2000). Retinoids exert their bioactivity by binding to retinoic acid receptors (RARs) (Garattini et al. 2007). Treatment of osteosarcoma and chondrosarcoma cell lines with all-trans retinoic acid (ATRA) resulted in reversible growth inhibition and decreased colony formation (Thein and Lotan 1982; Ng et al. 1985). Clinically, ATRA is successfully applied to treat acute promyelocytic leukemia (APL) with an aberrant chromosomal translocation (Waxman 2000). This translocation results from the fusion of the *PML* gene with the *RAR* gene (*PML-RARα*) (Spira and Carducci 2003). ATRA differentiates APL cells into mature neutrophils (Huang et al. 1987a, b). Retinoids reduced cell proliferation and induced myogenic differentiation in a variety of rhabdomyosarcoma (RMS) cell lines derived from either alveolar or embryonal RMS (Luo et al. 2010; Crouch and Helman 1991; Brodowicz et al. 1999; Barlow et al. 2006).

14.3.3 Histone Deacetylase Inhibitors

Histone deacetylation by histone deacetylases (HDACs) leads to chromatin compaction. Histone deacetylation is related to transcriptional repression of tumor suppressors involved in regulating cell growth and differentiation in various cancers (Mai et al. 2005; Cress and Seto 2000). DMS53 small cell lung carcinoma cells changed their morphology upon treatment with the histone deacetylase inhibitor, trichostatin A, and showed cellular differentiation (Platta et al. 2007). Five quinolone compounds, which inhibited HDAC activity in vitro, stimulated cell differentiation at growth inhibitory concentrations in MCF-7 breast carcinoma cells (Martirosyan et al. 2004). The morphology of MCF-7 cells was changed after treatment of suberoylanilide hydroxamic acid (SAHA) or vorinostat, suggesting the induction of epithelial mammary differentiation (Munster et al. 2001).

14.3.4 PPARγ Agonists

Peroxisome proliferator-activated receptor-γ (PPARγ) is an important regulator of cell proliferation, differentiation, and apoptosis in a variety of cell types such as hepatocytes, fibroblasts, myoblasts, and adipocytes (Grommes et al. 2004; Sertznig et al. 2007). Treatment of 3T3-L1 preadipocytes and murine fibroblast cells with the PPARγ agonist, troglitazone, induced the expression of cyclin-dependent kinase inhibitors (CDKIs) p18 and p21 allowing terminal adipogenic differentiation (Morrison and Farmer 1999). Activation of PPARγ with either endogenous PPARγ agonists or synthetic agonists induced cell cycle exit by terminal differentiation of preadipocytes and fibroblast cells (Morrison and Farmer 1999; Tontonoz et al. 1994; Wahli et al. 1995). Primary human liposarcoma (LPS) cells were effectively induced to undergo terminal adipocytic differentiation after treatment of the PPARγ agonist, pioglitazone (Tontonoz et al. 1997). Furthermore, promising preclinical results about the effects in differentiation of PPARγ agonist treatment in liposarcoma have been reported in a clinical phase II trial utilizing the PPARγ agonist, rosiglitazone (Debrock et al. 2003; Dusso et al. 2005).

14.3.5 Vitamin D

Vitamin D receptor (VDR) is expressed in many cell types and tissues. It is of a small intestine, kidney, and bone and is involved in the homeostasis of calcium and minerals (Dusso et al. 2005; Nagpal et al. 2005; Samuel and Sitrin 2008). Vitamin D alters cellular proliferation through multiple mechanisms such as cell cycle progression, apoptosis, and differentiation (Dusso et al. 2005; Nagpal et al. 2005; Samuel and Sitrin 2008; Masuda and Jones 2006; Banerjee and Chatterjee 2003). Vitamin D induces cell cycle arrest by inhibiting the transition from the G1 to the S phase of the cell cycle (Bohnsack and Hirschi 2004). By affecting multiple genes, multiple effects of vitamin D on this step of the cell cycle, including p21waf1, p27kip1, cyclin D1, and so on, were observed concerning their transcription and protein stability (Bohnsack and Hirschi 2004; Liu et al. 1996; Boyle et al. 2001; Hershberger et al. 2001; Inoue et al. 1999; Rots et al. 1998; Bettoun et al. 2002). Vitamin D induced maturation of HL-60 and U937 leukemia cells (Olsson et al. 1983; Rigby et al. 1984). It also induced CDKIs such as p27kip1 and perturbated the subcellular distribution of protein phosphatases (Wang et al. 1997; Song and Norman 1998).

14.3.6 Differentiation Therapy with Antihistaminic Drugs

A novel target for differentiation therapy is TCTP, because it was the most downregulated gene in tumor reversion experiments (Tuynder et al. 2004). Since TCTP encodes for a histamine-releasing factor, Tuynder et al. hypothesized that inhibitors of the histaminic pathway could be effective against tumor cells (Tuynder et al. 2004). Antihistaminics are widely used in cancer patients as antiallergics, antidepressants, or antiemetic agents (Tuynder et al. 2004). Therefore, it is also reasonable to test their possible antiproliferative effects. Moreover, some phenothiazines, including promethazine, thioridazine, perphenazine, and chlorpromazine, revealed antiproliferative effects (Strobl et al. 1990; Gil-Ad et al. 2004; Zhelev et al. 2004). Antihistaminic compounds decreased TCTP expression, killed cancer cells, and, eventually, led to strong reversion of the malignant phenotype (Tuynder et al. 2004). Hydroxyzine and promethazine as model drugs inhibited cell growth of human leukemia U937 cells and decreased TCTP expression of breast cancer MDA-MB-231 and monocytic leukemia U937 cells (Tuynder et al. 2004). These two drugs were also investigated in vivo. The volumes of MDA-MB-231 and U937 xenograft tumors were consistently reduced by treatment with hydroxyzine or promethazine, indicating that these drugs indeed inhibited tumor growth by targeting TCTP (Tuynder et al. 2004).

We investigated a series of antihistaminic drugs as new TCTP inhibitors in a systematic way (Seo and Efferth 2016). In our study, levomepromazine and buclizine showed higher in silico binding affinities to TCTP among 12 different antihistaminic compounds including the control drugs, promethazine and hydroxyzine, by using Autodock4 and AutodockTools-1.5.7.rc1. We found that levomepromazine and buclizine bound to the same sites at TCTP as promethazine and hydroxyzine, but with higher affinities. Recombinant human TCTP protein was obtained by codon optimization, heterogeneous expression in *E. coli*, and purification using chitin affinity chromatography. We were able to experimentally validate the binding of levomepromazine and buclizine to recombinant human TCTP using microscale thermophoresis. Furthermore, we explored the effects of two selected compounds on cell growth and TCTP protein and observed indeed that they inhibited cell growth and downregulated TCTP expression in MCF-7 breast cancer cells, indicating TCTP direct binding and downregulation as causative growth-inhibitory mechanism of levomepromazine and buclizine. We also investigated the cell cycle distribution of MCF-7 cells after drug treatment using flow cytometry and found that the percentage of G1 phase cells after levomepromazine or buclizine treatment increased without showing apoptosis. The mode of the action of two compounds was investigated using annexin V/PI staining. High concentrations (IC_{50} or $2 \times IC_{50}$) of both drugs for 72 h treatment did not increase the fraction of dead cells, and most of cells were annexin V/PI negative, demonstrating that the cells were alive after treatment of two drugs. These results indicated that these two antihistaminics cause neither necrosis nor apoptosis. Therefore, they were not cytotoxic. Our cell cycle analysis and annexin V/PI staining results strongly implied that levomepromazine and buclizine caused cell growth inhibition by G1 cell cycle arrest without induction of cell death. Moreover, trypan blue exclusion test showed that more than 90% of cells were living cells possessing intact cell membranes that excluded trypan blue staining upon treatment with IC_{50} or $2 \times IC_{50}$ concentrations of levomepromazine or buclizine for 72 h. This result is another proof that these two drugs inhibited cell growth without inducing cell death. Based on our results, we conclude that the interaction of TCTP with the apoptotic machinery was not of major mechanism for the antiproliferative effects of antihistaminic compounds. The effect of these two drugs on cell cycle arrest, annexin V/PI staining analysis, and cell viability using trypan blue staining demonstrated that cytostatic rather than cytotoxic mechanisms were operative. In order to confirm that two drugs really induced differentiation, lipid droplet staining was performed. Lipid droplets are a reliable market for functional differentiation of mammary tissues (Munster et al. 2001). Finally, we demonstrated that those two antihistaminics really induced differentiation in MCF-7 cells by increase of lipid droplets. Thus, we found that two antihistaminics, levomepromazine and buclizine, inhibited cancer cell growth by binding to TCTP and induction of cell differentiation.

On the basis of data of Tynder et al. and our study, TCTP is a novel target for anticancer differentiation therapy, and antihistaminics are promising to serve as lead compounds for cancer differentiation therapy by targeting TCTP (Tuynder et al. 2004; Seo and Efferth 2016).

14.4 Conclusions and Perspectives

Many cytotoxic agents against cancer reveal side effects such as bone marrow suppression, gastrointestinal tract lesions, hair loss, nausea, etc. because these agents are active on both malignant tumor and healthy normal cells (Thurston 2007; Jain et al. 2013). Therefore, these drugs induce cell death not only in tumors but also in normal cells (Thurston 2007; Jain et al. 2013). Since cytotoxic drugs lack sufficient tumor selectivity, they frequently cannot cure patients due to non-tolerable high side effects that prevent the application of drug doses high enough to sustainably kill all cells of a tumor.

Another treatment is differentiation therapy, which aims at reactivation of endogenous differentiation programs in cancer cells with subsequent cellular maturation and loss of the aggressive tumor phenotype (Pierce and Wallace 1971). This novel and potentially less toxic form of cancer therapy comprise agents that modify the state of differentiation and growth of cancer cells (Leszczyniecka et al. 2001). Although differentiation therapies such as retinoids, HDACI, PPARγ agonists and vitamin D have been investigated as described above, it is still relatively in its infancy.

TCTP represents an exquisite target for differentiation therapy, since downregulation of TCTP was responsible for the reprogramming of cancer cells into revertants (Tuynder et al. 2002, 2004). The antihistaminics, promethazine and hydroxyzine, showed the inhibition of TCTP indicating that antihistaminic drugs can be a suitable class of TCTP inhibitors (Tuynder et al. 2002, 2004). Furthermore, our study demonstrated that antihistaminic drugs, levomepromazine and buclizine, inhibited the breast cancer cell line MCF-7 growth by binding to TCTP and induce cell differentiation (Seo and Efferth 2016). Those studies showed the potential that differentiation therapy with higher tumor specificity and less side effects than cytotoxic therapy can be reached using antihistaminic TCTP inhibitors.

A synergistic effect was reported between the downregulation of TCTP by siRNA and antisense oligonucleotides in combination with docetaxel treatment of prostate cancer models in vitro and in vivo (Baylot et al. 2012). This result demonstrates that TCTP knockdown with docetaxel therapy could serve as a novel strategy to treat castration-resistant prostate cancer (Baylot et al. 2012).

Hence, the combination of the targeting of TCTP with classical chemotherapy is worth to be investigated, because it might reveal synergistic effects and the opportunity of treating cancer in a more effective way with higher response rates, lower risks of tumor resistance, and fewer side effects with treatment of less concentration of cytotoxic substances.

References

Acunzo J et al (2014) TCTP as therapeutic target in cancers. Cancer Treat Rev 40(6):760–769

Amit J et al (2009) Adamantinoma of tibial shaft. J Nepal Med Assoc 48(176):331–334

Amson RB et al (1996) Isolation of 10 differentially expressed cDNAs in p53-induced apoptosis: activation of the vertebrate homologue of the drosophila seven in absentia gene. Proc Natl Acad Sci U S A 93(9):3953–3957

Amson R et al (2011) Reciprocal repression between P53 and TCTP. Nat Med 18(1):91–99

Amson R, Karp JE, Telerman A (2013a) Lessons from tumor reversion for cancer treatment. Curr Opin Oncol 25(1):59–65

Amson R et al (2013b) TPT1/TCTP-regulated pathways in phenotypic reprogramming. Trends Cell Biol 23(1):37–46

Amzallag N et al (2004) TSAP6 facilitates the secretion of translationally controlled tumor protein/histamine-releasing factor via a nonclassical pathway. J Biol Chem 279 (44):46104–46112

Askanazy M (1907) Die Teratome nach ihrem Bau, ihrem Verlauf, ihrer Genese und im Vergleich zum experimentellen Teratoid. Verhandlungen der deutschen pathologischen Gesellschaft i 11:39–82

Banerjee P, Chatterjee M (2003) Antiproliferative role of vitamin D and its analogs--a brief overview. Mol Cell Biochem 253(1–2):247–254

Barlow JW et al (2006) Differentiation of rhabdomyosarcoma cell lines using retinoic acid. Pediatr Blood Cancer 47(6):773–784

Baylot V et al (2012) Targeting TCTP as a New Therapeutic Strategy in Castration-resistant Prostate Cancer. Mol Ther 20(12):2244–2256

Bettoun DJ et al (2002) A vitamin D receptor-Ser/Thr phosphatase-p70 S6 kinase complex and modulation of its enzymatic activities by the ligand. J Biol Chem 277(28):24847–24850

Bini L et al (1997) Protein expression profiles in human breast ductal carcinoma and histologically normal tissue. Electrophoresis 18(15):2832–2841

Bohnsack BL, Hirschi KK (2004) Nutrient regulation of cell cycle progression. Annu Rev Nutr 24:433–453

Bommer UA, Thiele BJ (2004) The translationally controlled tumour protein (TCTP). Int J Biochem Cell Biol 36(3):379–385

Boyle BJ et al (2001) Insulin-like growth factor binding protein-3 mediates 1 alpha,25-dihydroxyvitamin d(3) growth inhibition in the LNCaP prostate cancer cell line through p21/WAF1. J Urol 165(4):1319–1324

Brenner S et al (2000a) *In vitro* cloning of complex mixtures of DNA on microbeads: physical separation of differentially expressed cDNAs. Proc Natl Acad Sci U S A 97(4):1665–1670

Brenner S et al (2000b) Gene expression analysis by massively parallel signature sequencing (MPSS) on microbead arrays. Nat Biotechnol 18(6):630–634

Brioudes F et al (2010) Translationally controlled tumor protein is a conserved mitotic growth integrator in animals and plants. Proc Natl Acad Sci U S A 107(37):16384–16389

Brodowicz T et al (1999) Inhibition of proliferation and induction of apoptosis in soft tissue sarcoma cells by interferon-alpha and retinoids. Br J Cancer 80(9):1350–1358

Burgess A et al (2008) Chfr interacts and colocalizes with TCTP to the mitotic spindle. Oncogene 27(42):5554–5566

Cans C et al (2003) Translationally controlled tumor protein acts as a guanine nucleotide dissociation inhibitor on the translation elongation factor eEF1A. Proc Natl Acad Sci U S A 100(24):13892–13897

Chambers I et al (2003) Functional expression cloning of Nanog, a pluripotency sustaining factor in embryonic stem cells. Cell 113(5):643–655

Chen SH et al (2007a) A knockout mouse approach reveals that TCTP functions as an essential factor for cell proliferation and survival in a tissue- or cell type-specific manner. Mol Biol Cell 18(7):2525–2532

Chen Z et al (2007b) The expression of AmphiTCTP, a TCTP orthologous gene in amphioxus related to the development of notochord and somites. Comp Biochem Physiol B Biochem Mol Biol 147(3):460–465

Chou M et al (2016) A translationally controlled tumor protein gene Rpf41 is required for the nodulation of Robinia pseudoacacia. Plant Mol Biol 90(4–5):389–402

Cress WD, Seto E (2000) Histone deacetylases, transcriptional control, and cancer. J Cell Physiol 184(1):1–16

Crouch GD, Helman LJ (1991) All-trans-retinoic acid inhibits the growth of human rhabdomyosarcoma cell lines. Cancer Res 51(18):4882–4887

Debrock G et al (2003) A phase II trial with rosiglitazone in liposarcoma patients. Br J Cancer 89 (8):1409–1412

Donovan PJ, Gearhart J (2001) The end of the beginning for pluripotent stem cells. Nature 414 (6859):92–97

Dusso AS, Brown AJ, Slatopolsky E (2005) Vitamin D. Am J Physiol Ren Physiol 289(1):F8–28

Fiucci G et al (2004) Siah-1b is a direct transcriptional target of p53: identification of the functional p53 responsive element in the siah-1b promoter. Proc Natl Acad Sci U S A 101 (10):3510–3515

Funston G et al (2012) Binding of translationally controlled tumour protein to the N-terminal domain of HDM2 is inhibited by nutlin-3. PLoS One 7(8):e42642

Gachet Y et al (1999) The growth-related, translationally controlled protein P23 has properties of a tubulin binding protein and associates transiently with microtubules during the cell cycle. J Cell Sci 112(Pt 8):1257–1271

Garattini E, Gianni M, Terao M (2007) Cytodifferentiation by retinoids, a novel therapeutic option in oncology: rational combinations with other therapeutic agents. Vitam Horm 75:301–354

Gil-Ad I et al (2004) Characterization of phenothiazine-induced apoptosis in neuroblastoma and glioma cell lines: clinical relevance and possible application for brain-derived tumors. J Mol Neurosci 22(3):189–198

Ginis I et al (2004) Differences between human and mouse embryonic stem cells. Dev Biol 269 (2):360–380

Gnanasekar M, Dakshinamoorthy G, Ramaswamy K (2009) Translationally controlled tumor protein is a novel heat shock protein with chaperone-like activity. Biochem Biophys Res Commun 386(2):333–337

Graidist P, Phongdara A, Fujise K (2004) Antiapoptotic protein partners fortilin and MCL1 independently protect cells from 5-fluorouracil-induced cytotoxicity. J Biol Chem 279 (39):40868–40875

Graidist P et al (2007) Fortilin binds Ca2+ and blocks Ca2+-dependent apoptosis *in vivo*. Biochem J 408(2):181–191

Grommes C, Landreth GE, Heneka MT (2004) Antineoplastic effects of peroxisome proliferator-activated receptor gamma agonists. Lancet Oncol 5(7):419–429

Haghighat NG, Ruben L (1992) Purification of novel calcium binding proteins from Trypanosoma brucei: properties of 22-, 24- and 38-kilodalton proteins. Mol Biochem Parasitol 51(1):99–110

Hansen LA et al (2000) Retinoids in chemoprevention and differentiation therapy. Carcinogenesis 21(7):1271–1279

Hershberger PA et al (2001) Calcitriol (1,25-dihydroxycholecalciferol) enhances paclitaxel antitumor activity *in vitro* and *in vivo* and accelerates paclitaxel-induced apoptosis. Clin Cancer Res 7(4):1043–1051

Hsu YC et al (2007) *Drosophila* TCTP is essential for growth and proliferation through regulation of dRheb GTPase. Nature 445(7129):785–788

Huang ME, Ye YC, Zhao L (1987a) Treatment of acute promyelocytic leukemia by retinoic acid with or without low dose cytosine arabinoside: report of 4 cases. Zhonghua Nei Ke Za Zhi 26 (6):330–332. 380

Huang ME et al (1987b) All-trans retinoic acid with or without low dose cytosine arabinoside in acute promyelocytic leukemia. Report of 6 cases. Chin Med J 100(12):949–953

Inoue T, Kamiyama J, Sakai T (1999) Sp1 and NF-Y synergistically mediate the effect of vitamin D(3) in the p27(Kip1) gene promoter that lacks vitamin D response elements. J Biol Chem 274 (45):32309–32317

Israeli D et al (1997) A novel p53-inducible gene, PAG608, encodes a nuclear zinc finger protein whose overexpression promotes apoptosis. EMBO J 16(14):4384–4392

Jain MV et al (2013) Interconnections between apoptotic, autophagic and necrotic pathways: implications for cancer therapy development. J Cell Mol Med 17(1):12–29

Kirchhof N et al (2000) Expression pattern of Oct-4 in preimplantation embryos of different species. Biol Reprod 63(6):1698–1705

Koziol MJ, Garrett N, Gurdon JB (2007) Tpt1 activates transcription of oct4 and nanog in transplanted somatic nuclei. Curr Biol 17(9):801–807

Koziol MJ, Gurdon JB (2012) TCTP in development and cancer. Biochem Res Int 2012:105203

Kubiak JZ et al (2008) Temporal regulation of embryonic M-phases. Folia Histochem Cytobiol 46 (1):5–9

Lal S et al (1993) Levomepromazine receptor binding profile in human brain--implications for treatment-resistant schizophrenia. Acta Psychiatr Scand 87(6):380–383

Lespagnol A et al (2008) Exosome secretion, including the DNA damage-induced p53-dependent secretory pathway, is severely compromised in TSAP6/Steap3-null mice. Cell Death Differ 15 (11):1723–1733

Leszczyniecka M et al (2001) Differentiation therapy of human cancer: basic science and clinical applications. Pharmacol Ther 90(2–3):105–156

Li F, Zhang D, Fujise K (2001) Characterization of fortilin, a novel antiapoptotic protein. J Biol Chem 276(50):47542–47549

Liang P, Pardee AB (1992) Differential display of eukaryotic messenger RNA by means of the polymerase chain reaction. Science 257(5072):967–971

Liu M et al (1996) Transcriptional activation of the Cdk inhibitor p21 by vitamin D3 leads to the induced differentiation of the myelomonocytic cell line U937. Genes Dev 10(2):142–153

Liu H et al (2005) Stabilization and enhancement of the antiapoptotic activity of mcl-1 by TCTP. Mol Cell Biol 25(8):3117–3126

Luo P et al (2010) Retinoid-suppressed phosphorylation of RARalpha mediates the differentiation pathway of osteosarcoma cells. Oncogene 29(19):2772–2783

MacDonald SM et al (1995) Molecular identification of an IgE-dependent histamine-releasing factor. Science 269(5224):688–690

Mai A et al (2005) Histone deacetylation in epigenetics: an attractive target for anticancer therapy. Med Res Rev 25(3):261–309

Martirosyan AR et al (2004) Differentiation-inducing quinolines as experimental breast cancer agents in the MCF-7 human breast cancer cell model. Biochem Pharmacol 68(9):1729–1738

Masuda S, Jones G (2006) Promise of vitamin D analogues in the treatment of hyperproliferative conditions. Mol Cancer Ther 5(4):797–808

Masui S et al (2007) Pluripotency governed by Sox2 via regulation of Oct3/4 expression in mouse embryonic stem cells. Nat Cell Biol 9(6):625–635

Miao X et al (2013) TCTP overexpression is associated with the development and progression of glioma. Tumour Biol 34(6):3357–3361

Mitsui K et al (2003) The homeoprotein Nanog is required for maintenance of pluripotency in mouse epiblast and ES cells. Cell 113(5):631–642

Miyamoto K et al (2011) Nuclear actin polymerization is required for transcriptional reprogramming of Oct4 by oocytes. Genes Dev 25(9):946–958

Morrison RF, Farmer SR (1999) Role of PPARgamma in regulating a cascade expression of cyclin-dependent kinase inhibitors, p18(INK4c) and p21(Waf1/Cip1), during adipogenesis. J Biol Chem 274(24):17088–17097

Mousset S, Rommelaere J (1982) Minute virus of mice inhibits cell transformation by simian virus 40. Nature 300(5892):537–539

Munster PN et al (2001) The histone deacetylase inhibitor suberoylanilide hydroxamic acid induces differentiation of human breast cancer cells. Cancer Res 61(23):8492–8497

Nagano-Ito M, Ichikawa S (2012) Biological effects of mammalian translationally controlled tumor protein (TCTP) on cell death, proliferation, and tumorigenesis. Biochem Res Int 2012:204960

Nagpal S, Na S, Rathnachalam R (2005) Noncalcemic actions of vitamin D receptor ligands. Endocr Rev 26(5):662–687

Nemani M et al (1996) Activation of the human homologue of the Drosophila sina gene in apoptosis and tumor suppression. Proc Natl Acad Sci U S A 93(17):9039–9042

Ng KW et al (1985) Effect of retinoids on the growth, ultrastructure, and cytoskeletal structures of malignant rat osteoblasts. Cancer Res 45(10):5106–5113

Nishimoto M et al (1999) The gene for the embryonic stem cell coactivator UTF1 carries a regulatory element which selectively interacts with a complex composed of Oct-3/4 and Sox-2. Mol Cell Biol 19(8):5453–5465

Niwa H, Miyazaki J, Smith AG (2000) Quantitative expression of Oct-3/4 defines differentiation, dedifferentiation or self-renewal of ES cells. Nat Genet 24(4):372–376

Oh YT et al (2013) Aspergillus nidulans translationally controlled tumor protein has a role in the balance between asexual and sexual differentiation and normal hyphal branching. FEMS Microbiol Lett 343(1):20–25

Olsson I et al (1983) Induction of differentiation of the human histiocytic lymphoma cell line U-937 by 1 alpha,25-dihydroxycholecalciferol. Cancer Res 43(12 Pt 1):5862–5867

Pan GJ et al (2002) Stem cell pluripotency and transcription factor Oct4. Cell Res 12 (5–6):321–329

Passer BJ et al (2003) The p53-inducible TSAP6 gene product regulates apoptosis and the cell cycle and interacts with Nix and the Myt1 kinase. Proc Natl Acad Sci U S A 100(5):2284–2289

Pesce M, Schöler HR (2001) Oct-4: gatekeeper in the beginnings of mammalian development. Stem Cells 19(4):271–278

Pierce GB, Wallace C (1971) Differentiation of malignant to benign cells. Cancer Res 31 (2):127–134
Platta CS et al (2007) The HDAC inhibitor trichostatin A inhibits growth of small cell lung cancer cells. J Surg Res 142(2):219–226
Rho SB et al (2011) Anti-apoptotic protein TCTP controls the stability of the tumor suppressor p53. FEBS Lett 585(1):29–35
Rigby WF et al (1984) Differentiation of a human monocytic cell line by 1,25-dihydroxyvitamin D3 (calcitriol): a morphologic, phenotypic, and functional analysis. Blood 64(5):1110–1115
Riley T et al (2008) Transcriptional control of human p53-regulated genes. Nat Rev Mol Cell Biol 9(5):402–412
Roperch JP et al (1998) Inhibition of presenilin 1 expression is promoted by p53 and p21WAF-1 and results in apoptosis and tumor suppression. Nat Med 4(7):835–838
Roperch JP et al (1999) SIAH-1 promotes apoptosis and tumor suppression through a network involving the regulation of protein folding, unfolding, and trafficking: identification of common effectors with p53 and p21(Waf1). Proc Natl Acad Sci U S A 96(14):8070–8073
Rots NY et al (1998) A differential screen for ligand-regulated genes: identification of HoxA10 as a target of vitamin D3 induction in myeloid leukemic cells. Mol Cell Biol 18(4):1911–1918
Samuel S, Sitrin MD (2008) Vitamin D's role in cell proliferation and differentiation. Nutr Rev 66 (10 Suppl 2):S116–S124
Sanchez JC et al (1997) Translationally controlled tumor protein: a protein identified in several nontumoral cells including erythrocytes. Electrophoresis 18(1):150–155
Sato N et al (2003) Molecular signature of human embryonic stem cells and its comparison with the mouse. Dev Biol 260(2):404–413
Seo EJ, Efferth T (2016) Interaction of antihistaminic drugs with human translationally controlled tumor protein (TCTP) as novel approach for differentiation therapy. Oncotarget 7 (13):16818–16839
Sertznig P et al (2007) Present concepts and future outlook: function of peroxisome proliferator-activated receptors (PPARs) for pathogenesis, progression, and therapy of cancer. J Cell Physiol 212(1):1–12
Sinha P et al (2000) Identification of novel proteins associated with the development of chemoresistance in malignant melanoma using two-dimensional electrophoresis. Electrophoresis 21(14):3048–3057
Song X, Norman AW (1998) 1Alpha,25-dihydroxyvitamin D3 and phorbol ester mediate the expression of alkaline phosphatase in NB4 acute promyelocytic leukemia cells. Leuk Res 22 (1):69–76
Spira AI, Carducci MA (2003) Differentiation therapy. Curr Opin Pharmacol 3(4):338–343
Stierum R et al (2003) Proteome analysis reveals novel proteins associated with proliferation and differentiation of the colorectal cancer cell line Caco-2. Biochim Biophys Acta 1650 (1–2):73–91
Strobl JS et al (1990) Inhibition of human breast cancer cell proliferation in tissue culture by the neuroleptic agents pimozide and thioridazine. Cancer Res 50(17):5399–5405
Susini L et al (2008) TCTP protects from apoptotic cell death by antagonizing bax function. Cell Death Differ 15(8):1211–1220
Tani T et al (2007) Bovine oocytes with the potential to reprogram somatic cell nuclei have a unique 23-kDa protein, phosphorylated transcriptionally controlled tumor protein (TCTP). Cloning Stem Cells 9(2):267–280
Telerman A, Amson R (2009) The molecular programme of tumour reversion: the steps beyond malignant transformation. Nat Rev Cancer 9(3):206–216
Telerman A et al (1993) A model for tumor suppression using H-1 parvovirus. Proc Natl Acad Sci U S A 90(18):8702–8706
Thein R, Lotan R (1982) Sensitivity of cultured human osteosarcoma and chondrosarcoma cells to retinoic acid. Cancer Res 42(11):4771–4775

Thurston DE (2007) Chemistry and pharmacology of anticancer drugs. CRC Press/Taylor and Francis Group, Boca Raton

Tontonoz P, Hu E, Spiegelman BM (1994) Stimulation of adipogenesis in fibroblasts by PPAR gamma 2, a lipid-activated transcription factor. Cell 79(7):1147–1156

Tontonoz P et al (1997) Terminal differentiation of human liposarcoma cells induced by ligands for peroxisome proliferator-activated receptor gamma and the retinoid X receptor. Proc Natl Acad Sci U S A 94(1):237–241

Toolan HW (1967) Lack of oncogenic effect of the H-viruses for hamsters. Nature 214(5092):1036

Toscano-Morales R et al (2015) *AtTCTP2*, an *Arabidopsis thaliana* homolog of Translationally Controlled Tumor Protein, enhances *in vitro* plant regeneration. Front Plant Sci 6:468

Tsarova K, Yarmola EG, Bubb MR (2010) Identification of a cofilin-like actin-binding site on translationally controlled tumor protein (TCTP). FEBS Lett 584(23):4756–4760

Tuynder M et al (2002) Biological models and genes of tumor reversion: cellular reprogramming through tpt1/TCTP and SIAH-1. Proc Natl Acad Sci U S A 99(23):14976–14981

Tuynder M et al (2004) Translationally controlled tumor protein is a target of tumor reversion. Proc Natl Acad Sci U S A 101(43):15364–15369

Vitale AM et al (2007) Proteomic profiling of murine oocyte maturation. Mol Reprod Dev 74 (5):608–616

Wahli W, Braissant O, Desvergne B (1995) Peroxisome proliferator activated receptors: transcriptional regulators of adipogenesis, lipid metabolism and more. Chem Biol 2(5):261–266

Wang D, Gao L (2005) Proteomic analysis of neural differentiation of mouse embryonic stem cells. Proteomics 5(17):4414–4426

Wang QM, Luo X, Studzinski GP (1997) Cyclin-dependent kinase 6 is the principal target of p27/Kip1 regulation of the G1-phase traverse in 1,25-dihydroxyvitamin D3-treated HL60 cells. Cancer Res 57(14):2851–2855

Waxman S (2000) Differentiation therapy in acute myelogenous leukemia (non-APL). Leukemia 14(3):491–496

Williams AC et al (1996) *In vitro* models for studying colorectal carcinogenesis: cellular and molecular events including APC and Rb cleavage in the control of proliferation, differentiation and apoptosis. Biochim Biophys Acta 1288(1):F9–19

Xu A, Bellamy AR, Taylor JA (1999) Expression of translationally controlled tumour protein is regulated by calcium at both the transcriptional and post-transcriptional level. Biochem J 342 (Pt 3):683–689

Yan L et al (2000) A cnidarian homologue of translationally controlled tumor protein (P23/TCTP). Dev Genes Evol 210(10):507–511

Yang T, Kozopas KM, Craig RW (1995) The intracellular distribution and pattern of expression of Mcl-1 overlap with, but are not identical to, those of Bcl-2. J Cell Biol 128(6):1173–1184

Yang Y et al (2005) An N-terminal region of translationally controlled tumor protein is required for its antiapoptotic activity. Oncogene 24(30):4778–4788

Yarm FR (2002) Plk phosphorylation regulates the microtubule-stabilizing protein TCTP. Mol Cell Biol 22(17):6209–6221

Ying QL et al (2003) BMP induction of Id proteins suppresses differentiation and sustains embryonic stem cell self-renewal in collaboration with STAT3. Cell 115(3):281–292

Zhang D et al (2002) Physical and functional interaction between myeloid cell leukemia 1 protein (MCL1) and fortilin. The potential role of MCL1 as a fortilin chaperone. J Biol Chem 277 (40):37430–37438

Zhelev Z et al (2004) Phenothiazines suppress proliferation and induce apoptosis in cultured leukemic cells without any influence on the viability of normal lymphocytes. Phenothiazines and leukemia. Cancer Chemother Pharmacol 53(3):267–275

Chapter 15
Targeting TCTP with Sertraline and Thioridazine in Cancer Treatment

Robert Amson, Christian Auclair, Fabrice André, Judith Karp, and Adam Telerman

Abstract We have initially demonstrated in knocking down experiments that decreasing TCTP in cancer cells leads in some tissues to cell death while in others to a complete reorganization of the tumor into architectural structures reminiscent of normal ones. Based on these experiments and a series of other findings confirming the key role of TCTP in cancer, it became important to find pharmacological compounds to inhibit its function, and this became for us a priority. In the present text, we explain in detail the experiments that were performed and the perspectives of sertraline in cancer treatment, as this became today a reality with a clinical study that started in collaboration with Columbia University and Johns Hopkins University.

15.1 Introduction

The first indication that TCTP could be used as a pharmacologic target was based on the observation that it was the most downregulated gene in the revertant cells, compared to the parental malignant cells. Indeed a signal of 248 (standing for the amount of mRNA) was found in the parental cells and only 2 in the revertants (Tuynder et al. 2002). Using RNA interference against TCTP, we tested whether the

R. Amson (✉) • A. Telerman
Bâtiment B2M, Institut Gustave Roussy, Unité Inserm U981, 114 rue Édouard-Vaillant, 94805 Villejuif, France
e-mail: amson.robert@gmail.com

C. Auclair
CNRS-UMR 8113, LBPA, École Normale Supérieure, 61 avenue du Président Wilson, 94235 Cachan, France

F. André
Institut Gustave Roussy, Unité Inserm U981, Bâtiment B2M, 114 rue Édouard-Vaillant, 94805 Villejuif, France

J. Karp
Division of Hematologic Malignancies, The Johns Hopkins Sidney Kimmel Comprehensive Cancer Center, Baltimore, MD 21231-1000, USA

A. Telerman, R. Amson (eds.), *TCTP/tpt1 - Remodeling Signaling from Stem Cell to Disease*, Results and Problems in Cell Differentiation 64,
DOI 10.1007/978-3-319-67591-6_15

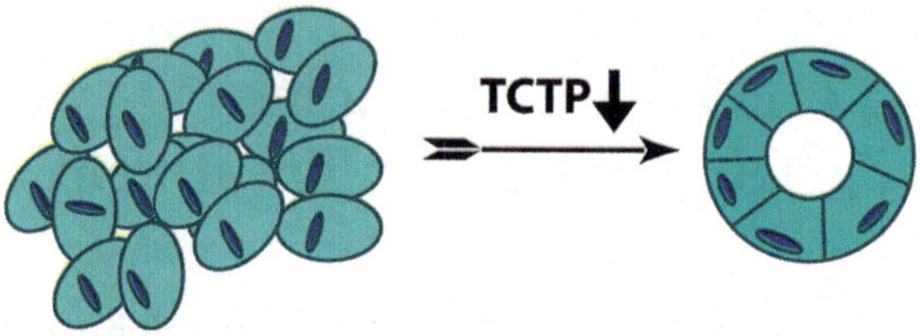

Fig. 15.1 Schematic representation of breast cancer cells growing randomly in a 3D Matrigel. Following inhibition of TCTP, these breast cancer cells are reprogrammed into ductal-like architectures

knocking down of TCTP would reprogram malignant cells into revertants. Breast cancer cells, MCF7 and T47D, grew in a disorganized fashion in 3D Matrigel (Petersen et al. 1992), (Tuynder et al. 2002) (Fig. 15.1). When TCTP expression was inhibited by siRNA (tpt1siRNA), the architecture of the breast cancer tumor cells was profoundly reorganized and reprogrammed, forming ductal structures reminiscent of normal breast cells (Fig. 15.1).

The second set of evidence came from inhibiting TCTP in v-SRC transformed NIH3T3 with antisense TCTP (as-TCTP), and this drastically increased the proportion of flat revertants up to 30% (Tuynder et al. 2004) of the culture instead of the 0.0001% for the spontaneous occurrence of flat revertants (Pollack et al. 1968; Tuynder et al. 2004).

These experiments indicated that inhibiting TCTP was overriding oncogenic transformation, suggesting that it was a promising pharmacologic target in the treatment of cancer (Fig. 15.2).

15.2 Experimental Approach

In order to find a shortcut for inhibiting TCTP pharmacologically, we based our rationale on results obtained by Susan MacDonald. Indeed, in the 1990 she identified a factor that induced the release of histamine and called it "histamine releasing factor (HRF)" (MacDonald et al. 1995). After purification and sequencing, she realized that it corresponded to TCTP. We thus hypothesized that some of the existing histamine antagonists might have an off-target activity by which TCTP could be influenced (Fig. 15.3).

We tested the only four antihistaminic drugs that were available in injectable form: hydroxyzine (Atarax®), brompheniramine (Dimegan®), promethazine (Phenergan®), and dexchlorpheniramine (Polaramine®) for their effect on cancer growth. While hydroxyzine and promethazine had a weak growth inhibitory effect at 100 and 50 μM, respectively, brompheniramine and dexchlorpheniramine were inactive (Tuynder et al. 2004) (Fig. 15.4). Importantly, intracellular levels of TCTP were decreased in hydroxyzine- and promethazine-treated cells but not with the two other drugs.

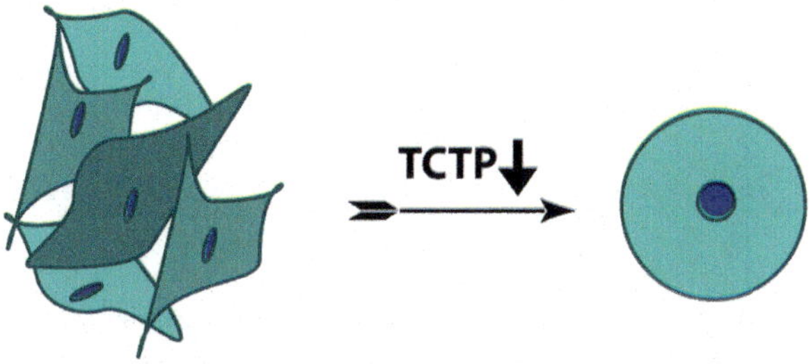

Fig. 15.2 Schematic representation of v-SRC transformed NIH3T3 cells. Decreasing TCTP with as-TCTP in these cells leads to a reprogramming into "flat revertants"

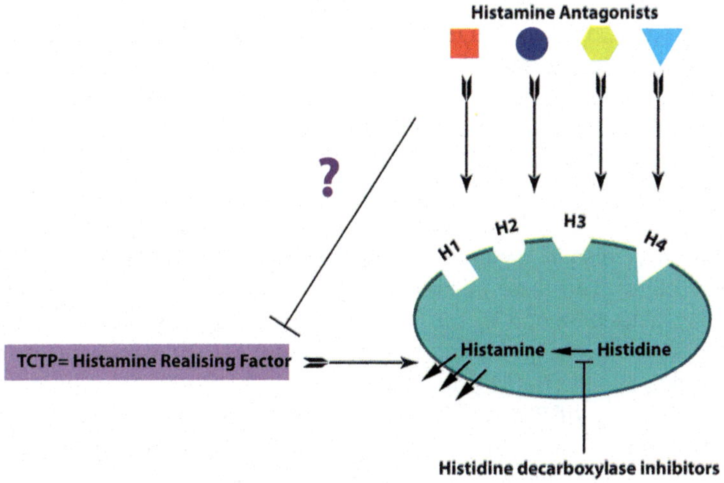

Fig. 15.3 The function of histamine can be inhibited by several known approaches. Histamine antagonists are able to target the H1, H2, H3, and H4 receptors of histamine and inhibit their respective functions. Histidine is decarboxylated by L-histidine decarboxylase to generate histamine. This reaction can be inhibited by histidine decarboxylase inhibitors. The *question mark* represents our hypothesis by which histamine antagonists may have an off-target activity on the inhibition of TCTP

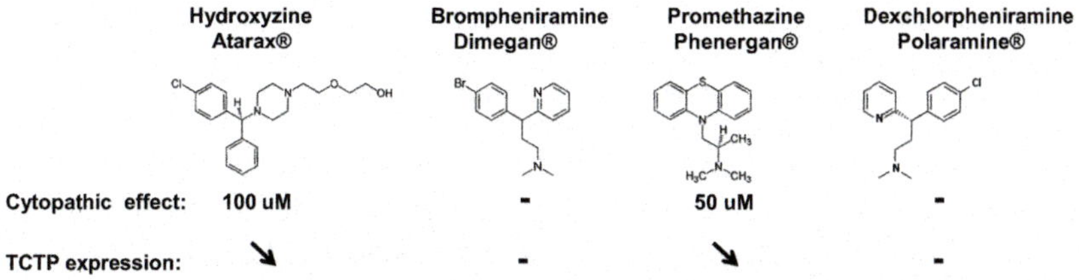

Fig. 15.4 The growth inhibitory effect of four antihistaminic drugs was tested on U937 cells and the expression of TCTP assessed by Western blot analysis (Tuynder et al. 2004)

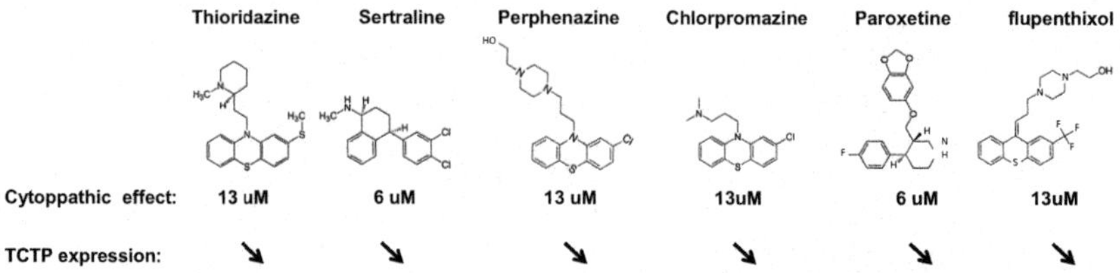

Fig. 15.5 The growth inhibitory effect of thioridazine, sertraline, perphenazine, chlorpromazine, paroxetine, and flupenthixol on U937 cells. TCTP expression was decreased (↘) after treatment with the abovementioned drugs (Tuynder et al. 2004)

This inhibitory effect on growth prompted us to search for other drugs with a similar chemical backbone such as thioridazine, sertraline, perphenazine, chlorpromazine, paroxetine, and flupenthixol (Tuynder et al. 2004) (Fig. 15.5). These drugs inhibited significantly the growth of U937 promyelomonocytic cells at concentrations ranging from 6 to 13 μM. Here again TCTP expression levels were decreased. This effect was confirmed in vivo for promethazine, sertraline, and thioridazine (Tuynder et al. 2004).

These μM concentrations were dependent on the biological assay performed. This is why we used a more sensitive one, consisting of long-term cultures (4 weeks) in semisolid medium (methylcellulose). When MCF7 breast cancer cells were cultured in these conditions, with sertraline or thioridazine added on the first day, a decrease of 30% and 20% for, respectively, sertraline or thioridazine at 1 nM was observed (Fig. 15.6a). At this concentration, the expression of TCTP is inhibited and p53 reactivated (Fig. 15.6b).

This reactivation of P53 was due to the reciprocal repression between P53 and TCTP (Amson et al. 2012). Several mechanisms underlying the anticancer effect of sertraline and thioridazine were suggested (Fig. 15.7):

1. Sertraline and thioridazine bind directly TCTP, inhibiting its function in a MDM2-dependent manner (Amson et al. 2012).
2. The reactivation of p53 by sertraline and thioridazine will have as a consequence the transcriptional inhibition of TCTP (Amson et al. 2012, 2013).
3. Sertraline and thioridazine induced cell death in a p53-independent manner. These cells had also elevated TCTP (Tuynder et al. 2004).
4. It was shown that sertraline is targeting the mTOR pathway (Lin et al. 2010). Whether this occurs in a TCTP-dependent or TCTP-independent manner remains to be elucidated.

This antiproliferative activity of sertraline (Tuynder et al. 2004) was confirmed by others. Indeed, sertraline was found in a large screening that aimed at identifying off-target effects and hidden phenotypes of drugs (MacDonald et al. 2006).

Previous reports indicated already that thioridazine restored doxorubicin responsiveness in leukemia cells (Ramu et al. 1984). Thioridazine was also described to

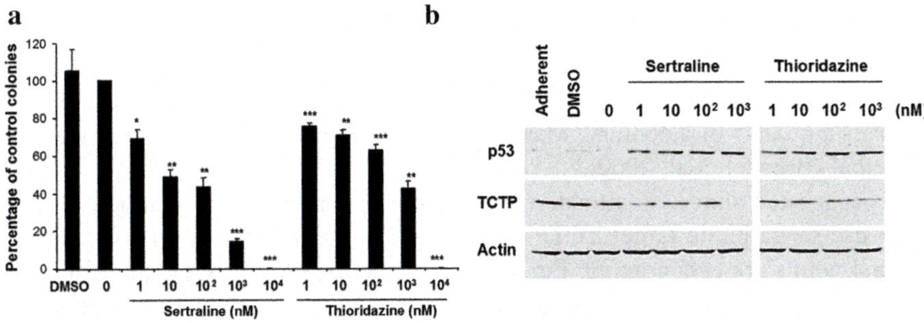

Fig. 15.6 (**a**) Breast cancer MCF7 cells cultured in methylcellulose during 4 weeks in the presence or absence of sertraline, thioridazine, or DMSO and the number of colonies were counted. (**b**) Western blot analysis of P53, TCTP, and actin expression

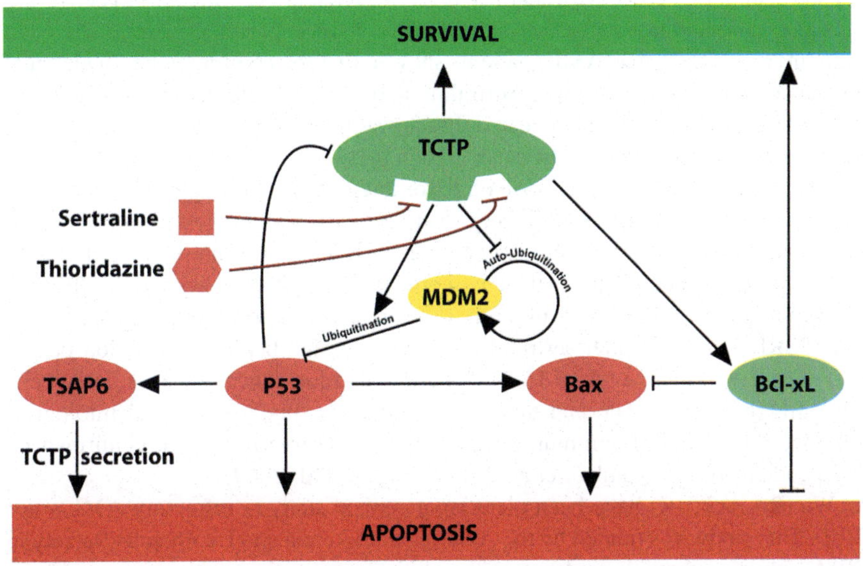

Fig. 15.7 Schematic representation of the interplay between TCTP and some of its protein partners leading to the control of apoptosis. Sertraline and thioridazine bind directly onto TCTP and inhibit its function on the degradation of P53, i.e., TCTP inhibits the auto-ubiquitination of MDM2 and activates the ubiquitination of P53. TCTP potentiated the anti-apoptotic Bcl-xL protein that inhibits the pro-apoptotic Bax protein (Susini et al. 2008; Thebault et al. 2016). P53 represses the transcription of TCTP but activates the transcription of Bax and TSAP6 (tumor suppressor-activated pathway6). TSAP6 favors the secretion of TCTP. TSAP6, P53, and Bax favor apoptosis. TCTP and Bcl-xL are pro-survival factors

induce apoptosis of a multidrug-resistant T-lymphoma cell line (Spengler et al. 2011).

The anticancer activity of these agents was "rediscovered" by others (Nagel et al. 2012; Sachlos et al. 2012).

15.3 Conclusion and Therapeutic Perspectives

Sertraline is a well-known FDA-approved drug used in the treatment of depression for over a quarter of a century. Massive intake of 8 g sertraline was described as nonlethal (Brendel et al. 2000).

As suggested above, an attractive therapeutic option in cancer treatment would be to use sertraline to target TCTP, as suggested by our tumor reversion studies. Several points are encouraging. First, TCTP is overexpressed in a wide variety of cancers (Amson et al. 2012; Arcuri et al. 2004; Baylot et al. 2012; Chung et al. 2000; Ma et al. 2010; Niforou et al. 2008; Stierum et al. 2003; Tuynder et al. 2002, 2004; Vercoutter-Edouart et al. 2001). Then, a series of in vitro and in vivo experiment indicated that sertraline was active on cancer cells including leukemic cells, breast cancer cells, and hepatocellular carcinoma cells (Tuynder et al. 2004; Amson et al. 2012, 2013; Kuwahara et al. 2015). Even breast cancer stem cells were sensitive to sertraline (Amson et al. 2012). Ex vivo experiment using cells from acute myeloid leukemia (AML) patients showed that the combination of a classical anticancer drug like Ara-C with sertraline had a synergistic effect (Amson et al. 2013). Other in vivo studies indicated that sertraline in combination with docetaxel inhibited tumor growth of breast cancer (Hallett et al. 2016). The use of sertraline in combination with a liposomal form of doxorubicin was also suggested to modulate cancer multidrug resistance (Drinberg et al. 2014). It was suggested to add sertraline to the CUSP9 treatment protocol for recurrent glioblastoma (Kast et al. 2014).

At the epidemiological level, it is worth to highlight the work of Collet that showed that patients treated with high doses of selective serotonin reuptake inhibitor (SSRI) (among which sertraline) do develop significantly less colon cancer (Xu et al. 2006). The work of Collet focused on colon cancer because colon cells have a high turnover rate and therefore suitable for this kind of epidemiological study (Collet personal communication). Similar observations were confirmed by independent studies (Chubak et al. 2011; Coogan et al. 2009).

We (RA, AT, JK) have been planning a clinical study in refractory AML since 2010. This study was funded by the American Leukemia and Lymphoma Society in 2014 (https://www.lls.org/content/the-clinical-application-of-tumor-reversion-a-phase-i-study-of-sertraline-zoloft-in-combination-with-timed-sequential-cytosine-arabinoside-ara-c-in) and is currently ongoing in collaboration with Mark Frattini and Ivana Gojo, respectively, at Columbia University and Johns Hopkins.

Acknowledgements We thank Manale Karam for precious help. This work was supported by grants from the French National Agency for Research (ANR, ANR-09-BLAN-0292) to AT, RA, and MV; European Union Network of Excellence CONTICANET to AT and RA; LabEx LERMIT to AT and RA; INCa Projets libres de Recherche 2013-1-PL BIO-10 "Biologie et Sciences du Cancer" to AT and RA; Ligue Nationale Contre le Cancer to AT and RA; Odyssea fund to RA.

References

Amson R, Pece S, Lespagnol A, Vyas R, Mazzarol G, Tosoni D, Colaluca I, Viale G, Rodrigues-Ferreira S, Wynendaele J et al (2012) Reciprocal repression between P53 and TCTP. Nat Med 18:91–99

Amson R, Karp JE, Telerman A (2013) Lessons from tumor reversion for cancer treatment. Curr Opin Oncol 25:59–65

Arcuri F, Papa S, Carducci A, Romagnoli R, Liberatori S, Riparbelli MG, Sanchez JC, Tosi P, del Vecchio MT (2004) Translationally controlled tumor protein (TCTP) in the human prostate and prostate cancer cells: expression, distribution, and calcium binding activity. Prostate 60:130–140

Baylot V, Katsogiannou M, Andrieu C, Taieb D, Acunzo J, Giusiano S, Fazli L, Gleave M, Garrido C, Rocchi P (2012) Targeting TCTP as a new therapeutic strategy in castration-resistant prostate cancer. Mol Ther 20(12):2244–2256

Brendel DH, Bodkin JA, Yang JM (2000) Massive sertraline overdose. Ann Emerg Med 36:524–526

Chubak J, Boudreau DM, Rulyak SJ, Mandelson MT (2011) Colorectal cancer risk in relation to antidepressant medication use. Int J Cancer 128:227–232

Chung S, Kim M, Choi W, Chung J, Lee K (2000) Expression of translationally controlled tumor protein mRNA in human colon cancer. Cancer Lett 156:185–190

Coogan PF, Strom BL, Rosenberg L (2009) Antidepressant use and colorectal cancer risk. Pharmacoepidemiol Drug Saf 18:1111–1114

Drinberg V, Bitcover R, Rajchenbach W, Peer D (2014) Modulating cancer multidrug resistance by sertraline in combination with a nanomedicine. Cancer Lett 354:290–298

Hallett RM, Girgis-Gabardo A, Gwynne WD, Giacomelli AO, Bisson JN, Jensen JE, Dvorkin-Gheva A, Hassell JA (2016) Serotonin transporter antagonists target tumor-initiating cells in a transgenic mouse model of breast cancer. Oncotarget 7:53137–53152

Kast RE, Karpel-Massler G, Halatsch ME (2014) CUSP9* treatment protocol for recurrent glioblastoma: aprepitant, artesunate, auranofin, captopril, celecoxib, disulfiram, itraconazole, ritonavir, sertraline augmenting continuous low dose temozolomide. Oncotarget 5:8052–8082

Kuwahara J, Yamada T, Egashira N, Ueda M, Zukeyama N, Ushio S, Masuda S (2015) Comparison of the anti-tumor effects of selective serotonin reuptake inhibitors as well as serotonin and norepinephrine reuptake inhibitors in human hepatocellular carcinoma cells. Biol Pharm Bull 38:1410–1414

Lin CJ, Robert F, Sukarieh R, Michnick S, Pelletier J (2010) The antidepressant sertraline inhibits translation initiation by curtailing mammalian target of rapamycin signaling. Cancer Res 70:3199–3208

Ma Q, Geng Y, Xu W, Wu Y, He F, Shu W, Huang M, Du H, Li M (2010) The role of translationally controlled tumor protein in tumor growth and metastasis of colon adenocarcinoma cells. J Proteome Res 9:40–49

MacDonald SM, Rafnar T, Langdon J, Lichtenstein LM (1995) Molecular identification of an IgE-dependent histamine-releasing factor. Science 269:688–690

MacDonald ML, Lamerdin J, Owens S, Keon BH, Bilter GK, Shang Z, Huang Z, Yu H, Dias J, Minami T et al (2006) Identifying off-target effects and hidden phenotypes of drugs in human cells. Nat Chem Biol 2:329–337

Nagel D, Spranger S, Vincendeau M, Grau M, Raffegerst S, Kloo B, Hlahla D, Neuenschwander M, Peter von Kries J, Hadian K et al (2012) Pharmacologic inhibition of MALT1 protease by phenothiazines as a therapeutic approach for the treatment of aggressive ABC-DLBCL. Cancer Cell 22:825–837

Niforou KM, Anagnostopoulos AK, Vougas K, Kittas C, Gorgoulis VG, Tsangaris GT (2008) The proteome profile of the human osteosarcoma U2OS cell line. Cancer Genomics Proteomics 5:63–78

Petersen OW, Ronnov-Jessen L, Howlett AR, Bissell MJ (1992) Interaction with basement membrane serves to rapidly distinguish growth and differentiation pattern of normal and malignant human breast epithelial cells. Proc Natl Acad Sci U S A 89:9064–9068

Pollack RE, Green H, Todaro GJ (1968) Growth control in cultured cells: selection of sublines with increased sensitivity to contact inhibition and decreased tumor-producing ability. Proc Natl Acad Sci U S A 60:126–133

Ramu A, Spanier R, Rahamimoff H, Fuks Z (1984) Restoration of doxorubicin responsiveness in doxorubicin-resistant P388 murine leukaemia cells. Br J Cancer 50:501–507

Sachlos E, Risueno RM, Laronde S, Shapovalova Z, Lee JH, Russell J, Malig M, McNicol JD, Fiebig-Comyn A, Graham M et al (2012) Identification of drugs including a dopamine receptor antagonist that selectively target cancer stem cells. Cell 149:1284–1297

Spengler G, Molnar J, Viveiros M, Amaral L (2011) Thioridazine induces apoptosis of multidrug-resistant mouse lymphoma cells transfected with the human ABCB1 and inhibits the expression of P-glycoprotein. Anticancer Res 31:4201–4205

Stierum R, Gaspari M, Dommels Y, Ouatas T, Pluk H, Jespersen S, Vogels J, Verhoeckx K, Groten J, van Ommen B (2003) Proteome analysis reveals novel proteins associated with proliferation and differentiation of the colorectal cancer cell line Caco-2. Biochim Biophys Acta 1650:73–91

Susini L, Besse S, Duflaut D, Lespagnol A, Beekman C, Fiucci G, Atkinson AR, Busso D, Poussin P, Marine JC et al (2008) TCTP protects from apoptotic cell death by antagonizing bax function. Cell Death Differ 15:1211–1220

Thebault S, Agez M, Chi X, Stojko J, Cura V, Telerman SB, Maillet L, Gautier F, Billas-Massobrio I, Birck C et al (2016) TCTP contains a BH3-like domain, which instead of inhibiting, activates Bcl-xL. Sci Rep 6:19725

Tuynder M, Susini L, Prieur S, Besse S, Fiucci G, Amson R, Telerman A (2002) Biological models and genes of tumor reversion: cellular reprogramming through tpt1/TCTP and SIAH-1. Proc Natl Acad Sci U S A 99:14976–14981

Tuynder M, Fiucci G, Prieur S, Lespagnol A, Geant A, Beaucourt S, Duflaut D, Besse S, Susini L, Cavarelli J et al (2004) Translationally controlled tumor protein is a target of tumor reversion. Proc Natl Acad Sci U S A 101:15364–15369

Vercoutter-Edouart AS, Czeszak X, Crepin M, Lemoine J, Boilly B, Le Bourhis X, Peyrat JP, Hondermarck H (2001) Proteomic detection of changes in protein synthesis induced by fibroblast growth factor-2 in MCF-7 human breast cancer cells. Exp Cell Res 262:59–68

Xu W, Tamim H, Shapiro S, Stang MR, Collet JP (2006) Use of antidepressants and risk of colorectal cancer: a nested case-control study. Lancet Oncol 7:301–308

Chapter 16
History of Histamine-Releasing Factor (HRF)/ Translationally Controlled Tumor Protein (TCTP) Including a Potential Therapeutic Target in Asthma and Allergy

Susan M. MacDonald

Abstract Histamine-releasing factor (HRF) also known as translationally controlled tumor protein (TCTP) is a highly conserved, ubiquitous protein that has both intracellular and extracellular functions. Here we will highlight the subcloning of the molecule, its clinical implications, as well as an inducible-transgenic mouse. Particular attention will be paid to its extracellular functioning and its potential role as a therapeutic target in asthma and allergy. The cells and the cytokines that are produced when stimulated or primed by HRF/TCTP will be detailed as well as the downstream signaling pathway that HRF/TCTP elicits. While it was originally thought that HRF/TCTP interacted with IgE, the finding that cells not binding IgE also respond to HRF/TCTP called this interaction into question. HRF/TCTP or at least its mouse counterpart appears to interact with some, but not all IgE and IgG molecules. HRF/TCTP has been shown to activate multiple human cells including basophils, eosinophils, T cells, and B cells. Since many of the cells that are activated by HRF/TCTP participate in the allergic response, the extracellular functions of HRF/TCTP could exacerbate the allergic, inflammatory cascade. Particularly exciting is that small molecule agonists of the phosphatase SHIP-1 have been shown to modulate the P13 kinase/AKT pathway and may control inflammatory disorders. This review discusses this possibility in light of HRF/TCTP.

S.M. MacDonald, M.D. (✉)
The Johns Hopkins Asthma and Allergy Center, 5501 Hopkins Bayview Circle, Room 3B.69, Baltimore, MD 21224, USA

Department of Medicine, Johns Hopkins University School of Medicine, Baltimore, MD, USA
e-mail: smacdona@jhmi.edu

A. Telerman, R. Amson (eds.), *TCTP/tpt1 - Remodeling Signaling from Stem Cell to Disease*, Results and Problems in Cell Differentiation 64,
DOI 10.1007/978-3-319-67591-6_16

16.1 Introduction/Cloning

Histamine-releasing factor (HRF) was originally classified as a tumor protein (translationally controlled tumor protein, TCTP) in both mouse acidic tumor and mouse erythroleukemia. Brawerman's group in the 1980s named the protein, but its function remained a mystery (Yenofsky et al. 1983; Chitpatima et al. 1988). We identified a histamine-releasing activity that was found in late-phase fluids from nasal lavages, bronchoalveolar lavage fluids (BAL), and skin blister fluids that directly induced histamine release from basophils isolated from a subpopulation of allergic donors (HRF-responders [HRF/TCTP-R]) (MacDonald et al. 1987b). By definition, donors with basophils who did not directly respond to HRF/TCTP were termed HRF-non-responders (HRF/TCTP-NR).

Although an IgE-dependent HRF can be detected in nasal lavages (MacDonald et al. 1987b), PBMC culture supernatants (Sampson et al. 1989), and fluids from human late-phase reactions (LPR) (MacDonald et al. 1987b), we used supernatants from overnight cultures of U937 cells, a human macrophage cell line (Sundstrom and Nilsson 1976), for the isolation and sequencing of the HRF. Fifty liters of these supernatants were concentrated, and the proteins were contained therein purified by Sephadex G75 gel filtration, MONO Q anion exchange, and repetitive Superdex chromatography. The basophil-releasing activity was concentrated, subjected to SDS-polyacrylamide gel electrophoresis (PAGE), blotted onto a polyvinylidene difluoride membrane, and stained with Coomassie blue, revealing four major protein bands (at 60 kDa and 29 kDa and a doublet at 23 kDa). The NH2-terminal sequences of each of these four bands were determined by protein sequencing. The 18 NH2-terminal amino acids of one of the 23-kDa components after a GenBank search revealed 94% homology to p2l, a predicted 21-kDa murine peptide whose complementary DNA (cDNA) was isolated from mouse tumor cells (MacDonald et al. 1987b), as well as identity to p23, the human homolog, described by Bohm et al. (1989). Both were cloned on the basis of their abundant expression in tumor cells, and no function has been ascribed to either molecule. Because there is a stop codon upstream from the initial methionine, it appears that p2l and p23 are not posttranslationally processed at their NH2-termini. p21 cDNA was subcloned into the pGEX-2T plasmid (Smith and Johnson 1988), expressed as a fusion protein with glutathione-S-transferase (GST), purified, and isolated from GST by cleavage with thrombin. Due to the homology between p2l and p23, the same synthetic primers, based on the mouse p2l sequence, were used to isolate the human p23 cDNA from the U937 cell line. This protein also was expressed in *Escherichia coli* as a GST fusion protein and was subsequently cleaved from GST with thrombin.

After purification and cloning, HRF was found to be identical to TCTP, which is also known as p23 (MacDonald et al. 1995). Our recombinant molecule was found to have the same properties and ability to induce histamine release from selected donors as did the originally described HRF/TCTP derived from nasal secretions. The protein is ubiquitously expressed as an intracellular protein, and homologs of HRF/TCTP are found in parasites including *Plasmodium falciparum*, *Wuchereria bancrofti*, *Brugia malayi*, and *Schistosoma Mansoni*. All of these parasites possess mast cell/basophil histamine-releasing activity (MacDonald et al. 2001; Gnanasekar et al. 2002; Rao et al. 2002). Our group, as well as another group, has identified the interaction between HRF and elongation factor-1δ, also known as eukaryotic elongation factor 1B-β (Langdon et al. 2004; Cans et al. 2003). Thus, HRF/TCTP may have an intracellular role in interfering with the elongation step of protein synthesis.

16.2 Clinical Relevance of HRF/TCTP

The history of HRFs dates back to 1979 when Thueson et al. (1979) first described a histamine-releasing activity produced by cultured peripheral blood mononuclear cells that had been stimulated with mitogens or antigens. This HRF was further characterized and found to be very heterogeneous, containing molecular weight species ranging from 15,000 to 50,000 kDa. A number of other groups confirmed this finding (reviewed in MacDonald et al. 1987b). HRFs are produced in vitro by a variety of cell types such as T and B lymphocytes, mononuclear cells, alveolar macrophages, platelets, vascular endothelial cells, and various cell lines, including the U937 monocyte/macrophage-like cell line and RPMI 8866 B-cell line. Not only is HRF found in vitro, but it is also found in vivo.

HRF/TCTP's link to human asthmatic, allergic disease has been well accepted. It has been found in human respiratory secretions (BAL) and skin blister fluids (MacDonald et al. 1987b). Since not all donors' basophils release histamine when exposed to HRF/TCTP, we undertook a study to define the responding population. Sixty-four ragweed allergic patients with a history of seasonal rhinitis and one or more positive skin tests were compared to 17 nonatopic controls who were skin test negative. Sensitivity to HRF/TCTP was restricted to a subpopulation of atopic individuals (MacDonald et al. 1987a). In a separate study of 55 ragweed allergic patients, there was a significant correlation between the intensity of symptoms in the late-phase reaction and basophil histamine release to HRF/TCTP (MacDonald 1993). In studies from another group, peripheral blood mononuclear cells from patients with asthma spontaneously produced HRF/TCTP (Alam et al. 1984; Alam and Rozniecki 1985). That production of HRF/TCTP not only correlated with bronchial hyperreactivity but the bronchial sensitivity to methacholine of the patient correlated with the magnitude of HRF/TCTP production (Alam et al. 1987). Sampson et al. have shown that the production of HRF/TCTP also is associated with clinical status of food allergy and atopic dermatitis (Sampson et al. 1989). Using blood from food allergic children with atopic dermatitis, they found that their basophils have a high spontaneous release of histamine and their cultured mononuclear cells spontaneously produce HRF/TCTP. When these children were placed on an avoidance diet, they improved clinically, their basophils no longer spontaneously secreted histamine, and their mononuclear cells no longer spontaneously produced HRF/TCTP. Two groups have reported the effects of immunotherapy on HRF/TCTP production. One group showed a striking correlation between the production of

HRF/TCTP by mononuclear cells and the change in bronchial sensitivity to histamine (PC20) after 2 years of immunotherapy (Kuna et al. 1989). Brunet et al. showed immunotherapy in allergic rhinitis patients without asthma improved symptoms and also avoided the seasonal increase of spontaneous and antigen-driven HRF/TCTP production from peripheral blood mononuclear cells (Brunet et al. 1992). Moreover, we have measured HRF in human BAL fluids of allergics following antigen challenge. While HRF/TCTP increases over baseline after antigen challenge, it is not significant with the number of patients ($n = 8$) we have investigated (unpublished observations).

With the availability of recombinant material, we examined the lymphocytes of allergic and nonallergic patients for the generation of HRF/TCTP mRNA and protein. Twelve patients (four HRF/TCTP-R, four HRF/TCTP-NR, and four nonallergic) were recruited. Blood was drawn for serum IgE measurements and for basophil histamine release in response to recombinant HRF/TCTP and anti-IgE. In addition, peripheral blood mononuclear cells were cultured for HRF/TCTP production and processed for mRNA extraction and subsequent reversed transcribed polymerase chain reaction for HRF/TCTP mRNA. The geometric mean serum IgE levels were 356 ng ml^{-1} in the HRF/TCTP-R group versus 52 μg ml^{-1} and 4.2 μg ml^{-1} in the HRF/TCTP-NR and nonallergic subjects, respectively. Histamine release in response to the recombinant HRF/TCTP paralleled that of our native HRF/TCTP preparation in that only the four HRF/TCTP-R patients released histamine to this stimulus. The quantity of mRNA for HRF/TCTP, when compared to that for beta-actin, the housekeeping gene, did not appear different among the groups. The bioactivity of the recombinant HRF/TCTP on lactic acid-treated cells passively sensitized with an IgE containing serum from a HRF/TCTP-R, however, was greater in the allergic, HRF/TCTP-R patients than in the nonallergic subjects (MacDonald 1996; Langdon et al. 1995). Thus, it appears that all individuals make mRNA for HRF/TCTP, but atopic subjects more effectively translate it to protein. In an abstract, the serum from some patients with atopic dermatitis, but not normals, demonstrated increased levels of HRF/TCTP-reactive IgE levels (Ando et al. 2012). These atopic dermatitis patients' sera could cause cytokine secretion from human mast cells (Ando et al. 2012).

16.3 HRF/TCTP Extracellular Functions

Secreted by an ER/Golgi-independent route, HRF/TCTP has no leader sequence, as documented by Amzallag et al. (2004). They discerned that secreted HRF/TCTP comes from an existing intracellular pool and co-distributes with TSAP6, a member of a family that is involved in vesicular trafficking and secretory processes (Amzallag et al. 2004; Moldes et al. 2001; Korkmaz et al. 2002). Our focus has been on the extracellular functions of HRF/TCTP. HRF was initially described as a complete secretagogue for histamine and IL-4 secretion from basophils of allergic donors (Schroeder et al. 1996). These donors were thought to have a certain type of

IgE that interacted with HRF/TCTP to induce secretion (MacDonald et al. 1995). However, it was subsequently demonstrated that HRF/TCTP primed all basophils for histamine release, IL-4 and IL-13 secretion regardless of the type of IgE (Schroeder et al. 1997). Additional studies demonstrated that HRF/TCTP did not appear to interact with IgE. Namely, pharmacologic agents that altered HRF/TCTP-induced histamine release, i.e., rottlerin, did not affect anti-IgE-induced histamine release (Bheekha-Escura et al. 1999). Rat basophilic leukemia cells transfected with the α, β, and γ chains of the human IgE receptor, FcεRI, did not release histamine to HRF/TCTP despite sensitization with IgE molecules from an HRF/TCTP-R donor (Wantke et al. 1999). HRF/TCTP was shown to stimulate eosinophils to produce IL-8 and induce an intracellular calcium response (Bheekha-Escura et al. 2000). This was also observed in the eosinophil cell line, AML-3D10, which does not express the α chain of the FcεR1 on the surface of the cell (Bheekha-Escura et al. 2000). Very recently, HRF/TCTP was found to have an inflammatory role in mouse models of asthma and allergy, whereby HRF/TCTP was found to exist as a dimer, bound to a subset of IgE and IgG antibodies by interacting by its N-terminus and some internal regions with the Fab region of immunoglobulins (Kashiwakura et al. 2012). These interactions were described with mouse HRF/TCTP and interacted on mouse mast cells.

At the level of gene transcription, HRF/TCTP has been shown to inhibit cytokine production from stimulated primary T cells and the Jurkat T-cell line (Vonakis et al. 2003). Thus, HRF/TCTP, in addition to functioning as a histamine-releasing factor, can modulate secretion of cytokines from human basophils, eosinophils, and T cells. It has also been identified as a B-cell growth factor by Kang et al. They demonstrated that HRF/TCTP bound to B cells and induced cytokine production (Kang et al. 2001). More recently, HRF/TCTP was shown to stimulate bronchial epithelial cells to produce IL-8 and GM-CSF (Yoneda et al. 2004). These effects of HRF/TCTP on different cell types are depicted in Fig. 16.1.

16.4 Other Functions of HRF/TCTP (Mainly Intracellular)

While this review focuses mainly on the extracellular functions of HRF/TCTP, it is important to discuss some of its broad spectrums of intracellular functions. HRF/TCTP is both transcriptionally and posttranscriptionally regulated by calcium (Xu et al. 1999). It is also a tubulin-binding protein and has been shown to transiently associate with microtubules during the cell cycle (Gachet et al. 1999). Also the vitamin D receptor, the NF-κβ regulatory subunit, Iκκγ(NEMO), the myeloid cell leukemia protein 1 (MCL1), and Bcl-XL have been demonstrated to interact with HRF/TCTP (Rid et al. 2010; Fenner et al. 2010; Zhang et al. 2002; Yang et al. 2005). High levels of HRF/TCTP have been associated with various cancers, such as prostate, breast, and colon cancer (Arcuri et al. 2004; Vercoutter-Edouart et al. 2001; Chung et al. 2000). Furthermore, the gene for HRF/TCTP was downregulated in tumor reversion, and more specifically, the level was significantly

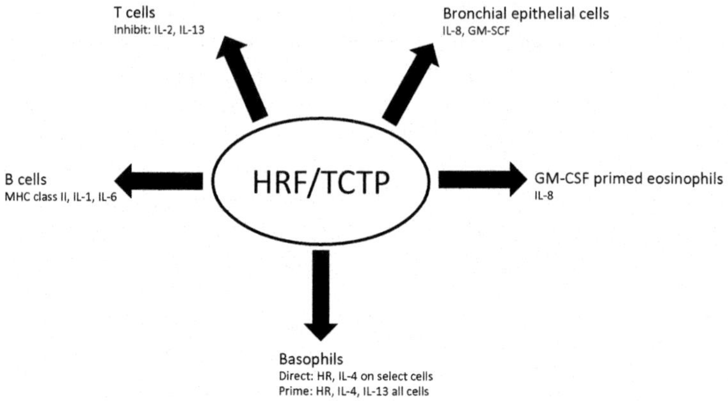

Fig. 16.1 Effects of HRF/TCTP on various cell types. HRF/TCTP either directly activates (direct) basophils producing HR and IL-4 on certain cells or primes (prime) anti-IgE-induced HR and IL-4 and IL-3. HRF/TCTP induces IL-8 from GM-CSF-primed eosinophils. Similarly, it produces IL-8 and GM-CSF from bronchial epithelial cells and MHC class II, IL-1, and IL-6 from B cells. Contrary to the enhanced interleukin production, HRF/TCTP inhibits IL-2 and IL-12 from T cells. Abbreviations: GM-CSF, granulocyte-macrophage colony-stimulating factor; HR, histamine release; HRF/TCTP, histamine-releasing factor/translationally controlled tumor protein; IL, interleukin; MHC, major histocompatibility complex

reduced in a lung cancer cell line, A549, and revertant cells (Tuynder et al. 2004). This was eloquently described by Telerman and Amson in a publication entitled *The Molecular Programmme of Tumors Reversion: The Steps Beyond Malignant Tumor Formation* (Telerman and Amson 2009). The role of HRF/TCTP in tumor development may be associated with its antiapoptotic activity (Yang et al. 2005; Li et al. 2001). This is further supported by reports of HRF/TCTP antagonizing Bax function and controlling the stability of the tumor suppressor p53 (Susini et al. 2008; Rho et al. 2011). In a very recent publication, HRF/TCTP promoted p53 degradation, and p53 directly repressed HRF/TCTP transcription (Amson et al. 2011). With this report of a previously unrecognized regulatory circuit, HRF/TCTP may be extremely relevant in cancer (Amson et al. 2011). As previously mentioned, our lab and others have shown involvement of HRF/TCTP in the elongation step of protein synthesis (Langdon et al. 2004; Cans et al. 2003). Thus HRF/TCTP's intracellular functions are wide ranging. The extracellular functions, however, seem to focus on inflammation.

16.5 An Inducible HRF/TCTP Transgenic Mouse

Although HRF/TCTP has been extensively investigated for many years, most studies have been carried out in cultured cells and pathologic samples. Until recently, there has been no established animal model available to explore the function of HRF/TCTP. Several groups generated HRF/TCTP knockout mice by targeted gene disruption, but these HRF/TCTP knockout mice were embryonically lethal (Susini et al. 2008; Chen et al. 2007). There has been a TCTP mouse generated by Telerman and colleagues (Thébault et al. 2016). Since HRF/TCTP is ubiquitous and highly conserved, our approach was to create an inducible HRF/TCTP mouse model using the tet-on system. We wanted to target HRF/TCTP to the lungs, so we used the CC10 promoter that is expressed in Clara cells of the lung epithelium to generate a transgenic TRE-HRF-EGFP mouse. The HRF transgenic plasmid was generated by the combination of three main components (Fig. 16.2). The first component is the pTRE-tight vector (provided by Dr. Zhu in our division), which contains a modified TRE (tetracycline response element) controlling the inducible expression of the gene of interest. The second component is human HRF cDNA, which was cloned from U937 cells by RT-PCR and further confirmed by sequencing. The third component is the pIRES2-EGFP vector, provided by Dr. Vonakis in our division. The IRES2 (internal ribosome entry site) allows the EGFP (enhanced green fluorescent protein) gene to be expressed individually as a reporter protein along with HRF in order to facilitate the recognition of expression of transgenic human HRF. Thus, the transgenic TRE-HRF-EGFP construct will express HRF and EGFP individually under the regulation of tetracycline or doxycycline. Using this model, we saw an enhanced asthmatic, allergic phenotype after OVA challenge (Yueh-Chiao et al. 2010). This enhancement is in the C57BL/6 mouse, not the traditional "allergic" BALB/c mouse. The development of an inducible-transgenic HRF/TCTP animal model will yield insights into its underlying pathophysiologic characteristics and provide a tool to define the mechanism of this enhanced or primed phenotype.

The mechanism of HRF/TCTP's enhanced response yielding increases in IL-4, IgE, and eosinophils after OVA challenge in our transgenic model is currently unknown. All of these events that could be attributed to the action of HRF/TCTP on the basophil is plausible considering the data on HRF/TCTP and the human basophil. We have shown that HRF/TCTP activates human basophils to produce IL-4 (Schroeder et al. 1996). It is well accepted that IL-4 is important for B-cell class switching and production of IgE. Furthermore, human basophils possess the β-1 integrin that is important for firm adhesion. The ligand for β-1 integrin, vascular cell adhesion molecule (VCAM-1) is upregulated by IL-4 and is important for the transendothelial migration of eosinophils and Th2 cells (Schroeder 2009). Therefore, the production of IL-4 by basophils could explain the enhanced asthmatic, allergic phenotype we see after overproduction of HRF/TCTP in our OVA-challenged model. This, of course, assumes that the mouse basophil acts in a similar manner as its human counterpart. While the existence of the mouse basophil dates back over two decades, this cell has reemerged in the last several years as an important initiator in mouse Th2 inflammation (Seder et al. 1991; Min and Paul 2008; Obata et al. 2007).

The multifactorial disease that is asthma makes it highly unlikely that one single cell such as the mouse basophil is solely responsible for HRF/TCTP's effect on asthmatic lung disease. Furthermore, one must consider eosinophils and T cells. It is well known that the activation, recruitment, and proliferation of the T cell are

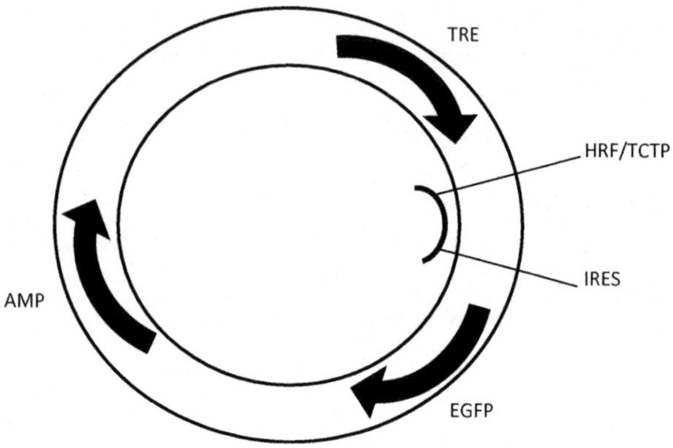

Fig. 16.2 Schema of the transgenic TRE-HRF/TCTP- EGFP plasmid construction. The human HRF gene was inserted between the TRE and IRES-EGFP elements. Abbreviations: TRE, tetracycline response element; HRF/TCTP, histamine-releasing factor; IRES, internal ribosome entry site; EGFP, enhanced green fluorescent protein

associated with asthmatic lung disease (Jacobsen et al. 2008). Given the HRF/TCTP's activation of eosinophils and the increased eosinophils we find in the BAL fluid of the OVA-challenged transgenic mouse, it is logical to examine the role of this cell in the mechanisms of action of HRF/TCTP in vivo (Bheekha-Escura et al. 2000; Yueh-Chiao et al. 2010). In the congenitally eosinophil-deficient PHIL mouse, there is a diminution of Th2 responses (Lee et al. 2004). Furthermore, eosinophils also secrete IL-4 and act as antigen-presenting cells yielding T-cell activation after allergen provocation in the lung (Obata et al. 2007; Mayr et al. 2002). Therefore, HRF/TCTP may exert additional enhancing effects through the eosinophil. Since an antibody to IL-5 has been shown to suppress eosinophil recruitment following OVA challenge in WT and FcεRα-/- mice (Mayr et al. 2002), giving anti-IL-5 to our OVA-challenged HRF/TCTP mice could help determine HRF/TCTP's mechanism upon eosinophil recruitment. Alternately, cross-breeding our HRF/TCTP transgenic mice with the eosinophil knockout PHIL mouse could ablate all HRF/TCTP-induced enhancing effects or just affect eosinophils. Future possibilities are many using this model as a tool.

16.6 The Importance of Ship-1 on HRF/TCTP Signaling

That human basophils are cells capable of being "primed" or having an enhanced functional response has long been appreciated. Some of the molecules that are known to prime human basophils include IL-3, NGF, HRF/TCTP, and the nonphysiologic stimulus D_2O (Schroeder 2009; MacDonald et al. 1989, 1991). In general, these substances show a greater *releasability* (as evidenced by histamine or IL-4 secretion) when stimulating basophils from allergic or allergic/asthmatic subjects. They do not generally activate basophils from normal subjects. The exception to this is the HRF/TCTP-R basophils. Basophils from these subjects are directly activated by D_2O, IL-3, and HRF/TCTP (MacDonald et al. 1989, 1991).

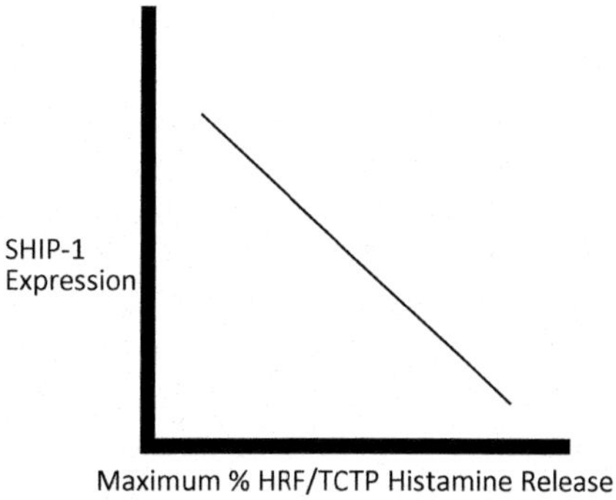

Fig. 16.3 Negative correlation between SHIP-1 protein expression and histamine release to HRF/TCTP. Histamine release was performed on the same day as SHIP-1 measurements

The molecular basis for the *releasability* of the HRF/TCTP's basophils remained elusive until relatively recently. It has become accepted that the term *releasability* (i.e., control of release of mediators from basophils in response to different stimuli) involves several biochemical events in addition to the surface density of IgE molecules. There have been reports of certain signaling molecule deficiencies in nonreleasing basophils (Kepley et al. 1999; Lavens-Phillips and MacGlashan 2000). While these deficiencies are documented, there is little variation of SHIP-1 in the general population (Vilariño et al. 2005). To date, we are the first group to show the negative association of the phosphatase, SHIP-1, with histamine release to HRF/TCTP in *hyperreleasing* basophils (Vonakis et al. 2001). See Fig. 16.3. Variation of SHIP-1 levels is also documented in a subset of patients with chronic idiopathic urticaria, where levels of SHIP-1 are increased and anti-IgE-induced histamine release is reduced (Vonakis et al. 2007). Thus, SHIP-1 levels appear to be altered in some human disease states.

A clue to the underlying mechanisms of increased *releasability* of basophils was demonstrated in a mouse knockout of SHIP-1 (Krystal 2000; Huber et al. 1998). Mast cells grown from the bone marrow (BMMC) of SHIP-1 knockout mice showed decreased hydrolysis of phosphatidylinositol (PI)-3,4,5,P_3 (PIP_3) (Huber et al. 1998). SHIP-1 participates in the pathway in which the lipid phosphatidylinositol 4,5 bisphosphate (PI4,5,P_2) is phosphorylated by PI3 kinase to produce PIP_3 which can be acted upon to produce PI(3,4)bisphosphate (PI3,4P_2) (Rohrschneider et al. 2000; Scharenberg and Kinet 1998). We have demonstrated that the compound, LY294002, an inhibitor of PI3 kinase, inhibits histamine release induced by HRF/TCTP in basophils from HRF/TCTP-R donors (Vonakis et al. 2001). The activity of PI3 kinase is central to many basophil functions, and SHIP-1 acts to oppose the function of PI3 kinase by removing the 5′phosphatase from PIP_3, making SHIP-1 an important regulator of these reactions. Mouse SHIP-1 knockout mast cells had an excess of PIP_3 that resulted in a sustained calcium signal that was critical for degranulation (Lioubin et al. 1996). Furthermore, SHIP is a suppressor of IgE plus antigen-induced degranulation of not only bone marrow-derived mast cells but also negatively regulates IgE plus antigen-induced degranulation of connective tissue and mucosal mast cells by repressing the P13 kinase pathway (Ruschmann et al. 2012). Additionally, PIP_3 recruits the serine tyrosine kinase, Akt, to the plasma membrane (Brauweiler et al. 2000), which is present in human basophils and transiently phosphorylated after anti-IgE stimulation (Miura et al. 2001). Akt is phosphorylated by HRF/TCTP in HRF/TCTP-R donors but not in HRF/TCTP-NR donors (Vonakis et al. 2008). Furthermore, we see prolonged Akt phosphorylation kinetics in HRF/TCTP-R (Vonakis et al. 2008), which is supportive of the involvement of this pathway in HRF-induced activation. Data from the SHIP-1 knockout mice and our own published data suggest that SHIP-1 may play a "gatekeeper role" in mouse and human basophils and mast cells. One would expect SHIP-1 to limit effector cell responsiveness in normal individuals, while a SHIP-1 deficiency would predispose an individual to excess inflammatory-mediator production and, hence, a *hyperreleasable* phenotype.

In order to address this more directly, we altered SHIP-1 levels in human basophils. These studies have been limited by the fact that the basophil is an end-stage nondividing cell and extremely difficult to transfect or transduce. Many attempts have been tried to transfect primary human basophils. These include lipid-based reagents, lentivirus, and nucleofection. Most failed either due to toxicity or very low transfection efficiency. Only nucleofection (Amaxa) gave a limited transfection efficiency that was useful only for single-cell analysis (Vilarino and MacGlashan 2005). There is one report that a TAT-fusion protein was used in transfecting human basophils (Didichenko et al. 2008). We set out to determine a more efficient method of altering signal transduction pathways in human basophils. To that end, we established a model of culturing human peripheral blood-derived basophils from CD34+ cells that have the morphologic and functional characteristics of human basophils (Langdon et al. 2008). We utilized this model to alter SHIP-1 levels using siRNA technology and demonstrated a decrease in SHIP-1 levels that was associated with an increase in histamine release to HRF/TCTP. Using CD34+ peripheral-derived basophils, it is possible to perform a more direct test of the hypothesis that SHIP-1 has a role in modulating basophil responsiveness, both to HRF/TCTP and IgE-mediated stimulation.

16.7 Additional Intracellular Signaling by HRF/TCTP

Another possible mechanism of action for HRF/TCTP may be IgE-dependent enhancement. Originally, HRF/TCTP was called the IgE-dependent HRF (MacDonald et al. 1995). This designation resulted from the fact that HRF seemed to act as a secretagogue for human basophils from a subpopulation of allergic donors. Moreover, passive sensitization of serum containing IgE from these responding donors rendered nonresponsive donors' basophils responsive to HRF/TCTP (MacDonald et al. 1995). HRF/TCTP was then shown to activate other cells that do not possess the high-affinity IgE receptor, FcεR1 (Bheekha-Escura et al. 2000; Vonakis et al. 2003). We have demonstrated that HRF/TCTP has signal transduction events that are similar, but not identical, to signaling through FcεR1 (Vonakis et al. 2008). With the availability of both the FcεR1α knockout mouse and the IgE knockout mouse (Dombrowicz et al. 1993; Oettgen et al. 1994), the question of whether HRF/TCTP is dependent on IgE can be definitively addressed. As mentioned, a manuscript has very recently been published that demonstrates mouse HRF/TCTP does bind to certain IgE and IgG molecules (Kashiwakura et al. 2012).

In order to address the molecular mechanisms of HRF/TCTP-induced secretion, we designed experiments to elucidate specific actions of HRF/TCTP on human basophils and to characterize the nature of intracellular signaling that follows stimulation with HRF/TCTP. Given the similarities in secretion kinetics following IgE-mediated stimulation, we hypothesized there would be some signaling characteristics similar to those previously found for IgE-mediated release. However, due to the differential sensitivity to treatment with rottlerin between HRF/TCTP and anti-IgE (Wantke et al. 1999), we also expected differences in signaling. We used human basophils from two donor populations, HRF/TCTP-R and HRF/TCTP-NR. Consistent with the ability of HRF/TCTP to either induce secretion directly from HRF/TCTP-R basophils or prime HRF/TCTP-NR basophils, we have shown binding of HRF/TCTP by flow cytometry to both donor populations (Vonakis et al. 2008). We demonstrated that HRF/TCTP induced activation of intracellular signal transduction events in basophils only from those donors who directly release histamine to HRF/TCTP, namely, HRF/TCTP-R. Specifically, we have been able to demonstrate increases in the arachidonic acid metabolite, LTC_4, from basophils of HRF/TCTP-R donors stimulated with anti-IgE. Additionally, we have demonstrated LTC_4 release from basophils stimulated with HRF/TCTP (Vonakis et al. 2008). One might predict that this might well be due to prolonged phosphorylation of MEK and ERK 1/2. Using human basophils isolated from leukopheresis packs, Miura et al. have demonstrated that the activation of ERK1/2 is linked to arachidonic acid metabolism but not to histamine or IL-4 release (Miura et al. 1999). Phosphorylation of ERK1/2 is transient, peaks at 5 min, and returns to baseline by 30 min. We have demonstrated that both MEK and ERK1/2 are phosphorylated by HRF/TCTP in basophils from HRF/TCTP-R donors but not from HRF/TCTP-NR donors (Vonakis et al. 2008). Thus, the characteristics of the signaling responses were very similar to those observed for stimulation with anti-IgE antibody or antigen with a couple of exceptions. Notably, there was no phosphorylation of FcεR1γ, and there was absolutely no phosphorylation of any downstream signal transduction molecules in the HRF/TCTP-NR basophils.

16.8 HRF/TCTP as a Therapeutic Target

Based on the above observations, we believe that HRF/TCTP may be an important element of the pathogenesis of asthmatic, allergic diseases. Since HRF/TCTP is present in late-phase reaction fluids in vivo, it may be contributing to mediator release that is found in the late response. Therefore, it is most reasonable to consider HRF as a therapeutic target. The most direct way to prove that HRF/TCTP is a therapeutic target would be to block its binding to its receptor. However, despite numerous attempts by different laboratories, the HRF/TCTP receptor has remained elusive. An HRF/TCTP-blocking antibody would prove useful in this approach. Unfortunately, no specific antibody exists. A recent publication does demonstrate that the extracellular actions of HRF/TCTP can be explained, at least in part by specific binding sequences on mouse HRF (mHRF) to some IgE and IgG molecules (Kashiwakura et al. 2012). In two regions, the N-terminal 19-residue peptide and residues 107–135, the H3 region, were found to be important for this binding to immunoglobulins (Kashiwakura et al. 2012). These regions overlapped only in part with the antigen binding site. Furthermore, only certain, but not all, IgE and IgE molecules supported or bounded to HRF/TCTP (Kashiwakura et al. 2012). Nevertheless, this observation warrants further investigation.

Two additional observations remain. The possibility exists that this HRF/TCTP-immunoglobulin interaction could be explained by nonspecific ionic interactions or interactions of different parts of the immunoglobulins. Moreover, BAL and sera from naïve mice contain HRF/TCTP that does not normally yield inflammation (Kashiwakura et al. 2012). This suggests there might be a suppressive mechanism of inflammation induced by endogenous HRF/TCTP. Our own SHIP data with the inverse correlation of levels of SHIP-1 protein with histamine release to HRF/TCTP would support this (Vonakis et al. 2001).

Miu, Ong, and colleagues have discovered small molecule agonists of SHIP-1 that inhibit the P13K pathway (Ong et al. 2007). These are potent and specific activators of SHIP-1. Initial mouse model studies suggested that these agonists might be useful therapeutically. Our laboratory received such an agonist and was able to demonstrate that anti-IgE-induced basophil histamine release was inhibited while F-met-leu-phe-induced release was not (data not shown). In fact, it has recently been reported at the American Thoracic Society Meeting in May 2012 in San Francisco that the SHIP-1 agonist, AQX-1125, from Aquinox Pharmaceuticals was tested in a three-part phase I study that included a single ascending dose, a multiple ascending dose, and a food-effect study in healthy human volunteers (Tam et al. 2012). The drug was well tolerated and had a half-life that supported a once-daily oral administration.

"Aquinox Pharmaceuticals is a clinical-stage pharmaceutical company discovering and developing targeted therapeutics in disease areas of inflammation and immune oncology. Our primary focus is anti-inflammatory product candidates targeting SH2-containing inositol-5′-phosphate 1, or SHIP1, which is a key regulator of an important cellular signaling pathway in immune cells, known as the P13K pathway. Our lead product candidate, AQX-1125, is a small molecule activator of SHIP1 suitable for oral, once daily dosing. Having successfully completed multiple preclinical studies and seven clinical trials with AQX-1125, they are now advancing towards pivotal Phase 3 trials with AQX-1125, in our lead indication of bladder pain syndrome/interstitial cystitis (BPS/IC). Aquinox has a broad intellectual property portfolio and pipeline of preclinical drug candidates that activate SHIP1 (www.aqxpharma.com)." It should be noted that AQX-1125 was used to determine its ability to reduce symptoms exacerbations in COPD. The results of that trial called

FLAGSHIP was reported in July 2015 and demonstrated no difference between AQX-1125 and placebo, and therefore further development of AQX-1125 for COPD has been suspended and redirected their efforts into bladder pain syndrome/intestinal cystitis (BAS/IC) (www.aqxpharma.com).

"AQX-1125, Aquinox's lead drug candidate, is a small molecule activator of SHIP1, which is a regulating component of the P13K cellular signaling pathway. By increasing SHIP1 activity, AQX-1125 accelerates a natural mechanism that has evolved to maintain homeostasis of the immune system and reduce the immune cell activation and migration to sites of inflammation. AQX-1125 has demonstrated preliminary safety and favorable drug properties for once daily oral administration in multiple preclinical studies and seven completed clinical trials. Aquinox completed a successful Phase 2 clinical trial with AQX-1125 for the treatment of IC/BPS in the third quarter of 2016 (www.aqxpharma.com)."

There have been three published papers, two in vitro and animal models and one in vivo in humans characterizing AQX-1125. The first describes the effects of cell activation and chemotaxis in vitro (Stenton et al. 2013b). This paper documents that this compound is suitable for testing in various models of inflammation. The second paper shows the effects of AQX-1125 in rodent models of pulmonary and allergy. The efficacy of AQX-1125 is dependent on the presence of SHIP-1 (Stenton et al. 2013a). Finally, AQX-1125 was demonstrated efficacious in humans with mild to moderate asthma (Leaker et al. 2014). In this manuscript AQX-1125 significantly reduced the late response to allergen challenge with a trend in reduction of inflammation. Clinical side effects were very mild and did not lead to discontinuation of therapy. This was performed in a randomized, double-blind placebo-controlled, two-way crossover study in 22 steroid-naïve mild-to-moderate asthmatic individuals with documented late-phase response to inhaled allergen. These results might suggest a role for this SHIP-1 agonist in HRF/TCTP- induced symptoms.

16.9 Summary

In conclusion, further defining the extracellular role of the mechanism of HRF/TCTP-induced priming in vivo using our HRF/TCTP inducible-transgenic mouse and in vitro using both peripheral blood-derived basophils and CD34+ peripheral-derived cultured basophils could yield additional insight into HRF/TCTP's participation in the propagation of the Th2 asthmatic allergic response. The successful completion of these studies could lead to an inhibition of the function of this unique cytokine and its amelioration of its role in the allergic, asthmatic diathesis. SHIP-1 agonists may well be useful as therapeutic targets for the actions of HRF/TCTP in allergic responses.

Acknowledgments/Disclosure Special thanks to Ms. Katherine Best for secretarial assistance.

References

Alam R, Rozniecki J (1985) A mononuclear cell-derived histamine releasing factor in asthmatic patients.II. Activity *in vivo*. Allergy 40:124–129

Alam R, Rozniecki J, Salmaj K (1984) A mononuclear cell derived histamine releasing factor (HRF) in asthmatic patients. Histamine release from basophils *in vitro*. Ann Allergy 53:66–69

Alam R, Kuna P, Rozniecki J, Kuzminska B (1987) The magnitude of the spontaneous production of histamine-releasing factor (HRF) by lymphocytes *in vitro* correlates with the state of bronchial hyperreactivity in patients with asthma. J Allergy Clin Immunol 79:103–108

Amson R, Pece S, Lespagnol A, Vyas R, Mazzarol G, Tosoni D, Colaluca I, Viale G, Rodrigues-Ferreira S, Wynendaele J, Chaloin O, Hoebeke J, Marine JC, Di Fiore PP, Telerman A (2011) Reciprocal repression between P53 and TCTP. Nat Med 18(1):91–99

Amzallag N, Passer BJ, Allanic D, Segura E, Théry C, Goud B, Amson R, Telerman A (2004) TSAP6 facilitates the secretion of translationally controlled tumor protein/histamine-releasing factor via a nonclassical pathway. J Biol Chem 279:46104–46112

Ando T, Kashiwakura J, Kawakami Y, Kawakami T (2012) Histamine releasing factor (HRF) in asthma and atopic dermatitis. J Immunol 188:177.16 (Abstract)

Arcuri F, Papa S, Carducci A, Romagnoli R, Liberatori S, Riparbelli MG, Sanchez JC, Tosi P, del Vecchio MT (2004) Translationally controlled tumor protein (TCTP) in the human prostate and prostate cancer cells: expression, distribution, and calcium binding activity. Prostate 60:130–140

Bheekha-Escura R, Chance SR, Langdon JM, MacGlashan DW Jr, MacDonald SM (1999) Pharmacologic regulation of histamine release by the human recombinant histamine-releasing factor. J Allergy Clin Immunol 103:937–943

Bheekha-Escura R, MacGlashan DW Jr, Langdon JM, MacDonald SM (2000) Human recombinant histamine-releasing factor activates human eosinophils and the eosinophilic cell line, AML14-3D10. Blood 96:2191–2198

Bohm H, Benndorf R, Gaestel M, Gross B, Nurnberg P, Kraft R, Otto A, Bielka H (1989) The growth-related protein P23 of the Ehrlich ascites tumor: translational control, cloning and primary structure. Biochem Int 19:277; Bohm H et al (1991) Biomed. Biochim Acta 12:1193

Brauweiler AM, Tamir I, Cambier JC (2000) Bilevel control of B-cell activation by the inositol 5-phosphatase SHIP. Immunol Rev 176:69–74

Brunet C, Bédard PM, Lavoie A, Jobin M, Hébert J (1992) Allergic rhinitis to ragweed pollen. II. Modulation of histamine-releasing factor production by specific immunotherapy. J Allergy Clin Immunol 89:87–94

Cans C, Passer BJ, Shalak V, Nancy-Portebois V, Crible V, Amzallag N, Allanic D, Tufino R, Argentini M, Moras D, Fiucci G, Goud B, Mirande M, Amson R, Telerman A (2003) Translationally controlled tumor protein acts as a guanine nucleotide dissociation inhibitor on the translation elongation factor eEF1A. Proc Natl Acad Sci U S A 100:13892–13897

Chen SH, Wu PS, Chou CH, Yan YT, Liu H, Weng SY, Yang-Yen HF (2007) A knockout mouse approach reveals that TCTP functions as an essential factor for cell proliferation and survival in a tissue-or cell type-specific manner. Mol Biol Cell 18:2525–2533

Chitpatima ST, Makrides S, Bandyopadhyay R, Brawerman G (1988) Nucleotide sequence of a major messenger RNA for a 21 kilodalton polypeptide that is under translational control in mouse tumor cells. Nucleic Acids Res 16:2350

Chung S, Kim M, Choi W, Chung J, Lee K (2000) Expression of translationally controlled tumor protein mRNA in human colon cancer. Cancer Lett 156:185–190

Didichenko SA, Spiegl N, Brunner T, Dahinden CA (2008) IL-3 induces a Pim1-dependent antiapototic pathway in primary human basophils. Blood 112:3949–3958

Dombrowicz D, Flamand V, Brigman KK, Koller BH, Kinet JP (1993) Abolition of anaphylaxis by targeted disruption of the high affinity immunoglobulin E receptor alpha chain gene. Cell 75:969–976

Fenner BJ, Scannell M, Prehn JH (2010) Expanding the substantial interactome of NEMO using protein microarrays. PLoS One 5:e8799

Gachet Y, Tournier S, Lee M, Lazaris-Karatzas A, Poulton T, Bommer UA (1999) The growth-related, translationally controlled protein P23 has properties of a tubulin binding protein and associates transiently with microtubules during the cell cycle. J Cell Sci 112:1257–1271

Gnanasekar M, Rao KV, Chen L, Narayanan RB, Geetha M, Scott AL, Ramaswamy K, Kaliraj P (2002) Molecular characterization of a calcium binding translationally controlled tumor protein homologue from the filarial parasites *Brugia malayi* and *Wuchereria bancrofti*. Mol Biochem Parasitol 121:107–118

Huber M, Helgason CD, Damen JE, Liu L, Humphries K, Krustal G (1998) The src homology 2 containing inositol phosphatase (SHIP) is the gatekeeper of mast cell degranulation. Proc Natl Acad Sci U S A 95:11330–11335

Jacobsen EA, Ochkur SI, Pero RS, Taranova AG, Protheroe CA, Colbert DC, Lee NA, Lee JJ (2008) Allergic pulmonary inflammation in mice is dependent on eosinophil-induced recruitment of effector T cells. J Exp Med 205:699–710

Kang HS, Lee MJ, Song H, Han SH, Kim YM, Im JY, Choi I (2001) Molecular identification of IgE-dependent histamine-releasing factor as a B cell growth factor. J Immunol 166:6545–6554

Kashiwakura J, Ando T, Matsumoto K, Kimura M, Kitaura J, Matho MH, Zajonc DM, Ozeki T, Ra C, MacDonald SM, Siraganian RP, Broide DH, Kawakami Y, Kawakami T (2012) Proinflammatory role of histamine-releasing factor in mouse models of asthma and allergy. J Clin Invest 122:218–228

Kepley CL, Youssef L, Andrews RP, Wilson BS, Oliver JM (1999) Syk deficiency in nonreleaser basophils. J Allergy Clin Immunol 104:279–284

Korkmaz KS, Elbi C, Korkmaz CG, Loda M, Hager GL, Saatcioglu F (2002) Molecular cloning and characterization of STAMP1, a highly prostate-specific six transmembrane protein that is overexpressed in prostate cancer. J Biol Chem 277:36689–36696

Krystal G (2000) Lipid phosphatases in the immune system. Immunology 12:397–403

Kuna P, Alam R, Kuzminska B, Rozniecki J (1989) The effect of preseasonal immunotherapy on the production of histamine-releasing factor (HRF) by mononuclear cells from patients with seasonal asthma: results of a double-blind, placebo-controlled, randomized study. J Allergy Clin Immunol 83:816–882

Langdon J, Anders K, Lichtenstein LM, MacDonald SM (1995) Atopics translate mRNA for the IgE-dependent histamine releasing factor (HRF) more effectively than normal. J Allergy Clin Immunol 95:336 (Abstract)

Langdon JM, Vonakis BM, MacDonald SM (2004) Identification of the interaction between the human recombinant histamine releasing factor/translationally controlled tumor protein and elongation factor-1 delta (also known as elongation factor-1B beta). Biochem Biophys Acta 1688:232–236

Langdon JM, Schroeder JT, Vonakis BM, Bieneman AP, Chichester K, MacDonald SM (2008) Histamine-releasing factor/translationally controlled tumor protein (HRF/TCTP)-induced histamine release is enhanced with SHIP-1 knockdown in cultured human mast cell and basophil models. J Leukoc Biol 84:1151–1158

Lavens-Phillips SE, MacGlashan DW Jr (2000) The tyrosine kinases p563/56lyn and p72syk are differentially expressed at the protein level but not at the messenger RNA level in nonreleasing human basophils. Am J Respir Cell Mol Biol 23:566–571

Leaker BR, Barnes PJ, O'Connor BJ, Ali FY, Tam P, Neville J, Mackenzie LF, MacRury T (2014) The effects of the novel SHIP1 activator AQX-1125 on allergen-induced responses in mild-to-moderate asthma. Clin Exp Allergy 44(9):1146–1153

Lee JJ, Dimina D, Macias MP, Ochkur SI, McGarry MP, O'Neill KR, Protheroe C, Pero R, Nguyen T, Cormier SA, Lenkiewicz E, Colbert D, Rinaldi L, Ackerman SJ, Irvin CG, Lee NA (2004) Defining a link with asthma in mice congenitally deficient in eosinophils. Science 305:1773–1776

Li F, Zhang D, Fujise K (2001) Characterization of fortilin, a novel antiapoptotic protein. J Biol Chem 276:47542–47549

Lioubin MN, Algate PA, Tsai S, Carlberg K, Aebersold A, Rohrschneider LR (1996) p150Ship, a signal transduction molecule with inositol polyphosphate-5-phosphatase activity. Genes Dev 10:1084–1095

MacDonald SM (1993) Histamine releasing factors and IgE heterogeneity. In: Middleton E, Reed CE, Ellis EF, Adkinson NF, Yunginger JW, Buss WW (eds) Allergy: principles and practice, 4th edn. Mosby-Year Book, St. Louis, MO, pp 1–11

MacDonald SM (1996) Histamine-releasing factors. Curr Opin Immunol 8:778–783

MacDonald SM, Kagey-Sobotk A, Proud D, Naclerio RM, Lichtenstein LM (1987a) Histamine-releasing factor: release mechanism and responding population. J Allergy Clin Immunol 79:248

MacDonald SM, Lichtenstein LM, Proud D, Plaut M, Kagey-Sobotka A (1987b) Studies of IgE-dependent histamine releasing factors: heterogeneity of IgE. J Immunol 139:506–512

MacDonald SM, Rafnar T, Langdon J, Lichtenstein LM (1995) Molecular identification of an IgE-dependent histamine releasing factor. Science 269:688–690

MacDonald SM, Schleimer RP, Kagey-Sobotka A, Gillis S, Lichtenstein LM (1989) Recombinant IL-3 induces histamine release from human basophils. J Immunol 142:3527–3532

MacDonald SM, White JM, Kagey-Sobotka A, MacGlashan DW Jr, Lichtenstein LM (1991) The heterogeneity of human IgE exemplified by the passive transfer of D_2O sensitivity. Clin Exp Allergy 21:133–138

MacDonald SM, Bhisutthibhan J, Shapiro TA, Rogerson SJ, Taylor TE, Tembo M, Langdon JM, Meshnick SR (2001) Immune mimicry in malaria: *Plasmodium falciparum* secretes a functional histamine-releasing factor homolog *in vitro* and *in vivo*. Proc Natl Acad Sci U S A 98:10829–10832

Mayr SI, Zuberi RI, Zhang M, de Sousa-Hitzler J, Ngo K, Kuwabara Y, Yu L, Fung-Leung WP, Liu FT (2002) IgE-dependent mast cell activation potentiates airway responses in murine asthma models. J Immunol 169:2061–2068

Min B, Paul W (2008) Basophils and type 2 immunity. Curr Opin Hematol 15:59–63

Miura K, Lavens-Phillips S, MacGlashan DW Jr (2001) Localizing a control region in the pathway to LTC_4 secretion following stimulation of human basophils with anti-IgE antibody. J Immunol 167:7027–7037

Miura K, Schroeder JT, Hubbard WC, MacGlashan DW Jr (1999) Extracellular signal-regulated kinases regulate leukotriene C4 generation, but not histamine release or IL-4 production from human basophils. J Immunol 162:4198–4206

Moldes M, Lasnier F, Gauthereau X, Klein C, Pairault J, Fève B, Chambaut-Guérin AM (2001) Tumor necrosis factor-alpha-induced adipose-related protein (TIARP), a cell-surface that is highly induced by tumor necrosis factor-alpha and adipose conversion. J Biol Chem 276:33938–33946

Obata K, Mukai K, Tsujimura Y, Ishiwata K, Kawano Y, Minegishi Y, Watanabe N, Karasuyama H (2007) Basophils are essential initiators of a novel type of chronic allergic inflammation. Blood 110:913–920

Oettgen HC, Martin TR, Wynshaw-Boris A, Deng C, Drazen JM, Leder P (1994) Active anaphylaxis in IgE-deficient mice. Nature 370:367–370

Ong CJ, Ming-Lum A, Nodwell M, Ghanipour A, Yang L, Williams DE, Kim J, Demirjian L, Qasimi P, Ruschmann J, Cao LP, Ma K, Chung SW, Duronio V, Andersen RJ, Krystal G, Mui AL (2007) Small-molecule agonists of SHIP1 inhibit the phosphoinositide 3-kinase pathway in hematopoietic cells. Blood 110:1942–1949

Rao KV, Chen L, Gnanasekar M, Ramaswamy K (2002) Cloning and characterization of a calcium-binding histamine-releasing protein from Schistosoma mansoni. J Biol Chem 277:31207–31213

Rho SB, Lee JH, Park MS, Byun HJ, Kang S, Seo SS, Kim JY, Park SY (2011) Anti-apoptotic protein TCTP controls the stability of the tumor suppressor p53. FEBS Lett 585:29–35

Rid R, Onder K, Trost A, Bauer J, Hintner H, Ritter M, Jakab M, Costa I, Reischl W, Richter K, MacDonald S, Jendrach M, Bereiter-Hahn J, Breitenbach M (2010) H2O2-dependent translocation of TCTP into the nucleus enables its interaction with VDR in human keratinocytes: TCTP as a further module in calcitriol signalling. J Steroid Biochem Mol Biol 118:29–40

Rohrschneider LR, Fuller JF, Wolf I, Liu Y, Lucas DM (2000) Structure, function and biology of SHIP proteins. Genes Dev 14:505–520

Ruschmann J, Antignano F, Lam V, Snyder K, Kim C, Essak M, Zhang A, Lin AH, Mali RS, Kapur R, Krystal G (2012) The role of SHIP in the development and activation of mouse mucosal and connective tissue mast cells. J Immunol 188:3839–3850

Sampson HA, Broadbent KR, Bernhisel-Broadbent J (1989) Spontaneous release of histamine from basophils and histamine-releasing factor in patients with atopic dermatitis and food hypersensitivity. N Engl J Med 321:228–232

Scharenberg AM, Kinet J-P (1998) PtdIns-3,4,5-P3: a regulatory nexus between tyrosine kinases and sustained calcium signals. Cell 94:5–8

Schroeder JT (2009) Basophils: beyond effector cells of allergic inflammation. Adv Immunol 101:123–161

Schroeder JT, Lichtenstein LM, MacDonald SM (1996) An immunoglobulin E-dependent recombinant histamine releasing factor induces IL-4 secretion from human basophils. J Exp Med 183:1265–1270

Schroeder JT, Lichtenstein LM, MacDonald SM (1997) Recombinant histamine-releasing factor enhances IgE dependent IL-4 and IL-13 secretion by human basophils. J Immunol 159:447–452

Seder RA, Paul WE, Dvorak AM, Sharkis SJ, Kagey-Sobotka A, Niv Y, Finkelman FD, Barbieri SA, Galli SJ, Plaut M (1991) Mouse splenic and bone marrow cell populations that express high-affinity Fc epsilon receptors and produce interleukin 4 are highly enriched in basophils. Proc Natl Acad Sci U S A 88:2835–2839

Smith DB, Johnson KS (1988) Single-step purification of polypeptides expressed in Escherichia coli as fusions with glutathione S-transferase. Gene 67:31–40

Stenton GR, Mackenzie LF, Tam P, Cross JL, Harwig C, Raymond J, Toews J, Chernoff D, MacRury T, Szabo C (2013a) Characterization of AQX-1125, a small-molecule SHIP1 activator: part 2. Efficacy studies in allergic and pulmonary inflammation models *in vivo*. Br J Pharmacol 168(6):1519–1529

Stenton GR, Mackenzie LF, Tam P, Cross JL, Harwig C, Raymond J, Toews J, Wu J, Ogden N, MacRury T, Szabo C (2013b) Characterization of AQX-1125, a small-molecule SHIP1 activator: part 1. Effects on inflammatory cell activation and chemotaxis *in vitro* and pharmacokinetic characterization *in vivo*. Br J Pharmacol 168(6):1506–1518

Sundstrom C, Nilsson K (1976) Establishment and characterization of a human histiocytic lymphoma cell line (U-937). Int J Cancer 17:565–577

Susini L, Besse S, Duflaut D, Lespagnol A, Beekman C, Fiucci G, Atkinson AR, Busso D, Poussin P, Marine JC, Martinou JC, Cavarelli J, Moras D, Amson R, Telerman A (2008) TCTP protects from apoptotic cell death by antagonizing bax function. Cell Death Differ 15:1211–1220

Tam P, Mackenzie LF, Toews J, Stenton GR, Harwig C, Raymond J, Szabo C, MacRury T, Chernoff D (2012) Absorption, pharmacokinetics, metabolism, distribution and tolerability of AQX-1125, a SHIP1 activator clinical development candidate, in healthy humans volunteers: comparison with dogs and rats. ATS Abstract, May

Telerman A, Amson R (2009) The molecular programme of tumour reversion: the steps beyond malignant transformation. Nat Rev Cancer 9(3):206–215

Thébault S, Agez M, Chi X, Stojko J, Cura V, Telerman SB, Maillet L, Gautier F, Billas-Massobrio I, Birck C, Troffer-Charlier N, Karafin T, Honoré J, Senff-Ribeiro A, Montessuit S, Johnson CM, Juin P, Cianférani S, Martinou JC, Andrews DW, Amson R, Telerman A, Cavarelli J (2016) TCTP contains a BH3-like domain, which instead of inhibiting, activates Bcl-xL. Sci Rep 6 (19725):1–11

Thueson DO, Speck LS, Lett-Brown MA, Grant JA (1979) Histamine releasing activity (HRA). I. Production by mitogen- or antigen-stimulated human mononuclear cells. J Immunol 123:628–631

Tuynder M, Fiucci G, Prieur S, Lespagnol A, Géant A, Beaucourt S, Duflaut D, Besse S, Susini L, Cavarelli J, Moras D, Amson R, Telerman A (2004) Translationally controlled tumor protein is a target of tumor reversion. Proc Natl Acad Sci U S A 101:15364–15369

Vercoutter-Edouart AS, Czeszak X, Crépin M, Lemoine J, Boilly B, Le Bourhis X, Peyrat JP, Hondermarck H (2001) Proteomic detection of changes in protein synthesis induced by fibroblast growth factor-2 in MCF-7 human breast cancer cells. Exp Cell Res 262:59–68

Vilarino N, MacGlashan D Jr (2005) Transient transfection of human peripheral blood basophils. J Immunol Methods 296:11–18

Vilariño N, Miura K, MacGlashan DW Jr (2005) Acute IL-3 priming up-regulates the stimulus-induced Raf-1-Mek-Erk cascade independently of IL-3-induced activation of Erk. J Immunol 175:3006–3014

Vonakis BM, Gibbons S Jr, Sora R, Langdon JM, MacDonald SM (2001) Src homology 2 domain-containing inositol 5' phosphatase is negatively associated with histamine release to human recombinant histamine releasing factor in human basophils. J Allergy Clin Immunol 108:822–831

Vonakis BM, MacGlashan DW Jr, Vilariño N, Langdon JM, Scott RS, MacDonald SM (2008) Distinct characteristics of signal transduction events by histamine releasing factor/translationally controlled tumor protein (HRF/TCTP)-Induced Priming and Activation of Human Basophils. Blood 111:1789–1796

Vonakis BM, Sora R, Langdon JM, Casolaro V, MacDonald SM (2003) Inhibition of cytokine gene transcription by the human recombinant histamine-releasing factor in human T lymphocytes. J Immunol 17:3742–3750

Vonakis BM, Vasagar K, Gibbons SP Jr, Gober L, Sterba PM, Chang H, Saini SS (2007) Basophil FcepsilonRI histamine release parallels expression of Src-homology 2-containing inositol phosphatases in chronic idiopathic urticaria. J Allergy Clin Immunol 119:441–448

Wantke F, MacGlashan DW Jr, Langdon JM, MacDonald SM (1999) The human recombinant histamine releasing factor: Functional evidence that it does not bind to the IgE molecule. J Allergy Clin Immunol 103:642–648

Xu A, Bellamy AR, Taylor JA (1999) Expression of translationally controlled tumour protein is regulated by calcium at both the transcriptional and post-transcriptional level. Biochemist 3:683–689

Yang Y, Yang F, Xiong Z, Yan Y, Wang X, Nishino M, Mirkovic D, Nguyen J, Wang H, Yang XF (2005) An N-terminal region of translationally controlled tumor protein is required for its antiapoptotic activity. Oncogene 24:4778–4788

Yenofsky R, Cereghini S, Krowczynska A, Brawerman G (1983) Regulation of mRNA utilization in mouse erythroleukemia cells induced to differentiate by exposure to dimethyl sulfoxide. Mol Cell Biol 3:1197–1203

Yoneda K, Rokutan K, Nakamura Y, Yanagawa H, Kondo-Teshima S, Sone S (2004) Stimulation of human bronchial epithelial cells by IgE-dependent histamine-releasing factor. Am J Phys Lung Cell Mol Phys 286:L174–L181

Yueh-Chiao Y, Xie L, Langdon JM, Myers AC, Oh SY, Zhu Z, Macdonald SM (2010) The effects of overespression of histamine releasing factor (HRF) in a transgenic mouse model. PLoS One 5:1–12

Zhang D, Li F, Weidner D, Mnjoyan ZH, Fujise K (2002) Physical and functional interaction between myeloid cell leukemia 1 protein (MCL1) and Fortilin. The potential role of MCL1 as a fortilin chaperone. J Biol Chem 277:37430–37438

Conclusion

It has been a privilege and an honor to work with all these scientists in order to put together a first book on TCTP/tpt1. It has been a discovery that editing a book could be so fascinating. After reading all these chapters, it will become clear why we have chosen this title: ***TCTP/tptl - Remodeling signaling from stem cell to disease.*** TCTP is already so complicated to understand, but what stands out is that, by interacting with such a wide variety of proteins in all subcellular compartments, it really fashions the way other proteins behave, communicate, and transduce signaling. For the moment, the most obvious diseases in which TCTP is directly implicated are asthma and cancer, but there is no doubt that with this large amount of interactions TCTP must be involved in other conditions. To be truthful, we do not understand yet much about how it functions, and there is still a long way to go with a bright future for TCTP. We just hope that this book has shed some light in our present understanding of it and, most importantly, that it will be a helpful tool in the future for the scientific community.

A. Telerman, R. Amson (eds.), *TCTP/tptl - Remodeling Signaling from Stem Cell to Disease*, Results and Problems in Cell Differentiation 64,
DOI 10.1007/978-3-319-67591-6

Printed by Printforce, the Netherlands